Comparative Pathobiology of Viral Diseases

Volume II

Editors

Richard G. Olsen, Ph.D.
Professor
Department of Veterinary Pathobiology
College of Veterinary Medicine
Ohio State University
Columbus, Ohio

Steven Krakowka, D.V.M., Ph.D.
Professor
Department of Veterinary Pathobiology
College of Veterinary Medicine
Ohio State University
Columbus, Ohio

James R. Blakeslee, Jr., Ph.D.
Associate Professor
Department of Veterinary Pathobiology
College of Veterinary Medicine
Ohio State University
Columbus, Ohio

CRC Press is an imprint of the
Taylor & Francis Group, an **informa** business

CRC Press
Taylor & Francis Group
6000 Broken Sound Parkway NW, Suite 300
Boca Raton, FL 33487-2742

Reissued 2019 by CRC Press

A Library of Congress record exists under LC control number:

ISBN 13: 978-0-367-25223-6 (hbk)
ISBN 13: 978-0-367-25224-3 (pbk)
ISBN 13: 978-0-429-28664-3 (ebk)

Visit the Taylor & Francis Web site at http://www.taylorandfrancis.com and the CRC Press Web site at http://www.crcpress.com

FOREWORD

Pathobiology has been introduced as a new expression in medical terminology within the last 3 decades. It reflects the expansion of pathology into related basic sciences (microbiology, immunology, biochemistry, molecular genetics etc.) in order to fulfill one of its traditional missions of redefining the ever changing concepts and principles of disease. Virchow, introducing the concepts of cellular pathology in the last century, already remarked in his keynote address to the Pathology Congress in Berlin that, unless methods are developed to reveal the function of the cellular structures discovered on the microscope, all our work will be in vain. These methods are now available. Modern pathologists (pathobiologists) employ a multitude of methods, ranging from gross observation to molecular genetics, to investigate disease-related problems.

A group of scientists from the Department of Veterinary Pathobiology, Ohio State University (a department combining pathology, microbiology, parasitology, and immunology) presents in 21 chapters of these volumes the current state of knowledge of a selected group of viral diseases in animals to which these authors made substantial contributions over the past decade. This publication reflects several years of cooperative research of a successful team of investigators trained in different disciplines but sharing common research interests. The book does not attempt to cover the whole field of virology but discusses selected viral diseases of broad interest to both veterinary and human medicine. It will fulfill an important need of all investigators involved in the study of viral diseases by providing them (in two volumes) with valuable information presently scattered in many national and international journals.

The title of the volumes, *Comparative Pathobiology of Viral Diseases* appropriately reflects the interdisciplinary team work of the authors. The range of interest is broader than the title indicates since most of the diseases treated in the volumes are excellent models of comparable human diseases. This two-volume set is greatly welcomed at this time.

Adalbert Koestner, D.V.M., Ph.D.
Michigan State University
East Lansing, Michigan
August, 1984

PREFACE

The pathobiology of viral diseases encompasses areas of knowledge of virology and the physiology of the host. These various parameters culminate in the discipline of "biology of viral disease". It was our intent in this treatise to select a few viral diseases of animal species to identify the various virus-host parameters that may be the basis of viral pathobiology. We feel that this level of understanding of viral diseases will be the basis of more effective control of viral diseases by prophylaxis and preventive medicine. Moreover, this approach may be the basis of understanding and developing innate disease resistance in animals of economic importance.

Richard G. Olsen
Steven Krakowka
James R. Blakeslee

ACKNOWLEDGMENTS

The senior editor wishes to acknowledge especially Dr. Eldon Davis, Dr. Lois Baumgartner, and Dr. William Beckenhauser of the Norden Company for their professional and personal friendship and for their many outstanding contributions to animal health research.

Dr. Blakeslee and Dr. Olsen had the distinct pleasure of working with Dr. Herold Cox during his tenure as chairman of the Department of Viral Oncology at Roswell Park Memorial Institute. They are indeed grateful to have been associated with this outstanding virologist, outdoorsman, and gentleman.

THE EDITORS

Richard G. Olsen, Ph.D., is currently Professor of Virology and Immunology in the Department of Veterinary Pathobiology, College of Veterinary Medicine, Microbiology, Department of Biological Sciences, and the Comprehensive Cancer Center, The Ohio State University. Professor Olsen is a native of Independence, Missouri and graduated with a B.A. from the University of Missouri and Kansas City. He obtained a M.S. degree from Atlanta University and a Ph.D. in Virology from the State University of New York (Roswell Park Memorial Cancer Institute Division), Buffalo, New York.

Professor Olsen joined the faculty at Ohio State University in 1969 and since then has developed a graduate program in pathobiology of viral diseases. He holds grants from the National Institutes of Health and the Department of Defense. He has published with his colleagues and graduate students over 200 papers in the fields of virology, immunology, immunopharmacology, and immunopathology. Professor Olsen has patented a unique procedure for the production of a feline leukemia vaccine. This patent reflects the pathobiologic approach of the immunotoxic effects of feline leukemia disease and indentifies the essential viral factor that cats immunologically recognize to resist disease. He is a member of many national and international associations. His current research interests are delineation of the mechanism of the retrovirus-induced acquired immune deficiency in cats, characterization of the preneoplastic events of retroviral disease, delineation of the biochemical mechanism of hydrazine-induced suppressor cell defects, and characterization of lichen planus dermatopathy as a presquamous cell carcinoma.

Steven Krakowka, D.V.M., Ph.D., received his D.V.M. degree from Washington State University in 1971 and a Ph.D. from the Ohio State University in 1974. He is currently a professor in the Department of Veterinary Pathobiology, College of Veterinary Medicine, The Ohio State University. His research interests are neuropathology, virology, and immunology. He is a co-author with Dr. Richard G. Olsen of a previous book entitled *Immunology and Immunopathology of Domestic Animals.*

James R. Blakeslee, Jr., Ph.D., received his B.S. in bacteriology in 1962 from the University of Pittsburgh and his M.S. and Ph.D. in Microbiology from the Roswell Park Memorial Institute Division of Microbiology of the State University of New York at Buffalo in 1971.

Dr. Blakeslee joined the faculty of the Department of Veterinary Pathobiology at the Ohio State University in 1973 and is an Associate Professor of virology and immunology in that department, the Department of Microbiology, and the Ohio State University Comprehensive Cancer Research Center. His research interests are mainly concerned with the interactions and effects of environmental chemicals on virus-induced neoplasias. Current research efforts are directed towards investigation of human and nonhuman primate T-cell lymphotropic viruses and co-factors that may interact with the infected host resulting in frank T-cell leukemias.

Dr. Blakeslee was a National Cancer Institute pre-doctoral fellow and, recently, an International Fellow of the Japan Society for the Promotion of Science while a Visiting Professor at the Institute for Virus Research at Kyoto University, Japan. He is a member of several national and international associations and has been co-editor of several books on comparative leukemia research and a contributor to a text on feline leukemia. He has over 50 publications on the subject of oncogenic viruses, modulators of virus expression, and host immunity.

CONTRIBUTORS

Michael K. Axthelm, D.V.M., Ph.D.
Postdoctoral Fellow
Department of Veterinary Pathobiology
Ohio State University
Columbus, Ohio

Wolfgang Baumgartner, Dr. Med. Vet.
Research Associate
Department of Veterinary Pathobiology
Columbus, Ohio

James R. Blakeslee, Ph.D.
Associate Professor
Department of Veterinary Pathobiology
Ohio State University
Columbus, Ohio

H Fred Clark, D.V.M., Ph.D.
Wistar Institute
Philadelphia, Pennsylvania

Robert M. Jacobs, D.V.M., Ph.D.
Assistant Professor
Department of Veterinary Pathobiology
Ohio State University
Columbus, Ohio

Gayle C. Johnson, D.V.M., Ph.D.
Assistant Professor
Department of Veterinary Pathobiology
Ohio State University
Columbus, Ohio

Steven Krakowka, D.V.M., Ph.D.
Professor
Department of Veterinary Pathobiology
Ohio State University
Columbus, Ohio

Lawrence E. Mathes, Ph.D.
Assistant Professor
Department of Veterinary Pathobiology
Ohio State University
Columbus, Ohio

Bellur S. Prabhakar, Ph.D.
NIH Laboratory of Oral Medicine
National Institute of Dental Research
Bethesda, Maryland

Anna-Lise Payne, Ph.D.
Molecular Biology Section
Veterinary Research Institute
Onderstepoort, Republic of South Africa

Niels C. Pedersen, D.V.M., Ph.D.
Associate Professor
Department of Medicine
School of Veterinary Medicine
University of California
Davis, California

R. Charles Povey, Ph.D., F.R.C.V.S.
Professor
Department of Clinical Studies
Ontario Veterinary College
Guelph, Ontario, Canada

Jennifer L. Rojko, D.V.M., Ph.D.
Assistant Professor
Department of Veterinary Pathobiology
Ohio State University
Columbus, Ohio

Royden C. Tustin, M. Med. Vet.
Professor and Head
Department of Pathology
Faculty of Veterinary Science
University of Pretoria
Onderstepoort, Republic of South Africa

D. W. Verwoerd, D.V.Sc., D.Sc.
Assistant Director and Head
Section of Molecular Biology
Veterinary Research Institute
Onderstepoort, Republic of South Africa

TABLE OF CONTENTS

Volume II

Chapter 1

FELINE LEUKEMIA

James R. Blakeslee, Jr. and Jennifer L. Rojko

TABLE OF CONTENTS

I. HISTORICAL PERSPECTIVES

Feline leukemia has been the subject of intense research over the last decade following its recognition as an infectious disease, and its potential significance as a model for the study of human disease. Feline leukemia provides a model of an outbred animal in close association with the human population which presents a unique opportunity to study aspects of neoplastic disease from a vantage point not possible with other animal systems. The evidence in favor of a causal relationship between a virus and the disease is in the epidemiologic data obtained in FeLV-positive leukemia cluster households.[1,2] One important finding was the identification of virus negative lymphomas in the cat. Although no virus was found in these cats, the virus was believed to be latent, in that feline oncornavirus-associated cell membrane antigen (FOCMA) was detected on the cell surface in these cats.[1,3-6] Under natural conditions cats become infected with the FeLV predominantly by contact with salivary and nasal secretions of cats with persistent, active FeLV infection. Within 6 weeks after infection of cats with FeLV, either of two major host-virus relationships usually develops: (1) persistent active infection, which is a progressive infection or (2) a selflimiting regressive infection. The cats may develop anemia, be immunosuppressed leading to other types of disorders, or they may develop leukemia. Thus, it is possible to divide diseases related to FeLV infection into two categories: neoplastic and non-neoplastic disease.

When FeLV infection results in neoplastic disease, classification is based on the cell type that has undergone malignant transformation. Subclassification is further based on the location of the primary lesion. The clinical manifestations are determined by the histologic type and location of the lesions.

A broad spectrum of non-neoplastic diseases has made diagnosis of many of these disorders quite difficult. The non-neoplastic diseases have been commonly confused with other feline diseases. Non-neoplastic diseases may include a nonregenerative anemia, a panleukopenia-like syndrome, thymic atrophy, reproductive disorders, glomerulonephritis, hemolytic anemia, and immunosuppression (acquired immune deficiency). In the case of neoplastic diseases, the spectrum includes lymphoproliferative disorders, including lymphocytic leukemia, lymphosarcoma, and/or myeloproliferative diseases such as myelosclerosis, granulocytic anemia, erythroleukemia, erythremic myelosis, myelofibrosis, and reticuloendotheliosis. Thus, the broad spectrum of neoplastic and non-neoplastic diseases induced by FeLV are manifold.

The FeLV model of non-neoplastic and neoplastic disease has taken on greater importance with the finding of human retroviruses that are involved in adult T cell leukemia/lymphoma and the acquired immunodeficiency syndrome (AIDS). Thus, it appears that the investigation of FeLV in cats and the responsible mechanism of leukemogenesis may be of great importance in the study of the pathogenesis of certain types of leukemia in man.

II. FELINE LEUKEMIA VIRUS

A. Nomenclature and Structure of FeLV

FeLV is an enveloped, oncogenic RNA virus containing a single-stranded 70s RNA composed of two identical 35s subunits and is classified in the family Retroviridae, subfamily Oncovirinae.[7] The FeLV core consists of tightly coiled single-stranded viral RNA surrounded by core proteins of 27,000, 15,000, and 10,000 daltons, and designated p27, p15(C), and p10, respectively,[8,9] and the RNA dependent DNA polymerase (RDDP).[10] The core complex is surrounded by acidic proteins of 12,000 daltons (p12) which constitute the inner coat. The viral envelope consists of a glycosylated protein of 70,000 daltons (gp70) arranged as spheres displayed on radiating spikes derived from the transmembrane 15,000 dalton protein designated p15(E).[11]

B. Molecular Biology of FeLV

The single-stranded viral RNA is composed of two identical 35s subunits. Each subunit has an extra cap nucleotide added at the 5′ end; 3′ end contains a tail of polyadenylic acid. Unique regulatory domains are located at the 5′ (U5) and 3′ (U3) ends, and each subunit is terminated by repeated (R) sequences shared by the 5′ and 3′ ends. The initiation site for reverse transcription is located at the 3′ boundary of the U5 domain where primer transfer RNA (tRNA) of host origin is attached by hydrogen bonds.[12] The coding domains for the virion structural proteins and the RDDP are organized from the 5′ to 3′ end in the following order: gag, pol, and env regions. The group-associated gene (gag) sequences encode the polyprotein precursors of the internal core proteins (p15(C), p12, p27, p10, the polymerase (pol) region codes for the RDDP, and the envelope (env) sequences encode the envelope polyprotein precursor which is glycosylated and cleaved post-translationally to yield gp70 and pl5(E).

C. Virus Replication

Following adsorption and penetration to susceptible cells, FeLV is uncoated and the RNA is copied into a single strand of complementary DNA (cDNA) by the virion RDDP which is primed by a small tRNA molecule.[12,13] The *de novo* synthesized DNA serves as a template for the formation of the double-stranded DNA provirus which circularizes and integrates to become part of the cell genome. Viral RNA (vRNA) is transcribed from the integrated provirus by DNA-dependent RNA polymerases (DDRP) and translated on host ribosomes to generate precursor structural and envelope polyproteins. These polyproteins migrate to the plasma membrane where "budding" occurs.[11,14] Peripherally, the envelope precursors undergo post-translational cleavage and the minor fragment (p15[E]) spans the plasma membrane as transmembranous spikes firmly embedded in the lipid bilayer.[11] Protruding knobs representing the major fragment are glycosylated (gp70) and are welded to the spikes of p15(E) by disulfide linkages.[11] Centrally, the structural precursors are processed and the individual proteins re-associate according to their biochemical and biophysical properties to encapsulate the ribonucleoprotein core.[8,14,15,16]

D. Biologic Properties of FeLV Proteins

The major group-specific antigen (GSA) which confers interspecies antigenic cross-reactivity to FeLV resides in the major core protein p17.[16] p17 Can be detected in the cytoplasm of circulating neutrophils and platelets by fixed cell immunofluorescence assay[76] and shows a high correlation with recovery of infectious virus[17] from plasma.[18] Serum from FeLV-infected cats contains soluble p27 generally detected by an enzyme-linked immunosorbent assay[19,20] which detects both viremic cats or occult FeLV infection.[19,29,21]

Three antigenically distinct forms of the major envelope glycoprotein (gp70) specify three subgroups (serotypes) of FeLV, designated $FeLV_A$, $FeLV_B$, and $FeLV_C$. These envelope gene products determine infectivity, interference, host range properties, and pathogenicity, and as subgroup-specific antigens evoke the virus neutralizing antibody responses important in the reversal of viremia in cats that regress FeLV infection.[22-26] Virus adsorption, and hence infectivity, is dependent upon the affinity of gp70 for host cellular FeLV receptors; virus neutralizing antibody blocks initial adsorption. Dessication, heat (56°C, 3 min), and ultraviolet (UV) light detach gp70 and inactivate FeLV.

The minor envelope protein (p15[E]) depresses T lymphocyte function that accompanies viremia.[27] The binding of p15(E) to the first component of complement[28] activates the classical pathway which causes complement depletion and results in ineffective virolysis in cats.[29]

E. FOCMA

An apparent virus-directed antigen on the surface of cat cells transformed by feline sarcoma virus was detected by membrane immunofluorescence using antisera from resistant cats.[5,30,31] This antigen was subsequently named Feline oncornavirus-associated cell membrane antigen (FOCMA). It was subsequently found on plasma membranes of cells infected in vitro with FeSV and on tumor cell surfaces from cats with fibrosarcomas, leukemias, and lymphomas.[5,32,33] An inverse relationship was noted in that antibody to FOCMA correlated inversely with tumor progression[5,32] and was therefore thought to be associated with the development of anti-FeLV/FeSV-induced tumor immunity, and has since been observed in vaccine studies.[33-35] FOCMA was shown to evoke a complement-dependent antibody response effective in prevention of tumor development.[32,36-44] While the exact origins and functions of FOCMA remain speculative, it is clear that antibody to FOCMA determinants effects antitumor immunosurveillance in these tumor-bearing or leukemia-infected animals. Several lines of evidence suggest a nonvirion nature of FOCMA. These include: (1) the inability to remove FOCMA-specific antibodies from appropriate sera by absorption with disrupted FeLV-purified gp70 or p27 proteins;[45] (2) the lack of correlation between the presence of antibody to FOCMA and antibody to gp70 or p27 in the sera of a large number of FeLV or FeSV exposed cats, as determined by membrane immunofluorescence and radioimmunoprecipitation;[46] (3) the failure to correlate the presence of anti-gp70 antibody with FOCMA antibody;[38,42,47,48] and (4) the presence of FOCMA in nonproducer mink cells transformed by FeSV.[49]

Further, FOCMA is present on preneoplastic lymphocytes in the bone marrow and mesentery lymph nodes.[20] Recently, an FeLV vaccine was developed using FOCMA and gp70. The details of this vaccine will be discussed later in this chapter.

FeLV-induced malignancies are distinguished by a common tumor-specific antigen, the feline oncornavirus-associated cell membrane antigen (FOCMA) which is present on transformed cells from cats with highly expressed (productive) FeLV infections and from cats with minimally expressed (nonproductive) FeLV infections.[50-53] Recent studies with monoclonal antibodies suggest that the expression of FOCMA and the partial expression of $FeLV_C$ envelope genes are interdependent events.[54]

F. FeLV Subgroups

$FeLV_A$ is the most common subgroup isolated from naturally infected pet cats. It replicates to high titers in infected cats, and always is present when mixtures of FeLV subgroups are isolated. $FeLV_A$ has a narrow host range in vitro,[22] being mostly restricted to cat cells. $FeLV_B$ is isolated always in conjunction with $FeLV_A$, replicates to lower titers, and $FeLV_B$ viremia is delayed in appearance relative to $FeLV_A$ viremia. $FeLV_B$ has an extended host range in vitro. $FeLV_C$ is only recovered if $FeLV_B$ and $FeLV_C$ also are present[22,25] and replicates in cat, human, and guinea pig cells.

Evidence has accumulated suggesting that FeLV subgroups induce different types of disease. Inoculation of susceptible kittens with the Rickard strain of FeLV-FeLV-R (subgroup AB) leads to a high incidence of viremia ($\geq$ 85%), severe immunosuppression, and thymic lymphomas in those animals surviving 17 to 30 weeks.[55,56] In contrast, exposure of cats to the Glasgow passaged Rickard strain of FeLV, which contains subgroup A only, generates a low incidence of viremia, hemorrhagic enteritis, and neutropenia, and occasional alimentary lymphomas or myelogenous leukemias after a very long latent period.[47,57] Although not associated with a specific disease state, $FeLV_B$ is intensely cytopathic for feline marrow cells in vitro.[58] Recent reports of homologous sequences shared by $FeLV_B$ and a murine retrovirus cytopathic for mink cells (mink cell focus-forming-MCF-virus) have raised speculations as to the origin and leukemogenicity of $FeLV_B$.[59] MCF viruses are spontaneous env gene recombinants between ecotropic and xenotropic retroviruses with high leukemogenicity in vivo, cytopathicity, and extended host range in vitro. $FeLV_B$ is the most frequent helper FeLV implicated in natural transduction of the fes proto-oncogene from cat fibroblasts and the resultant generation of acutely transforming feline sarcoma viruses (FeSVs).

The delay in the appearance of free, infectious $FeLV_C$ in plasma may result from its relative defectiveness for replication in vivo. Vedbrat et al.[54] report that even cells that produce only $FeLV_A$ or $FeLV_A$ and $FeLV_B$ as cell-free virus have many partially replicated, immature $FeLV_C$ buds embedded in the plasma membrane and suggest that the partial expression of $FeLV_C$ is equivalent to FOCMA-L expression. Based on its delay in appearance and defectiveness for replication of $FeLV_C$ and induction by $FeLV_C$-devoid viral stocks, many workers have considered $FeLV_C$ also a candidate recombinant virus.

The Kawakami-Theilen strain of FeLV(-FeLV-KT) is a mixture of $FeLV_{ABC}$, and causes profound erythrosuppression and death within 9 weeks of inoculation of neonatal kittens but is virtually apathogenic in weanlings and adults.[60,61,62] Passage of $FeLV_{ABC}$ through guinea pig cells in vitro selects for biologically cloned $FeLV_C$ (Sarma strain).[22] That this $FeLV_C$ induces viremia and erythroid aplasia in vivo and suppression of erythroid colony formation in vitro suggests a direct relationship between productive $FeLV_C$ infection and anemiagenesis.[62]

G. Pathogenesis of FeLV Infection

Domestic cats are exposed to FeLV following prolonged contact with the saliva or urine of naturally viremic cats.[18,63] High titers of infectious FeLV are excreted by pharyngeal, salivary, bladder, and intestinal epithelia but virus survival at room temperature or under conditions of dessication is less than 2 hr.[63] Therefore, efficient virus transmission appears to require either direct contact between cats, transfer of saliva on hands or feeding utensils, or exposure to recently voided urine in communal litter pans.

The presumed portal of entry following contact in nature[47,64] is the oronasal pharynx. Experimental oronasal challenge of cats with FeLV leads to viremia and disease or nonproductive infection and immunity (see below). Congenital transmission of FeLV has been proposed to account for the clinical observation that viremic queens may bear viremic kittens, but the question of intrauterine vs. lactation-associated vs. contact exposure of kittens has not been resolved.[65] Recent experiments document the transplacental passage of FeLV and its isolation from the embryonic hemolymphatic tissues of fetal kittens obtained by hysterectomy from viremic gravid queens.[66]

The induction of FeLV-related disease usually follows the onset of marrow origin viremia. Even though the leukemogenic event probably is random, high levels of viremia ensure that the damaging virus will be present when cells of the appropriate histogenesis and maturation reach a critical stage in their cycle. In the mouse, chronic antigenemia with MuLV gp71 incites chronic immunostimulation of uninfected T cells and renders them susceptible to the leukemogenic event. In cats, circulating virion proteins are immunosuppressive, erythrosuppressive, and embryosuppressive.[66] Other circulating virion proteins may be complexed to antibody to trigger immune complex disease. Even transient marrow origin viremia and probably mononuclear cell-associated viremia are important. It is true that cats with these usually develop immunity. However, the heavier the original FeLV burden, the smaller the likelihood that the cat will eliminate all FeLV-infected cells, and the larger the likelihood the cat will maintain persistent poorly expressed FeLV infections and risk nonproductive disease.

The pathogenesis of FeLV replication has six identifiable stages.[67] FeLV/host cell contact is initiated in the lymphoepithelia and follicular lymphocytes in the pharyngeal and palatine tonsils (stage 1) in the first 2 days following oronasal viral instillation.

Virus amplification occurs in the draining lymph node and infectious FeLV is transported by lymphocytes and macrophages (stage 2) to secondary sites (marrow, systemic and gut-associated lymphoid tissue (GALT) where massive viral amplification occurs (stages 3 and 4). The secondary virus amplification provides rapid, efficient distribution of FeLV to mitotically active cells in the bone marrow and GALT. Viral integration also protects FeLV from reticuloendothelial clearance and from inactivation by humoral lipid or protein moieties as described for feline, murine, and bovine retroviruses.[28,68-72]

Proliferation of FeLV in the spleen, lymph nodes, and GALT distant from the site of inoculation is evident in cats between 3 and 12 days postinfection (stage 3). FeLV p27 is most concentrated in the rapidly dividing lymphoid cells of the germinal centers of cortical follicles. Early viral tropism for B lymphocytes has been shown for certain MuLV strains early in infection.[73-75]

In the fourth stage (7 to 21 days postinfection), widespread infection of nonlymphoid hematopoietic cells in the bone marrow overlaps the systemic lymphoid phase. Megakaryocytes accumulate large amounts of cytoplasmic viral antigen, resulting in infection of developing platelets. The majority of the marrow cells containing p27 are developing myelomonocytic precursors. The concentration of p27 increases as the cell matures. However, p27 is rare in eosinophil precursors. In erythroid maturation, the intensity of FeLV replication is inversely related to cell maturation. Concomitant with marrow infection is the onset of multiple foci of productive infection in the crypt germinal epithelium of the small and large intestines. Viral antigen is confined to the basilar mitotic cell population and is absent from the mature distal absorptive epithelium lining the villi. Retroviruses generally are not cytopathic and replication of FeLV in the rapidly dividing cells of the marrow, lymphoid tissue, and intestine is not associated with overt cytopathic change (necrosis, polykaryocytosis, etc.). However, the presence of FeLV in lymphoid and marrow tissues may depress normal cell turnover and initiate atrophy.

p27-Positive neutrophils and platelets next appear in the circulation and are considered the fifth stage in the evolution of progressive FeLV infection. This phase directly reflects infection of marrow protenitors and is the onset of marrow-origin viremia[76] and the induction of fatal FeLV-related disease.[18,67,76]

The initial marrow release of FeLV-infected neutrophils and platelets coincides exactly with development of protracted neutropenia, lymphopenia, and thrombocytopenia 21 to 56 days postinfection.[67,77,78] These may be affected by atrophic or aplastic responses of hemolymphatic precursor cells, extravascular sequestration of FeLV-infected cells in myeloid or lymphoid tissue, or immunologic elimination of FeLV-infected cells by the host. Although the onset of marrow origin viremia usually signifies the establishment of progressive FeLV infection, some cats still are able to reverse this state by clearing FeLV-infected cells and producing both virus neutralizing (VN) and FOCMA antibody. This latter group of cats is at particular risk for reactivatable FeLV infections and p27 antigenemia.[19-21,78,79]

In cats that fail to develop VN and FOCMA antibody, FeLV infection extends to multiple mucosal and glandular epithelial tissues between 28 and 42 days postinfection and later. The earliest and most consistently infected epithelial tissues are those of the oropharynx, nasopharynx, larynx, trachea, stomach, salivary gland, pancreas, and urinary bladder. Replication begins in multiple foci in the mitotic layers and progresses to diffuse involvement of the mucosa of the bladder, oral and nasal pharynx, and trachea and the release of infectious FeLV into the secretions of persistently viremic cats.

In contrast is the tissue distribution of FeLV p27 in cats that develop immune (regressive) infection between 28 and 42 days postinfection. Immune cats develop FOCMA and VN antibody by 14 to 56 days[20,67] and are able to abort virus production prior to widespread marrow or epithelial infection. Recent evidence indicates that immune cats remain latently infected with FeLV.[78-81] The early cell-associated viremia and systemic lymphomyeloid viral replication must be eliminated in all cells with integrated FeLV proviruses. Most do not and rather develop a persistent nonproductive infection.

The pathogenesis of spontaneous viral infections often centers around a virus/host lymphomyeloid reciprocity in which certain cells are targets for productive or nonproductive (latent) infection, others are targets for cytosuppressive or cytoproliferative disease, and others are effectors of antiviral or antitumor resistance.[82-84] A sequence of lymphoreticular/virus interactions characterizes the infections of cats with FeLV: initial virus replication in hemolymphoreticular cells, and development of FeLV-related neoplasia or aplasia of the lymphoid or hematopoietic system. Available evidence suggests that the histogenesis and immunologic identities of the cells involved in early virus replication, virus containment and latency, chronic (preleukemic) virus replication, and eventual oncogenesis are divergent.

H. Acute, Persistent, and Latent FeLV Infection

Oronasal exposure of cats to the Rickard strain of FeLV simulates natural exposure and initiates FeLV replication in lymphoreticular cells in the tonsil, blood, germinal centers of lymphoid tissues, thymic medulla, and bone marrow, in that order.[67] Apparently, closely related events occur in other horizontally transmitted oncogenic viral infections, e.g., primary infections with the human EBV,[85] murine mammary tumor virus, and Marek's disease herpesvirus in chickens. It is believed that EBV enters the oropharynx, infects tonsillar and blood lymphocytes with complement receptors and specific EBV receptors, and is disseminated via an early lymphocyte-associated viremia. The rationale for FeLV infection of specific lymphoid subsets, whether due to distribution of FeLV receptors, capacity for spontaneous DNA synthesis, and distribution or migration patterns in vivo, currently is not understood.

In progressive FeLV infections, a persistent polyclonal infection of follicular lymphocytes and bone marrow precursors but not thymocytes in association with protracted lymphopenia and neutropenia[67,86] and limited anti-FeLV humoral responses during preleukemia precede the emergence of neoplastic T cells in the thymus and elsewhere.[67,87] The principal FeLV-infected cell in the lymph nodes of viremic cats is a nonadherent, CR-bearing B lymphocyte.[86] p27 Is diffuse in large (20 to 45 μm) lymphoblasts with eccentric, large, round to cleaved nuclei and prominent nucleoli. Plasma cells are densely stained and are identified by eccentric round nuclei with highly condensed chromatin, dense homogeneous cytoplasm at the rounded cell periphery, and a perinuclear (Golgi region) clear space. The pattern of increasing intensity of intracytoplasmic p27 staining with increasing differentiation in B-lineage cells is reminiscent of that described for myelomonocytic series cells in the bone marrow. Purified T cells and adherent macrophages are not particularly enriched for infectious FeLV nor do they contain p27.

The functional integrity of the B lymphocytes that replicate FeLV is unknown. In the mouse, splenic B lymphocytes that are concurrently infected with MuLV produce less antibody to sheep erythrocytes than do uninfected spleen cells, but this is most probably due to increased T-suppressor activity.[88] Naturally viremic pet cats have adequate IgM but ineffective and delayed IgG responses to a synthetic T-dependent antigen and a systemic T-helper defect has been postulated.[89] Experimentally induced viremic cats also produce IgM but not IgG antibody in response to FeLV-associated antigens.[90] These facts may signify altered B-cell function, altered regulation by T helper, T suppressor or accessory cells, or a changed microenvironment (FeLV-infected lymph node vs. control lymph node). Regarding the last, relative increases in nodal T cells and FcR cells have been reported to be simultaneous with the FeLV-associated peripheral blood lymphopenia in preleukemia.[86] Redistribution of lymphoid subpopulations also has been reported for preleukemic MuLV infections,[91] and increased numbers of CR- and Fc R-positive cells have been described in preneoplastic murine mammary tumor virus infections.[92]

A critical feature that distinguishes persistent productive FeLV infection from persistent nonproductive infection is viral replication by differentiating granulocytes and macrophages in the marrow and elsewhere. Furthermore, marrow origin viremia is accepted universally as the harbinger of fatal FeLV-associated disease.[18,64] In contrast, in cats that regress productive marrow infection and develop latent FeLV infection, refractoriness of myelomonocytic series cells to FeLV replication is correlated directly with increasing maturation. Even in viremic cats which experience extensive FeLV replication in myeloid progenitor cells, differentiated granulocytes, and marrow adherent macrophages, most mature peritoneal macrophages are spared.[93] It is speculated that these are derived from uninfected myeloid clones, or become refractory to, or abort FeLV infection during the process of differentiation in vivo.[79,93] This relationship between macrophage susceptibility to productive viral infection and disease progression also holds for mice infected with the leukemogenic Friend MuLV complex,[94] and was described originally for mice and monkeys infected with such nononcogenic viruses as herpes, corona, pox, and flavi viruses.[95-98] Susceptibility vs. resistance of macrophages to viral replication is central to the mechanism of age-related susceptibility vs. resistance to viral persistence.

FeLV replication in follicular (B) lymphocytes and myelomonocytic series cells is a constant feature of persistent productive FeLV infection. In contrast, replication in putative T-cell regions is limited to the recirculating lymphocyte pool and to the thymic medulla early in infection.[67] The disappearance of replicating FeLV from T-cell areas during preleukemia is associated with lymphopenia,[67,77,101,102] loss of circulating T-suppressor cell,[103] and T-cell mitogenic function, thymicolymphoid atrophy,[56,104] and redistribution of latently infected T cells to the mesenteric lymph nodes,[86] and precedes the emergence of productively infected T-lymphoma cells in the subcapsular cortical thymus and elsewhere.[67,87,101] Neoplastic lymphocytes from experimentally and naturally infected cats usually bear T-cell markers (E rosette forming capacity, surface thymocyte antigen[101]) but rarely do appear as SIg-bearing B cells or null cells.[52] The relative maturity of the transformed T-lymphoma cells is demonstrated by its lack of terminal deoxynucleotidyl transferase (TdT). Similar dissociation between lymphocyte tropisms for viral replication and virus-associated transformation have been observed in Moloney and AKR murine lymphomas.[105,106]

Cats that regress productive marrow and lymphoid infection and become immune must either eliminate all cells with integrated FeLV proviruses or risk persistent nonproductive infection. Cells most likely to escape immune elimination are those with a long interphase as retroviral antigens maximally expressed in mitotic cells. Based on this premise, candidate marrow and lymphoid cells would include the slow-to-cycle committed myelomonocytic precursor defined in the mouse by the in vitro spleen colony-forming unit assay (CFU_s of Till and McCulloch[107]), memory lymphocytes, and long-lived T lymphocytes. While none of these cells has been identified in the cat, it is known that the target cells for latent FeLV infection are compatible: marrow myelomonocytic precursor cells, macrophages, and Staphylococcal Protein A (SPA)-reactive T lymphocytes in the systemic lymphoid tissue.

I. Age-Related Susceptibility to FeLV Infection

Control of the FeLV dose and strain, coupled with variation of the host age at time of FeLV inoculation, will result in populations of cats destined to develop either viremic or immune FeLV infections. Even though the susceptibility of cats exposed to FeLV in nature may vary greatly, the rate of viremic disease in adults naturally exposed to FeLV in multi-cat households ranges from 15 to 28%. Experimentally, cats at risk for viremia include neonatal kittens (100% susceptible) and 8-week old weanlings (85% susceptible).[99]

Heightened susceptibility to FeLV in the neonate may be macrophage dependent in that macrophages from kittens are five times more sensitive to in vitro infection than macrophages from adults. This maturation-dependent event is abrogated by treatment with hydrocortisone[93] or prednisolone.[100]

Treatment with various doses of prednisolone acetate results in a sevenfold increase in susceptibility of adult cats to FeLV viremia. These animals eventually die of FeLV-related disease.

That corticosteroids are important in regulation of FeLV replication by bone marrow myelomonocytic precursors also may be inferred from their capacity to reactivate latent FeLV.

III. IMMUNOBIOLOGY OF FeLV

A. Humoral Antibody

Essex and colleagues were the first to promote the concept that antibodies are protective against neoplasms induced by feline retroviruses.[51] Protection is correlated to the development of IgG antibodies to FOCMA recognized by indirect membrane immunofluorescence (IMI) assay.

Actual clearance of virus-infected transformed cells probably is mediated by cytotoxic, complement-dependent antibody to both virion antigens and also to FOCMA.[108-112] In contrast, antibodies to nonvirion determinants of FL-74 cells (e.g., FOCMA) which activate feline complement arise later and persist throughout the cat's lifespan. It is likely that these antibodies serve to prevent emergence of producer or nonproducer lymphoma cells in vivo.[110-112] These same antibodies also inhibit the in vitro reactivation phenomenon.

The temporal expression of IgM vs. IgG antibody titers to FOCMA also has prognostic value.[90] In regressors, IgM anti-FOCMA appears soon after exposure and lasts 3 to 5 weeks. The decline in IgM anti-FOCMA is accompanied by a rapid rise in IgG anti-FOCMA. This IgM to IgG conversion is expected and is regarded as a T-helper-dependent response. Cats destined to become viremic differ in that this IgG anti-FOCMA response is low or absent, whereas the IgM anti-FOCMA persists at constant levels until death. The reasons for IgM persistence and failure of IgG conversion presently are not known, though it is postulated that this is due to defective T-helper function,[89] due to constant and recurrent antigen stimulation,[113] or to virus persistence in macrophages.

Limiting marrow-origin viremia in cats is essential. This appears to be mediated by virus neutralizing (VN) antibody to subgroup-specific envelope gp70.[18,21-24,26,70] Antibody to gp70 with high VN in vitro is passively protective in cats when transferred by colostrum[114] or systemic inoculation[70,115] if done early in the course of infection.[70] There is some suggestion that passive transfer of anti-gp70 induces partial remission or enhanced responses to chemotherapy in lymphoma.[115]

B. Cell-Mediated Immunity (CMI)

The first demonstrations that viremic cats have decreased CMI in vivo were those of Perryman et al.[104] and Hoover et al.[56] Allograft rejection responses in persistently infected kittens were delayed and this was correlated inversely with severe thymic atrophy and paracortical lymphoid depletion. In later studies, thymic depletion was attributed to altered traffic of thymocyte precursors from marrow to thymus, and/or altered exit of mature thymocytes to splenic and nodal paracortex.

Abrogation of immunity is T-cell specific. With the onset of viremia, cats also lose their capacity to respond to the T-lymphocyte mitogens and to the antigen keyhole limpet hemocyanin, whereas reactivity to Staphylococcal Protein A and to the B-cell mitogen lipopolysaccharide (LPS) are comparable to uninfected cats.[77,101,102,116]

T-cell specificity also is supported by the fact that viremic cats make ineffective IgG responses to a synthetic multichain T-cell specific polypeptide.[89] IgM antibodies were similar in infected and uninfected cats. This peptide indicates normal B-cell function and impaired T-helper cell function. FeLV-induced immunosuppression may provide a model for the specific impairment of OKT-4 positive T-helper cells observed following infection of human cells with the human T-cell leukemia virus (HTLV).[117-119]

T-cell suppression by FeLV is not, however, limited to the T-helper population. FeLV-related immunosuppression has been studied by evaluating suppressor cell function in vitro.[103] Viremic cats lack circulating suppressor cells or these cells are not functioning as those in normal cats.

C. Natural Killer Cells and Interferon

The evidence implicating natural killer (NK) cells and/or interferon in the biology of FeLV infection is fragmentary. Clinically, it has been reported that the administration of interferon to anemic, viremic cats has led to partial remission of viremia and recovery from anemia.[120-122] In vitro neoplastic T lymphoblasts are more susceptible to productive FeLV infection than are EBV-transformed lymphoblasts. FeLV buds from the surface of T lymphoblasts with HLA-A1, B12, but not HLA-A29, B8 determinants. Cycloheximide, an inhibitor of interferon synthesis, decreases NK lysis of FeLV-infected B cells but increases NK lysis of FeLV-infected T cells, indicating that interferon mediates enhanced NK lysis of FeLV-infected B cells only and that FeLV-infected, neoplastic T cells resist NK lysis.[123]

D. Complement and other Humoral Factors and FeLV

Circulating substances including proteins, glycoproteins, and lipoproteins are known to influence retrovirus replication. Complement components of humans, other primates, and cats but not guinea pigs, lyse retroviruses directly.[28,29,68,69] It appears that virion p15(E) binds to C1q thereby activating the classical pathway and thus virolysis. Surprisingly, there is no difference in FeLV lysis by normal vs. leukemic serum or by viremic vs. nonviremic serum.[29] Despite this, complement consumption in viremia is indicated by the facts that viremic cats are hypocomplemententemic, have circulating immune complexes containing gp70, p15(E), p27 and IgG,[123,124] and have deposits of FeLV, IgG and complement in renal glomeruli. Complement also mediates the antibody-dependent lysis of producer or nonproducer lymphoblasts transformed by FeLV. Other inhibitors with broad spectrum activity include the very low density lipoprotein of normal mouse serum origin which inactivates ecotropic mouse and feline viruses and broadly reacting antibodies directed against retroviral glycoproteins.[70,71]

E. Immunosuppression and FeLV

Many oncogenic retroviruses are associated with a rapid and sustained decline in immunocompetence soon after infection. In FeLV-infected cats, immunosuppression accompanies induction of marrow origin viremia, precedes detectable neoplastic transformation by months, and predisposes persistently infected cats to a variety of intercurrent, often opportunistic, pathogens. Viremic cats most commonly die of concurrent enteritis, gingivitis, pneumonia, or sepsis of bacterial origin; infectious peritonitis of coronaviral origin; or disease of hemotropic (*Hemobartonella felis*) or parasitic origin.[65] Clinical manifestations of immunomodulation include peripheral lymphopenia, thymicolymphoid atrophy, circulating immune complexes, hypocomplementemia, and membranous glomerulonephritis or periglomerular fibrosis.[29,36,123,125,126] Immunosuppression, therefore, is the most frequent and the most devastating, thus the most biologically prominent effect of FeLV in its natural host.

Early killed virus vaccine experiments suggested that immunosuppression did not necessarily require live virus.[33] Inactivated FeLV also interferes drastically with lymphocyte function in vitro. Incubation of feline leukocytes with serum from viremic cats or with UV-treated FeLV causes in vitro loss of lymphocyte reactivity to T-cell mitogens,[27,102,127] allogeneic leukocytes,[129] and depression of lymphocyte membrane lectin receptor mobility.[129,130] Similar membrane-related lymphocyte deficiencies accompany lymphoma in man, FeLV viremia in cats, and Friend MuLV infection in mice.

To delineate the FeLV component responsible for the immunosuppression, FeLV has been fractionated into its component polypeptides and each fraction was tested for suppression of mitogen-induced blastogenesis. Purification of the FeLV suppressive fraction has revealed a 15,000 dalton protein on polyacrylamide gel electrophoresis (FeLV p15(E)). Purified FeLV p15(E) has been shown to be suppressive to the mitogen-induced LBT at low concentrations.[27,127] Administration of FeLV p15(E) to cats reduces the subsequent response to FOCMA and also increases susceptibility to FeLV disease. The in vivo biologic effects of FeLV p15(E) are very similar to the effects of inactivated FeLV. In addition to tumor enhancement and decreased FOCMA antibody response, administration of FeLV p15(E) interferes with the apparent helper effect of T-lymphocytes and blocks the apparent conversion of IgM to IgG FOCMA antibody. Cats given FeLV p15(E) develop persistent IgM FOCMA antibody with only low levels of IgG.[99] This profile is similar to that observed in viremic cats.

FeLV-origin p15(E) does not appear to affect production of interleukin 1 (IL-1) by feline monocytes in vitro.[131] This contrasts with observation that antigenically similar, 15,000 dalton proteins derived from human and murine lymphomas inhibit monocyte (macrophage) functions. The mechanism of suppression is more likely related to the dramatic decrease in the secretion of IL-2 (T-cell growth factor) in FeLV- and p15(E)-treated leukocyte.[131] Similarly, avian retrovirus p15(E) has immunosuppressive action equivalent to FeLV p15(E) and recent studies by Wainberg et al.[132] demonstrate that blastogenesis can be restored to suppressed cultures with the addition of sufficient IL-2. Thus, retrovirus components, especially p15 envelope protein, disrupt recruited lymphocyte proliferation by eliminating secretion and action of IL-2.

FeLV p15(E) is highly hydrophobic and may bind readily to cell membrane lipids, thus interrupting normal membrane functions. The prostaglandin and cyclic nucleotide systems are logical candidates for effectors of the T-cell suppression induced by FeLV p15(E). Both are linked closely to the immune system, cell membrane-mediated events, mobility and expression of cell surface receptors and regulation of cell proliferation. Furthermore, colchicine, a microtubule disrupting agent, reverses FeLV suppression of lectin receptor mobility. Lewis et al.[133] have tested the putative involvement of cyclic nucleotides and prostaglandins in FeLV-related lymphorepression and have shown that only prostaglandins of the E series depress lymphagic blastogenesis alone or in conjunction with FeLV p15(E).

It is known that conA causes a rise in intracellular cyclic AMP levels in stimulated lymphocytes. Incorporation of FeLV p15(E) into the lymphocyte plasma membrane may interfere with the Ca^{+} transduction of the membrane signal which leads to the activation of adenylate cyclase.[133] Treatment of lymphocytes with FeLV p15(E) has no effect on cellular cyclic GMP but does inhibit cyclic AMP accumulation in the presence of mitogen. Indirect activation of adenylate cyclase by forskolin reverses FeLV suppression of lymphocyte function by raising intracellular cyclic AMP to normal levels. From this it is likely that the initial action of p15(E) at the cell membrane is to block the activation of adenylate cyclase. Failure to generate the second messenger cyclic AMP leads to failure of transmission of the message to proliferate or to undergo capping in response to the lectin signal.

F. Immunoprophylaxis

The basic virology of feline leukemia and the pathogenesis of the feline leukemia disease is well characterized and it is reasonable to assume that an appropriate vaccine to block this disease is possible. As discussed earlier, feline leukemia virus is the confirmed etiologic agent of feline leukemia disease and that this rough species of virus is responsible for the acquired immunodeficiency syndrome, lymphosarcoma, leukemia, thymic lymphoma, fibrosarcoma, nonregenerative anemia, and fetal absorption in cats. Not only are all these diseases thought to be induced by a single species of feline leukemia virus, but the neoantigen (FOCMA) is also common to all feline retrovirus diseases. Though the serotype of feline retrovirus may be associated with more than one particular form of feline leukemia disease, the feline retrovirus envelope antigen that elicits virus-neutralizing antibody has been more characterized. Most important, it is known that the pathogenesis of the feline leukemia virus diseases is initiated with the productive virus infection as a result of horizontal transmission of the feline leukemia agent. The ensuing disease culminates in the formation of the neoplastic disease.

The immune mechanisms that have been demonstrated to play a role in the protection from these diseases include the production of virus-neutralizing antibody to subgroup specific components that prevent viremia of the virus-envelope, and the induction of antibody towards FOCMA. Antibody to FOCMA has been demonstrated to bestow resistance to feline leukemia infected diseases. Thus, resistance to and recovery from feline leukemia disease, depends upon the development of at least two separate immunologic responses: one towards the infecting virus and one towards the surface antigen of the neoplastic cell. Because of the FeLV-gp70 (71,000 dalton glycoprotein), a feline leukemia virus is antigenically distinct from FOCMA. A feline vaccine would theoretically contain one or both of these components. Earlier attempts to develop a feline leukemia vaccine focused on the attempt to induce antiviral immunity as a means of prevention of feline leukemia disease. Vaccines composed of an activated feline leukemia virus-induced adequate virus neutralizing antibody in adult cats and, in fact, kittens born to dams immunized with said virus were apparently protected from disease. By contrast, kittens failed to produce significant virus neutralizing antibody to feline leukemia and these studies show that kittens less than 1 month of age lacked a vigorous immune response to feline leukemia and, in fact, were more susceptible to disease than nonimmunized cats. The relevance of these findings is still to be determined.

These studies and others suggest that conceivably the cat lacks the genetic determinant for complete response to many of the epitopes associated with FeLV-gp70; furthermore, it can be questioned whether the immune response to the glycosylated proteins with the feline leukemia envelope played the dominant role in protection against disease.

Recently, a subunit vaccine to feline leukemia has been developed and has been shown to be efficacious against FeLV disease. Investigators at Ohio State University developed a method of recovering FOCMA and other immunogens from spent-cell culture media from cells persistently infected with FeLV. It was found that actively growing FeLV infected cells when placed in serum free media maintain high cellular viability up to 96 hr of incubation. Moreover, 75% of these cells express FOCMA. Using a FOCMA-specific cytotoxicity inhibition assay and immunoblotting techniques, soluble FOCMA was detected in serum free media from cells grown and maintained in this incomplete media. The release of FOCMA and virion antigens from FOCMA-positive cells is not unexpected. It was found that synchronized cells expressed maximum FOCMA during the G1S phase of cell cycle. The amount of FOCMA diminished as the cell cycle passed through the S and G2 phases. It was logical to assume that membrane FOCMA and FeLV antigens were released into the media as the cells proceeded to the cell cycle.

Evaluation of soluble FOCMA and virion antigens as a vaccine demonstrated them to be a potent immunogen and cats immunized with these soluble factors demonstrated nearly complete protection against FeLV disease.

Western blot analysis demonstrated that all envelope and gag proteins, as well as a FOCMA-like substance were found in the soluble tumor vaccine, the difference being that many of these components were found in high molecular weight, and it's speculated that FOCMA and envelope and gag proteins may be associated with cell membrane components that influence their immunogenicity. It's interesting to speculate that conceivably these factors could be associated with Class I and Class II histocompatibility antigens which may influence both their immunogenicity as well as influencing the cell-mediated cytotoxicity response.

G. FeLV as a Model for Human Disease

FeLV is a unique retrovirus in that it is a horizontally transmitted infectious agent which infects an outbred population of cats. Recently, a family of T-lymphotropic human retroviruses designated human T cell leukemia/lymphoma viruses (HTLV) has been isolated from several lymphoproliferative diseases. HTLV-I virus originally was isolated from two patients with cutaneous T-cell lymphoma and leukemia (CTLC) after initiation of T-cell lines and recovery of the virus from the cell lines.[118,134] Subsequently, a new subgroup, designated HTLV-II virus, was isolated from cultured T-cells of a hairy cell leukemia[135] and recently a third subgroup, designated HTLV-III, was isolated from patients with acquired immunodeficiency syndrome (AIDS).[136,137] Both FeLV and HTLV viruses are T lymphotropic and in the human, the OKT4$^+$ subset ("helper") is the target cell.

Immunodeficiency in many FeLV infected cats is characterized by lymphopenia and neutropenia, cutaneous anergy, impaired macrophage function, impaired blastogenic responses to mitogens and antigens, and impaired humoral antibody responses.[93,102,104] Clinically infected cats may present with lymphadenopathy, pneumonias, gingivitis, skin sores, and susceptibility to viral diseases. Similarly, human AIDS patients also present with lymphadenopathy, particularly in the pre-AIDS period. The disease is manifested by opportunistic infection, predominantly *pneumocystis carinii* pneumonia, in addition to toxoplasmosis, candida and cryptococcus infections, and Kaposi's sarcoma.[136] Intercurrent viral infections are also found in AIDS. Thus, there are striking similarities between FeLV and the HTLV family of viruses.

REFERENCES

1. Essex, M., Feline leukemia and sarcoma viruses, in *Viral Oncology*, Klein, G., Ed., Raven Press, New York, 1980.
2. Essex, M., Grant, C. K., Cotter, S. M., Sliski, A. H., and Hardy, W. D., Jr., Leukemia specific antigens: FOCMA and immune surveillance, in *Modern Trends in Leukemia III*, Neth, R., Hofschneider, R., and Manweiler, K., Eds., Springer-Verlag, Basel, 1979, 453.

3. Essex, M., Cotter, S. M., Hardy, W. D., Jr., Hess, P., Jarrett, W., Jarrett, O., Mackey, L., Laird, H., Perryman, L., Olsen, R. G., and Yohn, D. S., Feline oncornavirus-associated cell membrane antigen. IV. Antibody titers in cats with naturally occurring leukemia, lymphoma and other diseases, *J. Natl. Cancer Inst.*, 55, 463, 1975.
4. Francis, D. P., Cotter, S. M., Hardy, W. D., Jr., and Essex, M., Feline leukemia and lymphoma: comparison of virus positive and virus negative cases, *Cancer Res.*, 39, 3866, 1979.
5. Essex, M., Klein, G., Snyder, S. P., and Harrold, J. B., Antibody to feline-oncornavirus-associated cell membrane antigen in neonatal cats, *Int. J. Cancer*, 8, 384, 1971.
6. Essex, M., Klein, G., Snyder, S. P., and Harrold, J. B., Feline sarcoma virus-induced tumors: correlation between humoral antibody and tumor regression, *Nature (London)*, 233, 195, 1971.
7. Fenner, F., The classification and nomenclature of viruses: the current position, *ASM News*, 42, 170, 1976.
8. Graves, D. C. and Velicer, L. F., Properties of feline leukemia virus. I. Chromatographic separation and analysis of the polypeptides, *J. Virol.*, 14, 349, 1974.
9. Schafer, W. and Bolognesi, D. P., Mammalian C-type oncornaviruses: relationships between viral structure and cell-surface antigens and their possible significance in immunological defence mechanisms, in *Contemporary Topics in Immunobiology*, Vol. 6, Plenum Press, New York, 1977, 127.
10. Scolnik, E., Rands, E., Aaronson, S. A., and Todaro, G., RNA-dependent DNA activity in five RNA viruses: divalent cation requirements, *Proc. Natl. Acad. Sci. U.S.A.*, 67, 1789, 1970.
11. Bolognesi, D. P., Montelaro, R. C., Frank, H., and Schafer, W., Assembly of type C oncornaviruses: a model, *Science*, 199, 183, 1978.
12. Litvak, S. and Araya, A., Primer transfer RNA in retroviruses, *Trends Biochem. Sci.*, 7, 361, 1982.
13. Varmus, H. E., Form and function of retroviral proviruses, *Science*, 216, 812, 1982.
14. Schafer, W. and Bolognesi, D. P., Mammalian C-type oncornaviruses: relationships between viral structure and cell surface antigens and their possible significance in immunological defense mechanisms, *Contemp. Top. Immunobiol.*, 6, 127, 1977.
15. Okasinki, G. F. and Velicer, L. F., Analysis of intracellular feline leukemia virus proteins. II. The generation of FeLV structural proteins from precursor polypeptides, *J. Virol.*, 22, 74, 1977.
16. Pinter, A. and Fleissner, E., Structural studies of retroviruses: characterization of oligomeric complexes of murine and feline leukemia virus envelope and core components formed upon cross-linking, *J. Virol.*, 30, 157, 1979.
17. Fischinger, P. J., Blevins, C. S., and Nomura, S., Simple quantitative assay for both xenotropic murine leukemia and ecotropic feline leukemia viruses, *J. Virol.*, 14, 177, 1974.
18. Hoover, E. A., Olsen, R. G., Mathes, L. E., and Schaller, J. P., Relationship between feline leukemia virus antigen expression in blood and bone marrow and viral infectivity in blood, bone marrow, and saliva of cats, *Cancer Res.*, 37, 3707, 1977.
19. Saxinger, C., Essex, M., Hardy, W., and Gallo, R., Detection of antigen related to feline leukemia virus in the sera of "virus-negative" cats, *Dev. Cancer Res. (Feline Leukemia Virus)*, 4, 489, 1980.
20. Rice, J. B. and Olsen, R. G., Feline oncovirus-associated cell membrane antigen and feline leukemia virus group specific antigen expression in bone marrow and serum, *J. Natl. Cancer Inst.*, 66, 737, 1981.
21. Lutz, H., Pedersen, N., Higgins, J., Hubscher, U., Troy, F. A., and Theilen, G. H., Humoral immune reactivity to feline leukemia virus and associated antigens in cats naturally infected with feline leukemia virus, *Cancer Res.*, 40, 3642, 1980.
22. Sarma, P. S. and Log, T., Subgroup classification of feline leukemia and sarcoma viruses by viral interference and neutralization tests, *Virology*, 54, 160, 1973.
23. Schaller, J. P. and Olsen, R. G., Determination of subgroup-specific feline oncorna virus neutralizing antibody, *Infect. Immun.*, 12, 1405, 1975.
24. Hardy, W. D., Jr., Hess, P. W., MacEwen, E. G., McClelland, A. J., Zuckerman, E. E., Essex, M., Cotter, S. M., and Jarrett, O., Biology of feline leukemia virus in the natural environment, *Cancer Res.*, 36, 582, 1976.
25. Jarrett, O. and Russell, P. H., Differential growth and transmission in cats of feline leukemia viruses of subgroups A and B, *Int. J. Cancer*, 21, 466, 1978.
26. Russell, P. H. and Jarrett, O., The specificity of neutralizing antibodies to feline leukemia viruses, *Intl. J. Cancer*, 21, 768, 1978.
27. Mathes, L. E., Olsen, R. G., Hebebrand, L. C., Hoover, E. A., and Schaller, J. P., Abrogation of lymphocyte blastogenesis by a feline leukemia virus protein, *Nature (London)*, 274, 687, 1978.
28. Bartholomew, R. M. and Esser, A. F., Differences in activation of human and guinea pig complement by retroviruses, *J. Immunol.*, 121, 1748, 1978.
29. Kobilinsky, L., Hardy, W. D., Jr., Ellis, R., and Day, N. K., Activation of feline complement by feline leukemia virus, *Fed. Proc.*, 38, 1089, 1979.

30. Essex, M. and Snyder, S. P., Feline oncornavirus-associated cell membrane antigen. I. Serological studies with kittens exposed to cell-free materials from various feline fibrosarcoma, *J. Natl. Cancer Inst.*, 51, 1007, 1973.
31. Essex, M., Stephenson, J. R., Hardy, W. D., Jr., Cotter, S. M., and Aaronson, S. A., Leukemia, lymphoma and fibrosarcoma of cats as models for similar diseases of man, *Cold Spring Harbor Symp. Cell Proliferation*, 4, 1197, 1977.
32. Essex, M., Jakowski, R. M., Hardy, W. D., Jr., Cotter, S. M., Hess, P., and Sliski, A., Feline oncornavirus-associated cell membrane antigen. III. Antibody titers in cats from leukemia cluster households, *J. Natl. Cancer Inst.*, 54, 637, 1975.
33. Schaller, J. P., Hoover, E. A., and Olsen, R. G., Active and passive immunization of cats with inactivated feline oncornaviruses, *J. Natl. Cancer Inst.*, 59, 1441, 1977.
34. Olsen, R. G., Hoover, E. A., Mathes, L. E., Heding, L., and Schaller, J. P., Immunization against feline oncornavirus disease using a killed tumor cell vaccine, *Cancer Res.*, 36, 3642, 1976.
35. Jarrett, W., Jarrett, O., Mackey, L., Laird, H., Hood, C., and Hay, D., Vaccination against feline leukemia virus using a cell membrane antigen system, *Int. J. Cancer*, 16, 134, 1975.
36. Essex, M., Sliski, A., Cotter, S. M., Jakowski, R. M., and Hardy, W. D., Jr., Immunosurveillance of naturally occurring feline leukemia, *Science*, 190, 790, 1975.
37. Charman, H. P., Kim, N., Gilden, R., Hardy, W. D., Jr., and Essex, M., Humoral immune responses of cats to feline leukemia virus: comparison of responses to the major structural protein, p30, and to a virus specific cell membrane antigen (FOCMA), *J. Natl. Cancer Inst.*, 56, 859, 1976.
38. Essex, M., Sliski, A., Hardy, W. D., Jr., and Cotter, S. M., The immune response to leukemia virus and tumor associated antigens in cats, *Cancer Res.*, 36, 640, 1976.
39. Stephenson, J. R., Khan, A. S., Sliski, A. H., and Essex, M., Feline oncornavirus-associated cell membrane antigen (FOCMA): identification of an immunologically cross-reactive feline sarcoma virus coded protein, *Proc. Natl. Acad. Sci.*, 74, 5608, 1977.
40. Francis, D. P., Essex, M., and Hardy, W. D., Jr., Excretion of feline leukemia virus by naturally infected pet cats, *Nature (London)*, 269, 252, 1977.
41. Hardy, W. D., Jr., McClelland, A. J., Zuckerman, E. E., Hess, P. W., Essex, M., Cotter, S. J., MacEwen, E. G., and Hayes, A. A., Prevention of the contagious spread of the feline leukemia virus and the development of leukemia in pet cats, *Nature (London)*, 263, 326, 1976.
42. Hardy, W. D., Jr., Hess, P. W., MacEwen, E. G., McClelland, A. J., Zuckerman, E. E., Essex, M., and Cotter, S. M., The biology of feline leukemia virus in the natural environment, *Cancer Res.*, 36, 582, 1976.
43. Grant, C. K. and Essex, M., Immunity to feline oncornavirus-induced tumors, in *Cell Mediated Immunity to Virus-Induced Tumors*, Blasecki, J., Ed., Marcel Dekker, New York, in press.
44. Mathes, L. E., Yohn, D. S., Hoover, E. A., Essex, M., Schaller, J. P., and Olsen, R. G., Feline oncornavirus-associated cell membrane antigen, *J. Natl. Cancer Inst.*, 56, 1197, 1976.
45. Stephenson, J. R., Essex, M., Hino, S., Aaronson, S. A., and Hardy, W. D., Jr., Feline oncornavirus-associated cell membrane antigen. VII. Relationship between FOCMA and virion glycoprotein gp70, *Proc. Natl. Acad. Sci. U.S.A.*, 74, 1219, 1977.
46. Essex, M., Cotter, S. M., Sliski, A. H., Hardy, W. D., Jr., Stephenson, J. R., Aaronson, S., and Jarrett, O., Horizontal transmission of feline leukemia virus under natural conditions in a feline leukemia cluster household, *Int. J. Cancer*, 19, 90, 1977.
47. Jarrett, W. F. H., Jarrett, O., Mackey, L., Laird, H. M., Hardy, W. D., Jr., and Essex, M., Horizontal transmission of leukemia virus and leukemia in the cat, *J. Natl. Cancer Inst.*, 51, 833, 1973.
48. Schaller, J., Essex, M., Olsen, R. G., and Yohn, D. S., Feline oncornavirus-associated cell membrane antigen. V. Humoral immune response to virus and cell membrane antigens in cats injected with Gardner-Arnstein feline sarcoma virus, *J. Natl. Cancer Inst.*, 55, 1373, 1975.
49. Sliski, A. H., Essex, M., Meyer, C., and Todaro, G. J., Feline oncornavirus-associated cell membrane antigen (FOCMA): expression on feline sarcoma virus transformed non-producer mink cells, *Science*, 196, 1336, 1977.
50. Essex, M., Klein, G., Snyder, S. P., and Harrold, J. B., Antibody to feline oncornavirus-associated cell membrane antigen in neonatal cats, *Int. J. Cancer*, 8, 384, 1971.
51. Essex, M., Klein, G., Snyder, S. P., and Harrold, J. B., Feline sarcoma virus-induced tumors: correlation between humoral antibody and tumor regression, *Nature (London)*, 233, 195, 1971.
52. Hardy, W. D., Jr., Zuckerman, E. E., MacEwen, E. G., Hayes, A. A., and Essex, M., A feline leukemia and sarcoma virus-induced tumor specific antigen, *Nature (London)*, 270, 249, 1977.
53. Hardy, W. D., Jr., McClelland, A. J., Zuckerman, E. E., Snyder, H. W., Jr., MacEwen, E. G., Francis, D., and Essex, M., Development of virus non-producer lymphosarcomas in pet cats exposed to FeLV, *Nature (London)*, 288, 90, 1980.
54. Vedbrat, S. S., Rasheed, S., Lutz, H., Gonda, M. A., Ruscetti, S., Gardner, M. B., and Prensky, W., Feline oncornavirus-associated cell membrane antigen: a viral and not a cellularly coded transformation-specific antigen of cat lymphomas, *Virology*, 124, 445, 1983.

55. Hoover, E. A., McCullough, B., and Griesemer, R. A., Intranasal transmission of feline leukemia, *J. Natl. Cancer Inst.*, 59, 973, 1972.
56. Hoover, E. A., Perryman, L. E., and Kociba, G. J., Early lesions in cats inoculated with feline leukemia virus, *Cancer Res.*, 33, 145, 1973.
57. Jarrett, W., Anderson, L. J., Jarrett, O., Laird, H. M., and Stewart, M. F., Myeloid leukaemia in a cat produced experimentally by feline leukaemia virus, *Res. Vet. Sci.*, 12, 385, 1971.
58. Onions, D., Testa, N., and Jarrett, O., Growth of FeLV in haemopoietic cells *in vitro*, *Dev. Cancer Res.*, 4, 513, 1980.
59. Elder, J. and Mullins, J. I., Nucleotide sequence of the envelope gene of Gardner-Arnstein feline leukemia virus B reveals unique sequence homologies with a murine MCF virus, *J. Virol.*, 46, 871, 1983.
60. Hoover, E. A., Kociba, G. J., Hardy, W. D., and Yohn, D. S., Erythroid hyperplasia in cats inoculated with feline leukemia virus, *J. Natl. Cancer Inst.*, 53, 1271, 1974.
61. Mackey, L. J., Jarrett, W., Jarrett, O., and Laird, H., Anemia associated with feline leukemia virus infection in cats, *J. Natl. Cancer Inst.*, 54, 1, 1975.
62. Onions, D., Jarrett, O., Testa, N., Frassoni, F., and Toth, S., Selective effect of feline leukemia virus on early erythroid precursors, *Nature (London)*, 296, 156, 1982.
63. Francis, D. P., Essex, M., and Hardy, W. D., Jr., Excretion of feline leukemia virus by naturally infected pet cats, *Nature (London)*, 269, 252, 1977.
64. Hardy, W. D., Jr., Old, L. J., Hess, P. W., Essex, M., and Cotter, S. M., Horizontal transmission of feline leukemia virus, *Nature (London)*, 244, 266, 1973.
65. Cotter, S. M., Hardy, W. D., Jr., and Essex, M., Association of feline leukemia virus with lymphosarcoma and other disorders in the cat, *J. Am. Vet. Med. Assoc.*, 166, 449, 1975.
66. Hoover, E. A., Rojko, J. L., and Quackenbush, S. L., Congenital feline leukemia virus infection, *Comp. Leuk. Res.*, 11, 7, 1983.
67. Rojko, J. L., Hoover, E. A., Mathees, L. E., Schaller, J. P., and Olsen, R. G., Pathogenesis of experimental feline leukemia virus infection, *J. Natl. Cancer Inst.*, 63, 759, 1979.
68. Welsh, R. M., Cooper, N. R., Jensen, F. C., and Oldstone, M. B. A., Human serum lyses RNA-tumor viruses, *Nature (London)*, 257, 612, 1975.
69. Sherwin, S., Benveniste, R., and Todaro, G., Complement-mediated lysis of type C viruses: effect of primate and human sera on various retroviruses, *Intl. J. Cancer*, 21, 6, 1978.
70. de Noronha, F., Schafer, W., Essex, M., and Bolognesi, D. P., Influence of antisera to oncornavirus glycoprotein (gp71) on infections of cats with feline leukemia virus, *Virology*, 85, 617, 1978.
71. Schwarz, H., Fischinger, P. J., Ihle, J. N., Thiel, J. J., Weiland, F., Bolognesi, D. P., and Schafer, W., Properties of mouse leukemia viruses. XVI. Suppression of spontaneous fatal leukemias in AKR mice by treatment with broadly reacting antibody against the viral glycoprotein gp71, *Virology*, 93, 159, 1979.
72. Gupta, P. and Ferrer, J., Expression of bovine leukemia virus genome is blocked by a nonimmunoglobulin protein in plasma from infected cattle, *Science*, 215, 405, 1982.
73. Hanna, M. G., Jr., Szakal, A. K., and Tyndall, R. L., Histoproliferative effect of Rauscher leukemia virus on lymphatic tissue: histological and ultrastructural studies of germinal centers and their relation to leukemogenesis, *Cancer Res.*, 30, 1748, 1970.
74. Ruddle, N. H., Armstrong, M. Y. K., and Richards, F. F., Replication of murine leukemia virus in bone marrow derived lymphocytes, *Proc. Natl. Acad. Sci. U.S.A.*, 73, 3714, 1976.
75. Isaak, D. D., Price, J. A., Reinish, C. L., and Cerny, J., Target cell heterogeneity in murine leukemia virus infection. I. Differences in susceptibility to infection with Friend leukemia virus between B lymphocytes from spleen, bone marrow and lymph node, *J. Immunol.*, 123, 1822, 1979.
76. Hardy, W. D., Jr., Hirshaut, Y., and Hess, P. W., Detection of the feline leukemia virus and other mammalian oncornaviruses by immunofluorescence, *Bibl. Haematol.*, 39, 778, 1973.
77. Cockerell, G. L., Hoover, E. A., and Krakowka, S., Lymphocyte mitogen reactivity and enumeration of circulating B- and T-cells during feline leukemia virus infection in the cat, *J. Natl. Cancer Inst.*, 57, 1095, 1976.
78. Pedersen, N. C., Theilen, G., Keane, M. A., Fairbanks, L., Mason, T., Orser, B., Chen, C. H., and Allison, C., Studies of naturally transmitted feline leukemia virus infection, *Am. J. Vet. Res.*, 38, 1523, 1983.
79. Rojko, J. L., Hoover, E. A., Quackenbush, S. L., and Olsen, R. G., Reactivation of latent feline leukaemia virus infection, *Nature (London)*, 298, 385, 1982.
80. Post, J. E. and Warren, L., Reactivation of latent feline leukemia virus, *Dev. Cancer Res. (Feline Leukemia Virus)*, 151, 1980.
81. Madewell, B. L. and Jarrett, O., Recovery of feline leukaemia virus from non-viraemic cats, *Vet. Rec.*, 112, 339, 1983.
82. Rouse, B. T. and Babiak, L. A., Mechanisms of recovery from herpesvirus infections — a review, *Can. J. Comp. Med.*, 42, 414, 1978.

83. Profitt, M. R., Ed., *Virus Lymphocyte Interactions: Implications for Disease*, Elsevier/North Holland, Amsterdam, 1979, 1.
84. Huddlestone, J. R. Lampert, P. W., and Oldstone, M. B. A., Virus-lymphocyte interactions: infection of T_G amd T_M subsets by measles virus, *Clin. Immunol. Immunopathol.*, 15, 502, 1980.
85. Epstein, M. A. and Achong, B. G., Recent progress in Epstein-Barr virus research, *Ann. Rev. Microbiol.*, 31, 421, 1977.
86. Rojko, J. L., Hoover, E. A., Finn, B. L., and Olsen, R. G., Determinants of susceptibility and resistance to feline leukemia virus infection. II. Susceptibility of feline lymphocytes to productive feline leukemia infection, *J. Natl. Cancer Inst.*, 67, 899, 1981.
87. Hoover, E. A., Krakowka, S., Cockerell, G. L., Mathes, L. E., and Olsen, R. G., Influence of thymectomy on the susceptibility of cats to feline leukemia virus and lymphosarcoma, *Am. J. Vet. Res.*, 39, 393, 1978.
88. Cerny, J. and Stiller, R. A., Immunosuppression by spleen cells from Moloney leukemia. Comparison of the suppressive effect on antibody response and on mitogen-induced response, *J. Immunol.*, 115, 943, 1975.
89. Trainin, Z., Wernicke, D., Ungar-Waron, H., and Essex, M., Suppression of the humoral antibody response in natural retrovirus infections, *Science*, 220, 858, 1983.
90. Mathes, L. E. and Olsen, R. G., Immunobiology of feline leukemia virus disease, in *Feline Leukemia*, Olsen, R. G., Ed., CRC Press, Boca Raton, Fla., 1981, 77.
91. Gillette, R. W. and Fox, A., Changes with age in homing properties and mitogen responses of lymphocytes from normal and leukemia prone mice, *Cell Immunol.*, 51, 32, 1980.
92. Epstein, R. S., Lopez, D. M., and Sigel, M. M., Divergent appearance of complement receptors and Fc receptors on T cells in a murine mammary tumor system, *Cell Immunol.*, 40, 154, 1978.
93. Hoover, E. A., Rojko, J. L., Wilson, P. L., and Olsen, R. G., Determinants of susceptibility and resistance to feline leukemia virus. I. Role of macrophages, *J. Natl. Cancer Inst.*, 67, 889, 1981.
94. Marcelletti, J. and Furmanski, P., Infection of macrophages with Friend virus: relationship to the spontaneous regression of viral erythroleukemia, *Cell*, 16, 649, 1979.
95. Mogensen, S. C., Role of macrophages in natural resistance to virus infections, *Microbiol. Rev.*, 43, 1, 1979.
96. Daughaday, C. C., Brandt, W. E., McCown, J. M., and Russell, P. K., Evidence for two mechanisms of dengue virus infection of adherent human monocytes: trypsin-sensitive virus receptors and trypsin-resistant immune complex receptors, *Infect. Immun.*, 32, 469, 1981.
97. Peiris, J. S. M., Gordon, S., Unkeless, J. C., and Porterfield, J. S., Monoclonal anti-Fc receptor IgG blocks antibody enhancement of viral replication in macrophages, *Nature (London)*, 289, 189, 1981.
98. Schlesinger, J. J. and Brandriss, M. W., Antibody-mediated infection of macrophages and macrophage-like cell lines with 17D-yellow fever virus, *J. Med. Virol.*, 8, 103, 1981.
99. Hoover, E. A., Olsen, R. G., Hardy, W. D., Jr., Schaller, J. P., and Mathes, L. E., Feline leukemia virus infection: age-related variation in response of cats to experimental infection, *J. Natl. Cancer Inst.*, 57, 365, 1976.
100. Rojko, J. L., Hoover, E. A., Mathes, L. E., Krakowka, S., and Olsen, R. G., Influence of adrenal corticosteroids on the susceptibility of cats to feline leukemia virus infection, *Cancer Res.*, 39, 3789, 1979.
101. Cockerell, G. L., Krakowka, S., and Hoover, E. A., Characterization of feline T- and B-lymphocytes and identification of an experimentally induced T-cell neoplasm in the cat, *J. Natl. Cancer Inst.*, 57, 907, 1976.
102. Cockerell, G. L. and Hoover, E. A., Inhibition of normal lymphocyte mitogenic reactivity by serum from feline leukemia virus infected cats, *Cancer Res.*, 37, 3985, 1976.
103. Stiff, M. I. and Olsen, R. G., Loss of the short-lived suppressive function of peripheral-infected cats, *J. Clin. Lab. Immunol.*, 7, 133, 1982.
104. Perryman, L. E., Hoover, E. A. and Yohn, D. S., Immunologic reactivity of the cat: immunosuppression in experimental feline leukemia, *J. Natl. Cancer Inst.*, 49, 1397, 1972.
105. Lee, J. C. and Ihle, J. N., Mechanisms of C-type viral leukemogenesis. I. Correlation of *in vitro* lymphocyte blastogenesis to viremia and leukemia, *J. Immunol.*, 123, 2351, 1979.
106. Isaak, D. and Cerny, J., T and B lymphocyte susceptibility to murine leukemia virus moloney, *Infect. Immun.*, 40, 977, 1983.
107. Till, J. E., McCulloch, E. A., and Siminovitch, L., A stochastic model of stem proliferation based on the growth of spleen colony-forming cells, *Proc. Natl. Acad. Sci. U.S.A.*, 51, 29, 1980.
108. Mathes, L. E., Yohn, D. S., Hoover, E. A., Essex, M., Schaller, J. P., and Olsen, R. G., Feline oncornavirus-associated cell membrane antigen. VI. Cytotoxic antibody in cats exposed to feline leukemia virus, *J. Natl. Cancer Inst.*, 56, 1197, 1976.

109. Grant, C. K., DeBoer, D. J., Essex, M., Worley, M. B., and Higgins, J., Antibodies from healthy cats exposed to feline leukemia virus lyse feline lymphoma cells slowly with cat complement, *J. Immunol.*, 119, 401, 1977.
110. Grant, C. K., de Noronha, F., Tusch, C., Michalek, M. T., and McLane, M. F., Protection of cats against progressive fibrosarcomas and persistent leukemia virus infection by vaccination with feline leukemia cells, *J. Natl. Cancer Inst.*, 65, 1285, 1980.
111. Grant, C. K., Harris, D., Essex, M. E., Pickard, D. K., Hardy, W. D., Jr., and de Noronha, F., Protection of cats against feline leukemia virus-positive and virus-negative tumors by complement dependent antibody, *J. Natl. Cancer Inst.*, 64, 1527, 1980.
112. Grant, C. K. and Michalek, M. T., Feline leukemia — unique and cross-reacting antigens on individual virus-producing tumors identified by complement-dependent antibody, *Int. J. Cancer*, 28, 209, 1981.
113. Cowan, K. M., Antibody response to viral antigens, *Adv. in Immunol.*, 17, 195, 1973.
114. Hoover, E. A., Schaller, J. P., Mathes, L. E., and Olsen, R. G., Passive immunity to feline leukemia: evaluation of immunity from dams naturally infected and experimentally vaccinated, *Infect. Immun.*, 16, 54, 1977.
115. Cotter, S. M., Essex, M., McLane, M. F., Grant, C. K., and Hardy, W. D. Jr., Chemotherapy and passive immunotherapy in naturally occurring feline mediastinal lymphoma, *Dev. Cancer Res.*, 4, 219, 1980.
116. Rojko, J. L., Hoover, E. A., Finn, B. L., and Olsen, R. G., Characterization and mitogenesis of feline lymphocyte populations, *Int. Arch. Allergy Appl. Immunol.*, 68, 226, 1982.
117. Miyoshi, I., Kubonishi, S., Yoshimoto, S., Akagi, T., Ohtsuki, Y., Shiraishi, Y., Nagatak, K., and Hinuma, Y., Type C virus particles in a cord T-cell line derived by co-cultivating normal human cord leukocytes and human leukaemic T-cells, *Nature (London)*, 394, 770, 1981.
118. Poiesz, B., Ruscetti, F. W., Reitz, M. S., et al., Isolation of a new type C retrovirus (HTLV) in primary uncultured cells of a patient with Sezary T-cell leukemia, *Nature (London)*, 293, 268, 1981.
119. Popovic, M., Kalyanaraman, V. S., Sarngadharan, M. G., Robert-Guroff, M., Nakgo, Y., Reitz, M. S., Miyoshi, Y., Ito, Y., Minowada, J., and Gallo, R. C., The virus of Japanese adult T-cell leukemia is a member of the human T-cell leukemia virus group, *Nature (London)*, 300, 63, 1982.
120. Tompkins, M. B. and Cummins, J. M., Response of feline leukemia virus-induced nonregenerative anemia to oral administration of an interferon-containing preparation, *Feline Pract.*, 12, 6, 1982.
121. Azocar, J. and Essex, M., Incorporation of HLA antigens into the envelope of RNA tumor viruses grown in human cells, *Cancer Res.*, 38, 3388, 1979.
122. Dubey, J. P., Staunton, D., Axocar, J., Stux, S., Essex, M., and Ynis, E. J., Natural killer activity against retrovirus infected cells: dichotomy in NK sensitivity of infected T & B cells, *Clin. Immunol. Immunopathol.*, 23, 215, 1982.
123. Jones, F. R., Yoshida, L. H., Ladiges, W. C., and Kenny, M. A., Treatment of feline leukemia and reversal of FeLV by *ex vivo* removal of IgG: a preliminary report, *Cancer*, 46, 675, 1980.
124. Snyder, H. W., Jones, F. R., Day, N. K., and Hardy, W. D., Jr., Isolation and characterization of circulating feline leukemia virus-immune complexes from plasma of persistently infected pet cats removed by *ex vivo* immunosorption, *J. Immunol.*, 124, 2726, 1982.
125. Anderson, L. J., Jarrett, W. F. H., Jarrett, O., and Laird, H. M., Feline leukemia virus infection of kittens: mortality associated with atrophy of the thymus and lymphoid depletion, *J. Natl. Cancer Inst.*, 47, 807, 1971.
126. Jakowski, R. M., Essex, M., Hardy, W. D., Jr., Stephenson, J. R., and Cotter, S. M., Membranous glomerulonephritis in a household cluster of cats persistently viremic with feline leukemia virus, *Dev. Cancer Res.*, 4, 141, 1980.
127. Hebebrand, L. C., Olsen, R. G., Mathes, L. E., and Nichols, W. S., Inhibition of a human lymphocyte mitogen and antigen response by a 15,000 dalton protein from feline leukemia virus, *Cancer Res.*, 39, 443, 1979.
128. Stiff, M. I. and Olsen, R. G., Effects of retrovirus protein on the feline one way mixed leukocyte reaction, *J. Gen. Virol.*, 64, 957, 1983.
129. Dunlap, J. E., Nichols, W. S., Hebebrand, L. C., Mathes, L. E., and Olsen, R. G., Mobility of lymphocyte surface membrane concanavalin A receptors of normal and feline leukemia virus-infected viremic felines, *Cancer Res.*, 39, 956, 1979.
130. Lewis, M. G. and Olsen, R. G., *Nature (London)*, submitted.
131. Copelan, E., Rinehart, J. J., Lewis, M. G., Olsen, R. G., and Sagone, A., The mechanism of feline leukemia virus suppression of human T cell proliferation *in vitro*, *J. Immunol.*, in press.
132. Wainberg, M. A., Vydelingum, S., and Margolese, R. G., Viral inhibition of lymphocyte mitogenesis: interference with the synthesis of functionally active T-cell growth factor (TCGF) activity and reversal of inhibition by the addition of same, *J. Immunol.*, 130, 2372, 1983.
133. Lewis, M. G., Fertel, R. H., and Olsen, R. G., *Nature (London)*, submitted.

134. Poiesz, B. J., Ruscetti, F., Gazdar, A., et al., Detection and isolation of type C retrovirus particles from fresh and cultured lymphocytes of a patient with cutaneous T-cell lymphoma, *Proc. Natl. Acad. Sci. U.S.A.*, 77, 7415, 1980.
135. Kalyanaraman, V., Sarngadharan, M., Robert-Guroff, M., et al., A new subtype of human t-cell leukemia virus (HTLV-II) associated with a T-cell variant of hairy cell leukemia, *Science*, 218, 571, 1982.
136. Popovic, M., Sarngadharan, M., Read, E., Gallo, R. C., Detection, isolation and continuous production of cytopathic retroviruses (HTLV-III) from patients with AIDS and pre-AIDS, *Science*, 224, 497, 1984.
137. Gallo, R. C., Salhuddin, S., Popovic, M., et al., Frequent detection and isolation of cytopathic retroviruses (HTLV-III) from patients with AIDS and at risk for AIDS, *Science*, 224, 500, 1984.

Chapter 2

BOVINE LYMPHOMA

Robert M. Jacobs

TABLE OF CONTENTS

I. INTRODUCTION

Lymphoma is a malignant lymphoproliferative neoplasm. It is characterized by the enlargement of lymphoid organs. The tumors are recognizable as solid tissue masses with effacement of normal architecture. This defines the disease as it occurs in cattle. True leukemia, indicative of primary bone marrow involvement, is an uncommon event in lymphoid neoplasia of cattle. For this reason the description, bovine leukemia, is inappropriate. Bovine lymphoma or lymphosarcoma are the preferred terms and are used interchangeably. When neoplastic cells appear in the circulation coexistent with a solid lymphoid tumor of the same cell type, the descriptive term is lymphosarcoma cell leukemia, indicating that secondary metastasis to the bone marrow has occurred. In bovine lymphoma this is usually a terminal event. The term leukosis refers to lymphoma as well as the benign lymphoproliferative response, persistent lymphocytosis (PL), and occult infection with the bovine leukemia virus (BLV). Since leukosis has several meanings, it should not be used to describe groups of cattle in experimental studies. Such use has contributed to much confusion in the literature regarding bovine lymphoma.

The objective of this chapter is to provide the reader with a succinct review of the current literature regarding bovine lymphoma. Historically important events are mentioned, but the reader must rely on the cited sources for comprehensive background information.

II. EPIZOOTIOLOGY OF LYMPHOMA

Lymphoma is the second most common malignancy in cattle, but it is the most frequent hemolymphatic neoplasm. The most frequent neoplasm is the orbital squamous cell carcinoma.[1] The frequency of bovine lymphoma has been difficult to determine, and this is reflected in the variability of incidence rates. Problems that make this determination difficult are the reporting efficiency on a national scale and accuracy of gross diagnoses by the inspectors at federally inspected slaughter plants, the nonuniform geographic distribution, and the fact that an ill-defined proportion of cattle with lymphoma never make it to slaughter. Reported incidence rates in the U.S. have been about 20 per 100,000 cattle slaughtered.[2,3] Recently the rate was estimated to be approximately 84 per 100,000 cattle slaughtered per year.[4] This latter figure is the mean value determined for the years 1959 through 1978 and was derived from beef and dairy cattle slaughtered at federally inspected plants. The authors suggested that only 50% of the affected cattle reach these plants and concluded that the real incidence rate is about 170 per 100,000 head slaughtered. The validity of this estimate is uncertain.

The age distribution of cattle with lymphoma is bimodal. There is an early peak, under 1 year of age, and a second peak between 5 and 8 years of age.[5,6] The incidence of lymphoma increases directly with age. Herds of more than 50 cattle have a higher frequency of lymphoma compared with those with less than 50.[7,8] A genetic predisposition to the development of lymphoma is evident. In a closed herd of Jersey cattle where there was a high incidence of BLV infection, cases of lymphoma clearly aggregated along familial lines.[9] This study substantiated a previous one in which a closed herd of Jersey cattle was followed for approximately 20 years. In this earlier study 7.9% of the herd died of lymphoma. When those cattle over 6 years of age were treated as a cohort, a striking 21.6% died of lymphoma. Cases of lymphoma clustered along sire groups and cow families.[10] There are data suggesting that the Guernsey and Brown Swiss breeds are more resistant, and grade cattle with Holstein-Friesian, Jersey, or Milking Shorthorn characteristics are more susceptible to the development of lymphoma. However, a breed predisposition has not been proven since the effects of man-

Table 1
DATE OF ADMISSION LABORATORY DATA IN ADULT BOVINE LYMPHOMA

	Group 1 (n = 33)	Group 2 (n = 38)
Age	5.84 (2—11)[a]	5.53 (2—10)
PCV	27 (15—37)	28 (10—38)
WBC	23,326 (3,300—146,000)	14,992 (3,900—37,700)
Segs	4,084 (0—13,986)	5,538 (192—12,2,85)
Lymphs	20,732 (1,350—146,000)	8,427 (1,768—31,668)

[a] Mean (range).

agement practices and genetic factors are difficult to separate.[7,11] Extensive epidemiologic surveys in Europe and North America have shown that cases of lymphoma in adult cattle aggregate in space and time in addition to the heredofamilial association.[5,12-15] When cattle from high incidence areas were introduced to low incidence areas, a gradual increase in occurrence resulted. Cattle inoculated with a whole blood piroplasmosis vaccine, produced in calves from herds with multiple cases of lymphoma, developed lymphoma and PL at an increased rate.[16-18] The epizootiologic evidence suggested an infectious agent, and the disease was termed enzootic leukosis. BLV was later detected in short-term cultures of peripheral blood lymphocytes from cattle with the adult form of lymphoma and is now considered the etiologic agent.[19] Other forms of the disease occurred randomly, at a very low rate, throughout the cattle population. These were termed the sporadic leukoses. They are generally seen in immature cattle and are unassociated with BLV. In addition to their sporadic occurrence in the cattle population, the results of molecular hybridization and serological studies suggest a separate etiology.[20-22] It has been hypothesized that an, as yet unidentified, bovine sarcoma virus requiring a BLV helper virus may be the cause of the sporadic forms.[23] BLV has been recovered from a cell culture originating from a case of lymphoma in a calf by treatment of the cell culture with 5′iodo-2′deoxyuridine.[24] The biological importance of this solitary event is unknown.

III. SYMPTOMS AND CLINICOPATHOLOGIC PROPERTIES

A. The Adult or Enzootic Form

This is the most common form. Ninety percent of lymphomas are in cattle 5 or more years of age and 98% are in cattle 2 or more years of age.[4] Peak incidence occurs between 5 and 8 years of age.[5] The pattern of organ invasion is multicentric. Due to the variety of tissues that may be affected, the clinical signs are variable. Among the most frequent symptoms are symmetrical or nonsymmetrical lymphadenopathy, weight loss, decreased milk production, exopthalmos, posterior paresis, cardiac failure, and melena.

The laboratory data presented in Table 1 show two groups of cattle with histologically confirmed lymphoma; group one accumulated at the Ontario Veterinary College (1976 through 1980) and group two accumulated at the Ohio State University (1977 through 1981.). All cattle were either near to or in the terminal stages of disease. Of the 71 cattle examined, there were 57 Holstein, 7 Jersey, 4 Angus, and 3 Guernsey. Only three were male. The mean packed cell volumes (PCV) were within normal limits. Group one had eight anemic cattle (24%) while group two had six anemic cattle (16%). The anemia was characterized as normocytic, normochromic in all but one cow that had a macrocytic hypochromic anemia. These rates of anemia are similar to a previous study that examined initial blood samples; however, on subsequent evaluation of the same cattle later in the disease, approximately 55% had developed anemia.[5] A recent study found that approximately 40% of adult cattle with lymphoma were anemic.[25] The anemia is due to chronic inflammation in addition to neoplastic proliferation in hematopoietic organs and blood loss in some cases. Immunohemolysis has not been reported in cattle with lymphoma. Lymphosarcoma cell leukemia was evident in 30% of the cattle with lymphoma presented in Table 1. In Table 1, the mean lymphocyte count in group one was markedly higher than in group two, but the proportion of cattle with leukemia was about the same in the two groups. The presence of neoplastic cells in the peripheral blood was not always reflected in the total lymphocyte count. These findings are similar to those of a previous study, although others have found leukemia in less than 10% of affected cattle.[5,26,27] This discrepancy may be accounted for by differences in the cattle under study. The cattle presented here were in the final stages of disease and lymphosarcoma cell leukemia is frequently a terminal event.

Persistent lymphocytosis (PL) is defined as an increase in the absolute lymphocyte count of three or more standard deviations above the normal mean for that respective age and breed for an extended period of time.[28] It is associated only with the enzootic form of the disease.[14] BLV can induce PL in about one third of experimentally inoculated sheep and calves.[29-32] The frequency of PL is similar in naturally infected cattle.[33,34] Approximately two thirds of cattle with lymphoma have histories of PL.[35] Most cattle with PL never develop lymphoma.[33,34,36] The phenomenon of PL has been exploited on a test and slaughter basis in an attempt to eradicate lymphoma from Denmark. This program was successful in reducing the prevalence of lymphoma, but did not eradicate it.[37] This result was likely due to one third of cattle with lymphoma and two thirds of BLV-infected cattle without PL.[35] PL and lymphoma may aggregate along different cow-families, suggesting that the genetic predispositions to develop PL or lymphoma are inherited as separate entities.[34,36,38] In a reciprocal foster nursing experiment, it was shown that PL was a function of the genetic composition of the calf and not influenced by the PL status of the foster dam. Indeed, there was no correlation between PL and BLV infection.[34]

Two subpopulations of B lymphocytes, one BLV-infected and the other noninfected that spontaneously incorporates thymidine, are expanded in the peripheral blood of some cattle with occult BLV infection and all cattle with BLV-induced PL.[39] In cattle with PL it is predominantly noninfected corticosteroid-sensitive B lymphocytes that comprise the expanded population. There is a strong correlation between numbers of B cells and total lymphocytes.[39,40-42] Although one subpopulation carries the BLV genome, it is not a neoplastic clone; PL is a benign lymphoproliferative response.[40,41,43] Thus PL is not a subclinical form of lymphoma nor is it an animal equivalent of chronic lymphocytic leukemia of man. It is suspected that in vitro production of BLV by the infected B cell subpopulation causes the increased spontaneous incorporation of thymidine by the remaining non BLV-infected subpopulation, the latter subset being memory cells. Supportive evidence for a BLV-driven proliferative response is that antiserum to BLV added to lymphocyte cultures abrogates the proliferative response by some noncytolytic mechanism.[44,45] Cattle with occult BLV infection or with BLV-induced PL have no deficits in cellular or humoral immunity.[46-48]

Cattle with the adult form of lymphoma may have a decrease in the size of the spontaneously proliferating B cell subset. The numbers of B lymphocytes carrying the BLV-genome are increased. Total numbers of peripheral blood B cells are intermediate between noninfected normal cattle and cattle with PL.[40,49] Their relative numbers of T cells are decreased in both peripheral blood and tumor tissue.[50] Neoplastic lymphocytes in the adult form of lymphoma have the surface receptor characteristics of B cells.[50-52]

Adult cattle in the terminal stages of lymphoma have a deficiency in cell-mediated immunity. Their responses to the intradermal injection of various recall antigens, phytohemagglutinin (PHA), and tissue extracts are significantly less than in normal adult cattle.[53] Peripheral blood lymphocytes from adult cattle with lymphoma have poor stimulation indexes to several nonspecific mitogens.[54-56] These lymphocytes may or may not have increased spontaneous proliferation. This may depend on where the affected cattle are in the disease process. Those cattle in the early stages of lymphoma have lymphocytes with poor stimulation indexes because of increased spontaneous proliferation while in the late stages, lymphocytes have poor responses due to absolute or relative unreactivity. A serum inhibitor of lymphocyte blastogenesis has been identified in approximately 60% of adult cattle in the terminal stages of lymphoma.[57] This inhibitor may account for refractoriness to stimulation of lymphocytes in vitro, and its presence in serum was highly correlated with cutaneous anergy seen in cases of lymphoma.[53]

The primary abnormality in the humoral immune system of cattle with lymphoma involves immunoglobulin M (IgM). IgM-producing cells are either absent or reduced in number in the spleen and lymph nodes of cattle with lymphoma.[58] Most of the neoplastic cells contain intracytoplasmic immunoglobulin and secrete IgG when cultured in vitro with pokeweed mitogen.[59] In most studies the serum concentration of IgM in cattle with lymphoma has been found to be reduced or absent.[58,60,61] It appears that the neoplastic lymphocytes fail to synthesize or secrete IgM or there is rapid catabolism.[59] Sensitization of cattle with lymphoma to human serum albumin and the synthetic polypeptide (T,G)-A-L, resulted in very little IgM production compared with normal cattle.[62] The antibody produced appears to have reduced binding affinity for antigen.[63] In those studies where serum IgM was present in normal concentration, IgM and IgG appeared heterogenous in molecular size and charge, indicating the presence of paraproteins.[64-66]

The microscopic and ultrastructural pathology have been dealt with in several studies and texts.[5,6,28,67-77] In decreasing frequency, the sites of involvement are lymph nodes, heart, abomasum, kidney, uterus, epidural fat, and intestine. Prior to recent modifications in the classification of human lymphoid tumors, most bovine lymphomas were described as being composed of immature cells (histocytic and stem-cell types) with a diffuse pattern, although 0.45% were identified as nodular lymphomas.[6,26,78] A recent study indicated that the major cell type had a noncleaved nucleus between one and two red cell diameters (small to medium size) with a diffuse pattern, while 11% were nodular.[79] Those cattle with cleaved cell lymphomas lived significantly longer than those with noncleaved tumors, as has been shown in man.[78] Using the Kiel classification, most bovine lymphomas were centroblastic-centrocytic, centroblastic, or immunoblastic.[80,81]

B. The Sporadic Forms

These are classified by pathoanatomic criteria into the calf or juvenile, thymic, and skin types.[13,25,28,82] Microscopically and ultrastructurally the sporadic forms are indistinguishable from the adult form of lymphoma. Among the peripheral blood lymphocytes there may be normal numbers of B cells.[49] Other studies report that most peripheral blood lymphocytes and tumor cells cannot be characterized as either T or B cells.[50,83]

1. *Calf type* — The age range is from birth to 2 years of age, but it is seen most commonly in calves 3 to 6 months of age.[84-88] It is the most frequent of the sporadic forms and accounts for approximately 4% of all bovine lymphomas.[6] There is usually generalized symmetrical lymphadenopathy. Anemia is common due to the high frequency of bone marrow involvement and leukemia. Other organs affected with some frequency are liver, spleen, and kidney.

2. *Thymic type* — This is most frequent in cattle between 6 months and 2 years of age.[25,82] Beef breeds are disproportionately affected.[11] The most common presenting sign is a large swelling in the brisket and/or the ventral neck. The masses are firm and surrounded by edema. Dysphagia and dyspnea may occur. Anemia is uncommon, occurring in approximately 7% of the cases.[25] Lymph nodes surrounding the thymus are frequently affected, but generalized lymphadenopathy is unusual.
3. *Skin type* — Affected cattle are between 1 and 3 years of age. The skin lesions are raised, circular, well-demarcated, and nonpruritic. Advanced lesions are alopecic. The lesions are more concentrated on the neck, shoulders, and perineum. A case observed by this author had large numbers of skin lesions, particularly numerous over the shoulders, and no lymphadenopathy at presentation. Historically, lesions had been present for 6 months. As the lesions regressed, hair regrew. At 6 months following admission, virtually all skin lesions had resolved and lymphadenopathy was then noted. The cow died 2 months later with multicentric lymphoma typical of the adult form. This unusual progression is similar to that described in other reports.[25,89-91]

IV. THE BOVINE LEUKEMIA VIRUS

A. Propagation of BLV in Cell Culture

BLV was first recognized in PHA-treated, short-term cultures of peripheral blood lymphocytes from cattle with lymphoma or PL.[19] It was later shown that in vitro expression of the virus was dependent on maintenance in culture and not PHA-treatment.[92] Concanavalin A was found to stimulate in vitro virus production in 28% of lymphocyte cultures while it inhibited production in 12%; the remainder were unaffected.[93] These short-term cultures did not yield sufficient quantity of virus for characterization. With the establishment of long-term cultures using heterologous cell lines, BLV was harvested in large quantities, allowing further study.[94-100] Two of the most widely used are cell lines of fetal lamb kidney (FLK)[73] and fetal lamb spleen (FLS).[94,96,97,100] BLV production appears to be cell cycle dependent. Using synchronized FLK cells, BLV production increased during S and G_2 phases and diminished during the M and G_1 phases. Release of virions took place during M phase.[10] Unfortunately, these cultures have become infected with the bovine virus diarrhea virus and some serological studies may have been influenced by the contaminating antigens.[102] Until recently, homologous cell lines established from neoplastic cells derived from cattle with lymphoma produced no virus.[103] A monolayer cell culture has now been established that produces as much virus as the heterologous cell lines.[104]

B. Biophysical Properties of BLV

BLV is similar in appearance to the C type oncoviruses of other species. In short-term cultures, budding from the cell surface is rarely observed.[35,105-108] Mature virus particles are found in intracytoplasmic vacuoles or in extracellular locations associated with cellular debris. Immunoferritin labeling using anti-BLV serum demonstrated heavy labeling of the virion but little labeling of the plasma membrane. These data support that the cell surface is not the preferred site of assembly and release in short-term cultures. Cell degeneration is apparently required for virion release from intracytoplasmic vacuoles. Development of the virion appears to be asynchronous since the nucleoid is condensed prior to release.[94,107,109] This is unlike the development of the typical mammalian retrovirus where particles with an electron-lucent core are released initially.[110] Viral budding from the plasma membrane is more frequently observed in long-term cultures.[94,109,111]

The diameter of the virion varies between 60 and 125 nm depending on the methods of preparation and fixation.[92,94,99,105-108,112,113] The nucleoid core ranges from 40 to 90 nm diameter and sometimes has a hexagonal shape suggesting an icosahedral structure.[113,114] The outer viral envelope has surface projections or spikes that are approximately 114 Å in length.[107] BLV has a buoyant density of 1.15 to 1.18 g/mℓ on sucrose gradients and a sedimentation coefficient of 600 to 700S. The virions contain a single strand of 60 to 70S RNA and reverse transcriptase.[114-116]

C. Biochemical Properties of BLV

1. BLV Structural Proteins

Structural proteins are primarily separated into the external envelope-associated, ether-sensitive glycoproteins and the nonglycosylated proteins that are internal and ether-resistant. Further classification is done by molecular weight determinations. Variation in molecular weights between laboratories is probably accounted for by technique differences and proteolysis. An international nomenclature is needed to prevent further confusion in this area.

Two major glycosylated proteins of the viral envelope are recognized. In this respect BLV is similar to the avian retroviruses, but unlike mammalian retroviruses that have only one glycosylated envelope protein.[117] The larger glycoprotein has a molecular weight between 48,000 and 70,000 daltons.[118-128] It has been purified to homogeneity and has an amino acid composition similar to the major glycoprotein of the mouse mammary tumor virus (MMTV) and the Rauscher murine leukemia virus (R-MuLV).[126,129,130] The smaller glycoprotein is between 30,000 and 48,000 daltons.[118-120] In a recent review, these glycoproteins have been designated gp^{51} and gp^{30}, respectively.[131] The same designation will be used in this manuscript.

When BLV is disrupted and electrophoresed under nonreducing conditions, glycoproteins are found with molecular weights of 162,000 and 85,000 daltons. These have been designated as viral glycoprotein I (VGP I) and viral glycoprotein II (VGP II), respectively, according to the nomenclature proposed for the avian system of tumor viruses.[132,133] When these glycoproteins were electrophoresed with reduced conditions, VGP I was shown to be composed of two molecules of gp^{60} (gp^{51}) and two molecules of gp^{30}, while VGP I contained one molecule of each.[132] Further evidence showed that the gp^{60}, (gp^{51}), and gp^{30} were linked by a disulfide bond and by a process of tryptic glycopeptide digestion; a partial relatedness between the two molecules was demonstrated. The biologic significance of the similarity of VGP I to the basic structure of the immunoglobulin molecule, in terms of the duplicity of heavy and light chains, is unclear at this time. It is hypothesized that it may function as a receptor at the cell surface.[132] There is evidence for another precursor polypeptide of gp^{51} and gp^{30} that has a molecular weight of 72,000 daltons.[134] The relation between this protein and VGP I or II has not been explored. Purified gp^{51} has agglutinating activity for mouse erythrocytes.[135] The equine infectious anemia virus and the murine leukemia virus are the only other mammalian retroviruses that share this property of hemagglutination.[136,137]

The major nonglycosylated protein of BLV is the 24,000 dalton ether-resistant antigen designated p24.[123,124,127,129,138-140] It has been purified to homogeneity, and the amino acid composition and partial sequencing have been determined. BLV p24 is similar in amino acid composition to the analogous proteins of the MuTV, R-MuLV, wild mouse leukemia virus (W-MuLV), baboon leukemia virus, feline leukemia virus (FeLV), and the avian sarcoma virus (ASV).[130,140] Like other mammalian type C virus p24, the BLV p24 has the amino-terminal proline and carboxy-terminal leucine. The prolylleucylarginine tripeptide and the amino-terminal conserved region are, however, lacking. This distinctive characteristic of BLV p24 has now been shown to be shared with the p24 of the human type C T-cell leukemia virus (HTLV). Amino acid sequencing has revealed a statistically significant sequence homology between BLV p24 — FeLV p27 and BLV p24 — HTLV p24. The FeLV p27 and HTLV p24 were unrelated.[140,141] These data suggest some common ancestry in evolution, however distant. The similarity in the primary structure of the p24 is not sufficient to allow for immunologic cross-reactivity with the analogous protein of other retroviruses. Until recently, no BLV antigen had been found to have cross-reactivity to those of other retroviruses as well as the bovine syncytial virus, parainfluenza-3, infectious bovine rhinotracheitis, bovine virus diarrhea, and the maedi-like bovine syncytial virus.[95,97,98,124,129,138,142] Antiserum to purified BLV p12 reacts with the analogous protein in HTLV, thus substantiating the relatedness of the two.[141] Results of molecular hybridization experiments fail to demonstrate any similarity of BLV with mammalian, including HTLV, and avian retroviruses.[20,116,131,143,144] Current evidence suggests that the ovine leukemia virus is identical to BLV.[118,127,145,146] The two are antigenically undistinguishable, but amino acid sequencing of the major proteins has not been done. It is assumed here that the BLV was naturally transmitted to sheep. Thus, among the mammalian C type viruses, BLV appears to be a distant relative of HTLV and FeLV, but perhaps somewhat closer to the former.

There are six minor nonglycosylated proteins with molecular weights of 10,000 to 12,000, 15,000, 18,000, 45,000, 65,000 to 70,000, and 145,000 daltons. These are designated as p10(12), p15, p18, p45, p65(70), and p145, respectively. Only p15 and p10(12) have been purified to homogeneity.[141,147] Evidence for their viral origin is provided by in vitro protein biosynthesis utilizing *Xenopus laevis* oocytes micro injected with BLV 38S RNA, pulse-chase experiments in BLV-infected fetal lamb kidney and bat lung cells, quantitative immunoprecipitation using hyperimmune serum, tryptic peptide digestion, and gel chromatography of disrupted BLV.[134,148-150] These studies found p145, p65(70), and p45 to be precursor polypeptides. The p45 is not derived from p65(70) by proteolysis, yet the sequence of p45 is entirely included in p70. Why there are two similar precursors is unknown. Posttranslational modification of these precursors results in the formation of the structural proteins p24, p15, and p10(12). The p145 contains the entire sequence of p65(70) and is thought to include the reverse transcriptase. It may be similar to the 180,000 dalton polymerase read-through precursor found in avian and murine retroviruses.[150] The p18 is unrelated antigenically and has no common sequences by tryptic peptide analysis with p45, p65(70), or p145 and may be derived from a subgenomic fragment or from cellular RNA.[150] The locations of the structural proteins within the virion are unknown, but, like other retroviruses, are probably associated with the core.

2. BLV Reverse Transcriptase

BLV reverse transcriptase (RT) responds well in the reaction using the synthetic template primer poly rA-oligo dT but poorly with poly dA-oligo dT.[116,138,151,152] This is typical of C type virus RT activity. The RT activity exists in a high molecular weight RNA-RT complex, has a density of 1.155 g/m*l* in sucrose gradients, and greater than 80% of complementary DNA back hybridizes to BLV RNA, confirming that BLV is indeed a retrovirus.[143] BLV-RT is antigenically distinct from the RT of the murine leukemia virus, simian sarcoma virus, and the avian myeloblastosis virus. BLV differs from the prototypic type C virus by its failure to antigenically cross-react with other C type viruses, its maturation sequence in short-term cultures (*vide supra*), and its divalent cation requirements for RT activity. BLV and 3 Type-B viruses, Mason-Pfizer monkey virus, mouse mammary tumor virus, and the Guinea pig virus have optimal RT activity in the presence of magnesium ions, while other mammalian Type-C virus RTs are most active with manganese.[138,143] It is clear that BLV is a nonprototypic member of the family Retraviridae. It is not decided whether it is a Type C, B, or in a separate subtype, although most authors refer to it as a Type C retrovirus.

3. BLV RNA Genome

Like other retroviruses, BLV has a high molecular weight 60 to 70S RNA that is made of a dimer of identical 30 to 40S RNA molecules.[134,149,150,153] The dimers have a molecular weight of about 2.95 × 10^6 daltons that corresponds to approximately 10 kilobases. A map of the RNA genome, aligned relative to the 5′-to-3′ orientation, has been prepared using nine restriction endonucleases. The order of the fragments, in terms of their kilobase weights, is 5′-4.00, 1.25, 2.61, 2.08-3′, and there is suggestive evidence for long terminal repeat (LTR) sequences on either side of the BLV genome. All of the genetic information for BLV is in about 9.2 kilobases and each LTR is about 0.75 kilobases in length.[153,154] These data and data from the in vitro translation of BLV RNA suggest that the BLV genome is likely organized in a similar fashion to the non-defective retroviruses, such as that from the 5′-to-3′ termii there are LTR, gag, pol, env, and LTR genes.[134,150,155] The LTR segments promote translation of adjacent genes. The gag, pol, and env genes code for internal viral structural proteins, RT, and envelope glycoproteins, respectively. The gene products are the precursor polypeptides (*vide supra*). The gag gene codes for p65(70) that is then cleaved and results in the formation of p24 and perhaps p15 and p10(12). The p145 is likely the product of both the gag and pol genes. It is not clear whether the env gene product is the VGP I with a molecular weight of 85,000 daltons or the 72,000 dalton protein. Like other chronic leukemia viruses, BLV lacks the onc or transforming gene, but may acquire this gene when cultured for long periods in the appropriate cell lines.[150,154] Evidence for this acquisition is provided by the observation of spontaneous transformation of BLV-infected FLK cells and a cell line derived from a case of lymphoma in calf from which BLV was rescued by treatment with 5′-iodo-2′-deoxyuridine treatment.[156] The virus from these cell lines when inoculated onto sheep fibroblasts also resulted in phenotypic transformation.[156] Restriction analysis has revealed that, among BLV proviruses isolated from cattle in different geographic locations, there are genetic mutants.[153,158] Mutants appear to be conserved within a geographic region.[159] This also raises the possibility that certain strains of BLV may be more lymphomagenic than others. There are no reports of serologic differences between isolates of BLV despite genetic differences.

Epizootiologic evidence and transmission experiments suggest that BLV infection is contagious. Along with molecular hybridization experiments, BLV has been proven to be an exogenous RNA virus.[20,21,143,154] BLV sequences are only found in neoplastic lymphocytes from cases with the adult form of lymphoma, lymphocytes from cattle with PL, and lymphocytes from cattle with occult BLV infection. The DNA from uninvolved tissues from cattle with lymphoma, neoplastic cells from the sporadic forms of bovine lymphoma, tissues from normal cattle, sheep, goats, mice, rats, dogs, deer, elk, cats, chickens, and humans, and tumor tissue from various human leukemias and sarcomas show no evidence of having relatedness to BLV sequences in their genomes; thus BLV is not related to the endogenous viruses of these species.[20,21,116,143,154]

The BLV provirus in neoplastic lymphocytes is integrated in the G + C-rich fragments having a buoyant density of 1.708 g/mℓ in CsCl gradients. There appears to be from less than one to three copies of the proviral DNA per haploid genome, but there are many possible integration sites.[158,160] This is similar to other species' neoplasms induced by exogenous retroviruses.[161] Defective proviral copies have been detected in about 25% (4 of 17) of bovine lymphoid tumors. The deleted sequences were from 5′ half of the provirus.[160] More than one variant complete provirus or defective provirus may be isolated from individual tumor cases.[160] With respect to the site of integration of the BLV provirus, the bovine lymphoid tumors are of mono- or oligoclonal origin. The site of integration also varies among cases of lymphoma.[158,160,162,163] These data are generally compatible with the results of karyotypic analysis showing a monoclonal origin within an animal, but no consistent abnormality between cattle with lymphoid tumors.[164]

In cattle with BLV-induced PL, the provirus is integrated at many sites; thus, the origin of these lymphocytes is polyclonal.[162,163] It was estimated that 25 to 40% of peripheral blood leukocytes (PBL) of cattle with PL have integrated BLV provirus.[162] The PBL from cattle that are BLV-infected, without PL, fail to demonstrate hybridization with BLV probes. This result was attributed to less than 5% of the PBL containing BLV provirus, placing it beyond the sensitivity of the assay.[163] It has been shown that a small percentage of lymphocytes in these cattle do harbor the provirus since, when cocultured with permissible cells, BLV-induced syncytia result.[48,98]

In each instance, whether one considers BLV-induced lymphoma, BLV-induced PL, or occult BLV infection, there is no transcription of viral genes in vivo.[160,163] Infected lymphocytes must be maintained in vitro for at least 12 to 24 hr so that they are rescued from a resting phase and enter the cell cycle.[92,93,101,165] Viremia and/or antigenemia has not been detected.[130,165-167] Once infected, cattle remain infected for life despite a persistent immune response to all of the BLV antigens.[168] A possible exception to this has recently been reported.[169] In this study, BLV was transiently recovered from six bulls with undetectable or low levels of antibody out of a group of 432 cattle. The significance of this finding is unknown at this time.

Virus neutralizing (VN) antibody in the serum of BLV-infected cattle is cytolytic to in vitro-cultured BLV-infected lymphocytes and FLK cells chronically infected with BLV. The antibody is in the IgG_1 subclass, complement-dependent, and is directed to the gp51 expressed on cultured cells.[170-172] It may play some role in preventing viremia, perhaps by eliminating lymphocyte clones producing virus in vivo, giving nonpermissive clones a selective growth advantage. However, in vitro, lymphocytes from BLV-infected cattle do produce BLV showing that the lymphocytes do not have an inherent defect. It may be that only a small percentage of BLV-infected cells produce virus at one point in time and that VN antibody kills these cells through complement-mediated cytotoxicity or antibody-dependent cellular cytotoxicity. Alternatively, the combination of antibody with membrane-bound BLV antigen may signal some cytosolic messenger to repress genomic expression. These hypotheses are untested. A nonimmunoglobulin plasma protein has been identified in BLV-infected cattle that inhibited production of p24 presumably by preventing translation. The authors suggested that the effect is mediated through a membrane receptor and hypothesized that the genome is modulated by a process similar to the mechanism exerted by hormones.[173]

There must be some expression of the BLV genome since antibody to BLV is persistent for life. It is possible that there is expression of the BLV genome at a very low rate, a rate too low for detection by the assay used to demonstrate failure of translation.[160] Fluctuation in the concentrations of VN antibody and/or the nonimmunoglobulin plasma protein may allow for the temporary expression of BLV proteins resulting in immune stimulation. Privileged sites within the body not accessible to mediators of immunity or regulators of protein synthesis may harbor infected lymphocytes that produce BLV antigens, which diffuse away from the site and stimulate immunity. Tumors may represent such an immunologically privileged site since, once tumor development begins in a BLV-infected animal, the antibody titer to BLV generally increases. Hopefully, study into the pathogenesis of BLV infection will shed some light onto the interactions at play here.

The mechanism of lymphomagenesis is unknown, but the data support the following concepts: (1) defective proviral sequences with intact 3′ regions are adequate to at least maintain the neoplastic condition; (2) gene expression of the BLV provirus is not required for the maintenance of the neoplastic condition; (3) the site of integration of the BLV provirus is variable between cattle with lymphoma and the tumors in different organs within a single animal with lymphoma; this may indicate that there are many sites in the bovine genome that can potentially be turned on by the BLV provirus resulting in transformation or that the BLV provirus can turn those genes on at a distance; (4) the BLV provirus does not contain transforming genes; the transforming genes are assumed to be part of the host cell genome; and (5) there are a large number of BLV-infected cattle but only a small proportion develop lymphoma (*vide infra*); this is attributable to a long incubation period and the interaction of BLV with other uncharacterized events in the environment (carcinogens and other infectious agents) and the genetic background of the animal.

D. Tumor-Associated Antigen

There have been several preliminary investigations that have dealt with tumor-associated antigens (TAA) and bovine lymphoma.[53,174-180] The putative antigen has been termed the bovine oncornavirus cell-membrane-associated antigen (BOCMA). Direct immunofluorescence, using antisera prepared in calves inoculated with tumor cell suspensions, localized antigen to the cytoplasm of acetone-fixed frozen sections of tumor cells. Similar preparations of normal bovine lymphoid tissue failed to react.[180] Later studies repeated these findings and demonstrated a similar antigen in cells from cases of ovine lymphoma. Using tumor cell suspensions, the TAA was localized to the cell membrane on both bovine and ovine lymphoid tumor cells.[175,178] Cattle with the calf and thymic form had few (less than 3%) tumor cells that contained TAA. Of 15 cattle, 3 with occult BLV infection but no tumor also had TAA in lymphocytes. The authors associated the appearance of the TAA with increased numbers of B cells that occurred in the 3 BLV-infected cattle and in the cattle with lymphoma. Cells from normal cattle, with no BLV or antibody to BLV, had no detectable TAA.[175] Using the technique of antibody-dependent cell-mediated cytotoxicity (ADCC), it was shown that two transformed cell lines derived from FLK cells, and sheep fibroblasts and cells from cases of adult lymphoma and a case of skin lymphoma all had distinct neoantigens unrelated to BLV structural proteins. Sheep that had been experimentally infected with BLV derived from the transformed sheep fibroblasts and had remained healthy had antibody to the neoantigen. In 26 of 51 cattle that were BLV-infected, there was serum antibody also directed to the respective neoantigen detected by ADCC, while only 1 of 23 sera from cattle with the adult form of lymphoma had the antibody, the implication here being that antibody played some protective role against tumor development.[179]

Semipurified membrane preparations from bovine lymphoid tumors caused delayed cutaneous hypersensitivity (DH) reactions in six cattle with lymphoma, while three controls failed to react. Although the control cattle did not have reactions to normal lymph node or tumor extracts, the cattle with lymphoma did have reactions to normal lymph node extracts.[174] In a subsequent study, with a larger battery of test antigens and a more homogeneous population of cattle with lymphoma, it was found that roughly two thirds of adult cattle in the terminal stage of lymphoma had cutaneous anergy and thus had a poor DH reaction to all antigens including the nonspecific mitogen PHA-M.[53] Normal cattle did respond to both normal lymphoid and tumor cell membrane extracts. In an in vitro test of cell-mediated immunity, it was found that approximately 70% of adult cattle with lymphoma and 4% of normal cattle (37% of which had serological evidence of BLV infection) had significant lymphocyte stimulation to a partially purified pooled tumor cell membrane extract. The lymphocytes from three cases of the sporadic form of lymphoma failed to respond. An antiserum raised in rabbits failed to cross-react with normal lymphoid cell membrane extracts and BLV proteins. The antigen was localized to the cytoplasm and cell membranes of tumor cells.[176] Circulating TAA was not detected in the serum of cattle with lymphoma or normal cattle.[177] Undoubtedly, further development in this area will lead to a better understanding of the process of lymphomagenesis, particularly in the area of tumor immunity.

Table 2
TESTS FOR THE DETECTION OF ANTIBODY TO BLV

Test	Comments	Ref.
Immunodiffusion using gp51 (ID gp51)	Termed the agar gel immunodiffusion (AGID) test or Leukassay B test, widely used as test for export of cattle, may be less sensitive than RIA, false negatives in periparturient period, Ab is directed to the carbohydrate moiety of gp51	111, 119, 123—125, 139, 182—196
Immunodiffusion using p24 (ID p27)	Made of ether disrupted BLV less sensitive than IDgp51	22, 33, 95, 111, 122, 123, 138—140, 142, 148, 185, 189, 190
Immunodiffusion using p15 (ID p15)	Less sensitive due to low MW and poor precipitability in agar	147
Immunofluorescence (IFA)	Target cells are virus-producing cell lines, indirect tests including anticomplement types, detects intracytoplasmic p24 and p15, greater sensitivity than ID tests	33, 95, 97, 98, 111, 165, 166, 184, 198—201
Immunoperoxidase (IP)	Indirect method, target cells are virus-producing cell lines, as sensitive as IFA	202
Complement fixation (CF)	gp51/p24 or gp51 used alone, bovine sera are frequently anticomplementary, as sensitive as IDp24 but less sensitive than IDgp51	122, 185, 202—211
Enzyme linked immunosorbent assay (ELISA)	gp51 or crude viral preparations used, detects antibody in serum and milk, as sensitive as ID gp51	212—217
Particle counting immunoassay (PACIA)	Crude viral preparations fixed to latex beads, totally automated system	218
Radioimmunoassay using p24 (RIA p24)	Double antibody methods, more sensitive than ID gp51 and IFA in some studies	124, 129, 130, 198, 219—227
Radioimmunoassay using gp51 (RIA gp51)	Double antibody methods, most sensitive assay	124, 126, 130, 210, 219, 222, 225, 227
Counter immunoelectrophoresis (CIEP) using gp51	As sensitive as ID gp51, results in 2—4 hr	228
Reverse transcriptase inhibition (RTI)	50% cattle with lymphoma have antibody to RT, BLV-infected cattle with no tumors are negative	229
Virus neutralization (VN) syncytia inhibition (SI), early polykaryocytosis inhibition (EPI)	Detects antibody in BLV-infected cattle that inhibits BLV, BLV-structural proteins or fusion factors from inducing giant cells or syncytia in permissible cell cultures, as sensitive as ID gp51, addition of complement enhances sensitivity	170—172, 184, 219, 226, 230—236
Pseudotype inhibition (PI)	Pseudotype particles with rhabdo-virus genome and BLV envelope, more sensitive than ID gp51 but as sensitive as RIA p24	238, 239

V. DETECTION OF BLV INFECTION

The methods for the detection of antibody to BLV and for the recovery of BLV from infected cattle have been discussed in detail elsewhere.[43,131,149,181] The various tests with appropriate references and comments are presented in Tables 2 and 3.

VI. EPIZOOTIOLOGY OF BLV

The world-wide distribution of BLV in the cattle population has been discussed in other recent reviews, and it will not be dealt with in detail here.[131,149] A summary of the salient features follows:

1. About 66% of dairy herds in north central U.S. and 10 to 30% (depending on the sensitivity of the assay) of dairy cattle are BLV-infected. In beef herds from

Table 3
TESTS FOR THE DETECTION OF BLV OR BLV ANTIGENS

Test	Comments	Ref.
Electron microscopy	Requires in vitro cultivation of lymphocytes as do all direct tests of infectivity, poor sensitivity	19,92
Immunofluorescence infectivity assay (IFIA)	Less sensitive than competitive RIA for detection of p24 in culture supernatants	95,165,166,198,220
Immunoperoxidase infectivity assay (IPIA)	Greater sensitivity than syncytia infectivity assay (SIA), detects p24	240
Immunodiffusion (ID)	Single radial and double ID used for both p24 and gp51	31,165,241
Radioimmunoassay using p24	As sensitive as SIA, competitive technique	198,220,242
Reverse transcriptase	Not reliable	115
Syncytia infectivity assay (SIA) early and late polykaryocytosis induction	Detects BLV, BLV structural protein or BLV-associated fusion factor by induction of syncytia or giant cells in permissible cell cultures, good agreement with serological results	115,184,226,243—250

the same area, the proportions are 14 and 1%, respectively.[183,251] Management practices and differences in genetic background between beef and dairy cattle probably account for the marked difference in prevalence. Variation in geographical location also influences prevalence rates. For example, in Florida, about 50% of dairy and 7% of beef cattle are BLV-infected.[252]

2. There is no sex predilection among BLV-infected cattle.[251]
3. Herds with a history of lymphoma have between 25 and 100% (depending on the sensitivity of the assay as well as herd prevalence) BLV-infected cattle. Herds without histories of cases of lymphoma have only 1 to 15% BLV-infected cattle.[182] This suggests some dose-related effect; the higher the concentration of virus within a herd, the higher the probability of finding cases of lymphoma.
4. All cattle with the adult form of lymphoma are BLV-infected.[33,129]
5. Once cattle become infected with BLV, they remain infected for life. Those cattle that are infected develop a persistent antibody response to many BLV antigens.[184]
6. Cattle, at least in herds with a history of lymphoma, begin to seroconvert when 1 to 2 years of age, and thereafter the prevalence increases with increasing age, but may remain relatively constant after 5 years of age.[168,251-255] The effect of introducing BLV-negative cattle into an infected herd may result in a gradual decline in prevalence. This effect is probably due to the low annual rate of seroconversion, 0.9 to 2.1%, observed in the study.[256] Hence, BLV is apparently not a very contagious infection.[257]
7. BLV-infected cattle may or may not develop PL. Generally, about one third of infected cattle have PL.[9,33,34] In BLV-infected herds almost all cattle with PL are BLV-infected.[9]
8. BLV is apparently unrelated to the sporadic forms of lymphoma.[22] These forms are not spread as an infectious disease.[14] It is conceivable that some of these cattle may harbor BLV and have serological evidence of infection since calves may become infected in utero.[168] Molecular hybridization studies show that in tumor cells derived from cases of sporadic lymphoma, there are no BLV sequences.[2,21,143]
9. The rate of conversion from BLV-infected to lymphoid tumors has been estimated at 1 in 1000 to 1 in 250 BLV-infected cattle per year.[131] Over their lifetimes, 5 to 10% of BLV-infected cattle will develop lymphoma.[9,184,230] This latter estimate may be somewhat high, reflecting the genetic predisposition for the development of lymphoma in the herds studied.

10. BLV infection results in (1) production of antibody, (2) antibody and PL, (3) antibody and adult lymphoma, and (4) antibody with adult lymphoma and PL. The only disease associated with BLV infection is lymphoma. Cattle with occult BLV infection (antibody alone) or BLV infection with PL show no signs of any illness. Production variables and reproductive performance in BLV-infected cattle are not decreased compared with uninfected cattle.[256,258]
11. The prevalence of BLV in a herd is apparently unrelated to herd size.[182,251]
12. Dairy breeds appear to be more susceptible to BLV infection than beef breeds and, among dairy breeds, Jersey cattle appear to be particularly susceptible.[252,255] These same authors estimated the heritability to be 0.48 ± 0.22, suggesting a marked effect of genetic background on an individual's susceptibility to BLV infection.

VII. NATURAL TRANSMISSION OF BLV

Since BLV does not infect germinal cells and there is essentially no virus production in vivo, congenital infection must result from the transfer of infected lymphocytes from dam to offspring.[166-168,259] In all probability, there is a temporary break in the placental barrier allowing this transfer to occur. Since BLV infection is most often detected by a positive serological test prior to colostral ingestion, only calves that were immunocompetent at the time of this transfer are detected. In those studies using tests of virus isolation, there was no evidence for immuno-tolerant BLV-infected individuals. The existence of such an individual, however, has not been specifically looked for. The reported frequency of congenital infection has ranged from 3 to 25%.[168,184,259-261] Those figures in the high end of this range may be due to the genetic characteristics of the herd that was bred for a high incidence of lymphoma. Caesarean-derived calves may be infected as a result of exposure to maternal blood. This may account for a study in which two thirds of calves delivered by Caesarean section later developed PL.[262] It is not known whether congenitally infected calves are more susceptible to the development of the adult form of lymphoma.

As with the congenital infection, postnatal transmission of BLV involves the transfer of infected lymphocytes. Most cattle become infected through some event after birth. In most herds this event is close contact with infected adult cattle, resulting in horizontal transmission.[168,254,259] It is not a highly contagious infection; consequently it spreads slowly within a herd.[257] Genetic background and age at exposure influence the outcome. Susceptibility to BLV has a high degree of heritability, apparently greater than the heritability of milk production. Few cattle under 1 year of age are BLV-infected. Most seroconvert between 1 to 4 years of age and after 5 years of age the rate of conversion decreases. Thus a plot of prevalence vs. age would resemble a sigmoidal curve.[131,168,259,263,264]

Transmission by blood-sucking vectors is supported by increased rates of seroconversion during the summer months and the recovery of BLV-infected lymphocytes from horse flies fed on BLV-infected cattle.[265-267] The heads and mouth parts of mosquitoes fed on blood from a BLV-infected cow with PL caused infection in sheep when inoculated subcutaneously.[268] The intradermal injection of as few as 2500 lymphocytes from a BLV-infected cow can result in seroconversion.[269] Other studies have reported that transmission was more frequent in the winter months.[270,271] This may have resulted from using a less sensitive test to detect antibody to BLV, thus apparently extending the incubation prior to seroconversion. Alternatively, insect vectors may not be important for transmission of BLV in these herds.

Iatrogenic transfer may also occur. Vaccination of cattle and sheep for babesiosis with a whole blood vaccine derived from a presumably BLV-infected individual resulted in increases in the rates of PL and lymphoma.[16,272] Contaminated syringes and needles may also contribute to the spread of infection.[273] Veterinarians should observe the appropriate precautions when performing mass operations such as castration or dehorning. Intradermal tuberculin testing has not been incriminated. As long as there was no contamination of the needle with blood, transmission did not occur.[274]

The inoculation of milk, milk cells, or colostrum from BLV-infected cows into sheep may result in infection demonstrating the presence of infected viable cells or virions.[275,276] Ingestion of milk or colostrum, containing BLV-infected lymphocytes, by susceptible calves is not an important route of transmission. Approximately 15% of calves that ingested colostrum and milk from the BLV-infected dams became infected.[168,277,278] Presumably, BLV-infected lymphocytes or virions released intraluminally are able to enter the calf's circulation during the first 24 hr of life, prior to intestinal closure. Early closure may be protective in some cases but, of course, will leave the calf hypogammaglobulinemic. Any enteric infection that may enhance the permeability of the mucosa may increase the chances of infection. The relative resistance of calves to this route of transmission is attributed to the presence of virus-neutralizing antibody in milk.[168,275-279] It is possible that dilution of colostrum with milk or the pooling of colostrum between infected and noninfected cows may dilute the virus-neutralizing antibody and abrogate its protective effects. Pooled sour colostrum has a low enough pH that BLV is inactivated.[279] The infectivity of BLV is totally destroyed by pasteurization.[280] Heat-inactivated colostrum may be of use in controlling BLV infection.[278,281]

Virus was not detected in semen, nasal secretions, saliva, or urine of BLV-infected cattle using sheep infectivity as a bioassay.[275] In other studies using the same assay, however, BLV was recovered from semen and saliva, but not from feces and urine.[282,283] A radioimmunoassay detected p24 in whole urine specimens from BLV-infected cattle. It was not clear whether the p24 was associated with infectious BLV or with membranes of infected cells present in urine. The study found the urinary excretion of p24 to be intermittent and suggested that the concentration of BLV may be below the threshold needed for a positive result in the sheep bioassay.[284] Artificial insemination and natural breeding do not appear to influence the BLV status of the offspring although semen may contain infectious BLV, and the intrauterine inoculation of BLV-infected lymphocytes may lead to the seroconversion of the recipient cows.[269,282,285,286] Fresh semen may contain an inhibitory factor that prevents BLV infection.[287]

BLV may also be transmitted to other animal species. Serological evidence of infection was detected in Zebu cattle (*Bos indicus),* sheep (*Ovis ovis),* water buffalo (*Bubolis bubolis),* and the capybara (*Hydrochoerus hydrochaeris).*[288] All of these animals species were either raised with or shared the same environment with domestic cattle. The high prevalence of insect vectors in Venezuela may be responsible for this unique finding. No serologic evidence for BLV infection was found in pigs, goats, sheep, ponies, 17 species of wild animals, and 5 species of wild birds in the U.S.[289] Current evidence shows that the BLV and the ovine leukemia virus (OLV) are identical. The putative OLV does not appear to be transmitted by contact among sheep as is BLV.[290] There are, however, herds in which infection has become endemic, probably through vertical (congenital) transmission.[185,197] Experimentally infected ewes can pass the infection to their lambs in utero.[291]

VIII. EXPERIMENTAL TRANSMISSION OF BLV

BLV can infect cattle, sheep, goats, pigs, deer, domestic rabbits, cats, and chimpanzees. These mammals all develop persistent antibody titers.[292-295] The dog, cottontail rabbit, and rat had transient titers.[294] In earlier studies, the same strain of rats and many other mammals and birds were found to be resistant to infection.[296,297] The discrepancies between investigators can be attributed to differences in techniques used for detecting antibody or virus and the source or nature of the inoculum. Intradermal or subcutaneous injection of the infective material (cells or cell-free, culture supernatant) appears to be the most efficacious route.[269,287] In a preliminary study of the pathogenesis of BLV infection in calves, the virus was first detected in the spleen 8 days postinoculation. Six days later it was detected in peripheral blood. There was no serological evidence of infection until 6 weeks postinoculation.[298]

The lymphomagenicity of BLV has been conclusively demonstrated only in sheep.[290] Methodological deficiencies discount reports of induction of lymphoma in cattle. Control cattle were not given a similarly treated inoculum from BLV-negative sources. Experimental animals were not housed separately allowing for contact transmission, and occasionally the induction of PL in the recipients was interpreted as successful transmission. These studies have been reviewed elsewhere.[43] Failure to induce the tumor in cattle has been attributed to the long incubation period, and the lack of a permissive genetic constitution of experimental cattle. Up to 50% of lambs inoculated with BLV during the first week of life will die of lymphoma within 5 years.[30,299-302] Recently, one of five goats that were inoculated with BLV-infected lymphocytes as neonates developed lymphoma after 8 years.[303] Other long-term studies using goats have found no tumor development.[30,149] These studies also suffered from lack of appropriate controls. Two of six newborn chimpanzees fed milk from a BLV-infected cow developed erythroleukemia.[304] Chimpanzees in a subsequent study failed to develop tumors, but did have persistent antibody titers to BLV although BLV was not recovered.[293] None of the small mammals or birds inoculated with BLV developed tumors. Cell cultures from a bovine lymphoid tumor and fresh lymphoma cells have been transplanted to nude mice and immunocompromised (radiation and antilymphocyte serum treated) calves.[305-308]

IX. PREVENTION AND CONTROL OF BLV INFECTION

The real economic loss associated with BLV infection is not through the death of cattle with lymphoma, although some herds may suffer significant losses, but results from the loss of export markets. These markets have restricted their imports to BLV-negative cattle. Semen must be from BLV-negative bulls. Considering that about 20% of all dairy cattle in the U.S. are BLV-infected, research into the mechanisms that will result in the prevention and control of BLV infection is of the utmost importance.

A. Vaccination

Preliminary studies have met with mixed results. Virus inactivated with 0.05% formalin resulted in loss of the antigenicity of gp51. *N*-acetylethylenimine (AEI) or binary ethylenimine (BEI) used at a concentration of 0.2% for 8 hr at 37°C caused complete inactivation of virus and preserved the antigenicity of gp51. Four calves were injected i.m. with AEI-inactivated virus in a 10% aluminum hydroxide gel. Two calves required two injections for seroconversions, while another two calves required a third injection. Each injection contained 340 μg of gp51. Challenge was done with BLV-infected lymphocytes injected intradermally. The dose of virus was approximately one to four minimum infective doses. Three of the four calves were protected.[309] The lymphocytes from protected calves were negative in a sheep bioassay, and the antibody titers decreased during the year following challenge, showing that there was no persistent infection. In a second study, 18 adult cattle were vaccinated three times using BEI-inactivated virus in a 50% aluminum hydroxide gel. Eight nonvaccinated controls and 14 vaccinated cattle were challenged with 100 minimum infective doses. Of these cattle, all of the nonvaccinated and 12 of 14 vaccinated cattle became infected. One of four cattle vaccinated but nonchallenged also became infected, showing that there was either infective virus remaining in the BEI-treated vaccine or the individual was infected by contact transmission. Although the vaccine did not prevent infection, vaccinated cattle produced less virus in vitro.[310] The authors related this reduced infectivity to virus-neutralizing antibody or other uncharacterized immune mechanisms induced by vaccination. Further study is needed to determine if infectivity can be reduced sufficiently to prevent contact transmission.

Sheep vaccinated with formalin-inactivated virus in aluminum hydroxide were reportedly protected against challenge with BLV produced in FLK cells. Details of the experiment were not reported. Complement-dependent cytotoxic antibody was not produced by the vaccinated sheep in contrast to cattle and sheep naturally infected with BLV.[311] The significance of this finding is not clear.

In another attempt at vaccination, BL-3 cells, derived from a case of lymphoma in a calf, were used to vaccinate 6-month-old beef calves. The cells grown in suspension culture were given alone once and twice with a 10% aluminum hydroxide gel. Two of three nonvaccinated controls became infected following challenge with the intradermal injection of 2500 lymphocytes from a BLV-infected animal. None of the four vaccinated calves became infected nor did they seroconvert. The authors hypothesized that the vaccinated calves developed immunity to TAA, which prevented infection by altering viral receptor sites on host lymphocytes.[312] If this were specific immunity to TAA, the results would support a common etiology of the calf and adult forms of lymphoma. Results of molecular hybridization, epizootiologic, and other studies concerned with TAA indicate that the sporadic lymphomas have a different etiology (*vide supra*). Perhaps the protective effect was due to adjuvant alone resulting in generalized immunostimulation (this control group was not included) or immunity against bovine leukocyte antigens (BOLA) causing death of the inoculated lymphocytes prior to virus replication. These studies do show potential and indicate that this will be an active area of research in the future.

B. Control

The policy of slaughtering cattle with PL, as an indication of BLV infection, has been successful in reducing the frequency of cases of lymphoma, but has not eradicated the disease.[37] It is possible that the prevalence of BLV infection has remained constant or not decreased in proportion to the prevalence of PL because non-PL, BLV-infected cattle were selected. With the advent of simple serological tests, there have been several successful attempts at reducing or eliminating BLV-infected cattle from herds.[186,187,278,281,313,314,315-318] The radioimmunoassay (RIA) may detect reactors earlier in infection and those with antibody titers too low for identification in the agar gel immunodiffusion (AGID) test. The RIA test may, therefore, be more efficient and perhaps critical in eliminating BLV from a herd.[281] The eradication programs have been based on either one or more of the following: (1) the isolation of BLV-free calves born to infected dams after 4 to 6 weeks of nursing, (2) separating adult cattle into infected and noninfected groups, (3) withholding colostrum and milk from infected dams, (4) use of noninfected bulls for natural service, (5) milking noninfected cows first, and (6) removal of BLV-infected cattle from the herd. Colostrum-fed calves from BLV-infected dams will be serologically positive until 6 to 8 months of age. If infected in the interim, they will remain serologically positive.[281,319] Depending on the sensitivity of the tests used to identify reactors, the tests should be repeated at least two times and preferably three times with less sensitive assays, separated by 3-month intervals to ensure that an individual is noninfected. All new additions to a herd should pass through this regimen during an isolation period prior to entering the herd.

X. PUBLIC HEALTH SIGNIFICANCE

Concern for BLV having zoontic potential has arisen since: (1) dairy products comprise a significant proportion of our diet and the dairy industry is a large employer resulting in ample opportunity for transmission, (2) BLV can grow in human tissue cultures,[98] (3) chimpanzees developed persistent antibody titers to BLV although virus was not recovered and two out of six chimpanzees fed raw milk developed erythroleukemia,[293,304] (4) BLV can cross species boundaries naturally and through experimental transmission,[288,292] (5) there are reports of isolated clusters of human leukemia cases that were suggested but not proven to be related to lymphoma in cattle or the dairy industry,[43,320] (6) antibody in sera from human patients with lymphoid malignancies and people exposed to cases of bovine lymphoma bound to bovine neoplastic lymphoid cells in an indirect membrane immunofluorescent test. Adequate controls were not done but the result remains of interest.[321]

The relation between BLV and neoplasms in the human population has been dealt with in several recent reviews and studies.[98,320,322-326] There is no hybridization of the BLV genome to the DNA from a small number of human neoplasms.[21] Large seroepidemiologic studies that included farmers drinking raw milk, veterinarians, and human leukemia patients report that there is no serologic evidence of infection in people. This, of course, does not infer that BLV cannot infect people and remain in a covert state. The results of epidemiologic studies remain inconclusive.[43,320,326] The studies suffered from lack of adequate controls, and trends were not apparent between different geographic locations. A recent study found a significant increase in mortality from leukemia and Hodgkin's disease in veterinarians, particularly those in clinical practice.[327] An excess of acute lymphoid leukemia in males living in rural counties in Iowa was found to be positively correlated with cattle density and dairy herds with cases of lymphoma.[322] Independent confirmation of these findings is awaited.

Although the current data suggest that BLV is not infectious to people, they certainly do not eliminate the possibility. Perhaps once the sensitivities of our testing systems in molecular hybridization, serology, and epidemiology are increased, a more firm conclusion can be reached.

XI. CONCLUSIONS

Infection with BLV is widespread in domestic cattle. Fortunately, the only disease associated with BLV infection is lymphoma and this is a relatively rare event. The protection of export markets gives economic justification for the development of vaccines and control programs. This area should receive vigorous support since the feasibility has been shown.

BLV induces lymphoma in sheep with a relatively short incubation period. This model of lymphomagenesis will undoubtedly be used to elucidate the pathogenesis of the disease in cattle. There has only been a single preliminary study into the early events of BLV infection. The evolutionary relationship between BLV and HTLV may stimulate interest in BLV-infected sheep as a model for retroviral infection in people.

Further study is needed to confirm the role of blood-sucking insects in the transmission of BLV. Also, the large contribution of heritability to the development of BLV infection and lymphoma indicates that investigators should attempt to correlate these propensities with biological markers, perhaps bovine leukocyte antigens. The relationship of environmental influences on the development of lymphoma remains uninvestigated. As a result of the remarkable advances in the molecular biological aspects of BLV, it is likely that the onc gene in the bovine genome will soon be identified.

In summary, although the research in BLV and BLV-induced lymphoma has seen dramatic advances, particularly in the field of molecular biology, there remain numerous gaps in our fundamental knowledge of the infection and tumors. The next few years should be an exciting period in BLV-related research.

REFERENCES

1. Preister, W. A. and McKay, F. W., *Occurrence of Tumors in Domestic Animals,* National Institutes of Health, Bethesda, Md., November 1980, 132.
2. Migaki, G., Hematopoietic neoplasms of slaughter animals, in *Comparative Morphology of Hematopoietic Neoplasms,* Lingeman, C. H. and Garner, F. M., Eds., Natl. Cancer Inst. Monogr. 32, National Institutes of Health, Bethesda, Md., 1969, 121.
3. Food Safety and Quality Service, Meat and Poultry Inspection Program: Statistical Summary — Federal Meat and Poultry Inspection for Calendar Year 1978, U.S. Department of Agriculture, Washington, D.C., 1978, 2.
4. Sorensen, D. K. and Beal, V. C., Prevalence and economics of bovine leukosis in the United States, in *Proc. Bovine Leukosis Symp.,* May 22, U.S. Department of Agriculture, Washington, D.C., 1979, 33.
5. Marshak, R. R., Coriell, L. L., Lawrence, W. C., Croshaw, J. E., Schryver, H. F., Altera, K. P., and Nichols, W. W., Studies on bovine lymphosarcoma. I. Clinical aspects, pathological alterations and herd studies, *Cancer Res.,* 22, 202, 1962.
6. Smith, H. A., The pathology of malignant lymphoma in cattle, *Pathol. Vet.,* 2, 68, 1965.
7. Anderson, R. K., Sorensen, I. K., Perman, V., Dirks, A., Snyder, M. M., and Bearman, J. E., Selected epizootiologic aspects of bovine leukemia in Minnesota (1961—1965), *Am. J. Vet. Res.,* 32, 563, 1971.
8. Conner, G. H., LaBelle, J. A., Langdon, R. F., and Crittenden, M., Studies on the epidemiology of bovine leukemia, *J. Natl. Cancer Inst.,* 36, 383, 1966.
9. Ferrer, J. F., Marshak, R. R., Abt, D. A., and Kenyon, S. J., Relationship between lymphosarcoma and persistent lymphocytosis in cattle: a review, *J. Am. Vet. Med. Assoc.,* 175, 705, 1979.
10. Cypress, R. H., Waller, J. H., Redmond, C. K., Tashjian, R. J., and Hurvitz, A. I., Epidemiologic and pedigree study of the occurrence of lymphosarcoma from 1953 to 1971 in a closed herd of Jersey cows, *Am. J. Epidemiol.,* 99, 37, 1974.
11. Theilen, G. H., Dungworth, D. L., Lengyel, J., and Rosenblatt, L. S., Bovine lymphosarcoma in California. I. Epizootiologic and hematologic aspects, *Health Lab. Sci.,* 1, 96, 1964.
12. Bendixen, H. J., Preventive measures in cattle leukemia: leukosis enzootica bovis, *Ann. N.Y. Acad. Sci.,* 108, 1241, 1963.
13. Bendixen, H. J., Bovine enzootic leukosis, in *Advances in Veterinary Science and Comparative Medicine,* Vol. 10, Brandly, C. A. and Cornelius, C., Eds., Academic Press, New York, 1965, 129.
14. Bendixen, H. J., *Studies of Leukosis Enzootica Bovis with Special Regard to Diagnosis, Epidemiology and Eradication, Public Health Service Publi. No. 1422, U.S. Government Printing Office, Washington, D.C., 1965.*
15. Crowshaw, J. E., Abt, D. A., Marshak, R. R., Hare, W. C. D., Switzer, J., Ipsen, J., and Dutcher, R. M., Pedigree studies in bovine lymphosarcoma, *Ann. N.Y. Acad. Sci.,* 108, 1193, 1963.
16. Olson, H., Studien über das auftreten und die verbreitung der rinderleukose in Schweden, *Acta Vet. Scand.,* 2(Suppl. 2), 13, 1961.
17. Bodin, S., Enhorning, G., Olson, H., and Wingvist, G., Die anzahl der lymphozyten im blut von rhindern bei lymphatischer leukose und piroplasmose, *Acta Vet. Scand.,* 2(Suppl. 2), 47, 1961.
18. Hugosen, G., Vennstrom, R., and Henriksson, K., The occurrence of bovine leukosis following the introduction of babesiosis vaccination, *Bibl. Haematol.,* 30, 157, 1968.
19. Miller, J. M., Miller, L. D., Olson, C., and Gillette, K. G., Virus-like particles in phytohemagglutinin-stimulated lymphocyte cultures with reference to bovine lymphosarcoma, *J. Natl. Cancer Inst.,* 43, 1297, 1969.
20. Callahan, C., Lieber, M., Todaro, J., Graves, C., and Ferrer, F., Bovine leukemia virus genes in the DNA of leukemic cattle, *Science,* 192, 1005, 1976.
21. Kettmann, R., Burny, A., Cleuter, Y., Ghysdael, J., and Mammerickx, M., Distribution of bovine leukemia virus proviral DNA sequences in tissues of animals with enzootic bovine leukosis, *Leukemia Res.,* 2, 23, 1978.
22. Onuma, M. and Olson, C., Bovine leukemia virus antigen in bovine lymphosarcoma cell cultures, in *Bovine Leucosis: Various Methods of Molecular Virology,* Burny, A., Ed., Comm. European Commun., Luxembourg, 1977, 95.

23. Ressang, A., Preliminary communications on results of studies on juvenile bovine leukaemia, *Vet. Microbiol.*, 1, 393, 1976.
24. Onuma, M., Okada, K., Yamazaki, Y., Fujinaga, K., Fujimoto, Y., and Mikami, T., Induction of C-type virus in cell lines derived from calf-form bovine lymphosarcoma, *Microbiol. Immunol.*, 22, 683, 1978.
25. Grimshaw, W. T. R., Wiseman, A., Petrie, L., and Selman, I. E., Bovine leucosis (lymphosarcoma): a clinical study of 60 pathologically confirmed cases, *Vet. Rec.*, 105, 267, 1979.
26. Anderson, L. J., Jarrett, W. F. H., and Crighton, G. W., A Classification of Lymphoid Neoplasms of Domestic Mammals, *Natl. Cancer Inst. Monogr.*, 32, National Institutes of Health, Bethesda, Md., 1968, 343.
27. Weber, W. T., Hematologic aspects of bovine lymphosarcoma, *Ann. N.Y. Acad. Sci.*, 108, 1270, 1963.
28. Marshak, R. R., Criteria for the determination of the normal and leukotic state in cattle, *J. Natl. Cancer Inst.*, 41, 243, 1968.
29. Van der Maaten, M. J. and Miller, J. M., Induction of lymphoid tumors in sheep with cell-free preparations of BLV, *Bibl. Haematol.*, 43, 277, 1976.
30. Ressang, A. A., Baars, J. C., Calafat, J., Mastenbroek, N., and Quak, T., Studies on bovine leukemia. III. The hematological and serological responses of sheep and goats to infection with whole blood from leukaemic cattle, *Zentralbl. Vet. Med. B.*, 23, 662, 1976.
31. Schmidt, F. W., Mitscherlich, E., Garcia de Lima, E., Milczewski, K. E. V., and Lembke, A., Cultivation of bovine C-type leukemia virus, its transmission to calves and the development of leukemia as determined by hematological, electron microscopical and immunodiffusion test, *Vet. Microbiol.*, 1, 231, 1976.
32. Miller, L. D., Miller, J. M., and Olson, C., Inoculation of calves with particles resembling C-type virus from cultures of bovine lymphosarcoma, *J. Natl. Cancer Inst.*, 48, 423, 1972.
33. Ferrer, J. F., Abt, D. A., Bhatt, D. M., and Marshak, R. R., Studies on the relationship between infection with bovine C-type virus, leukemia and presistent lymphocytosis, *Cancer Res.*, 34, 893, 1974.
34. Abt, D. A., Marshak, R. R., Ferrer, J. F., Piper, C. E., and Bhatt, D. M., Studies on the development of persistent lymphocytosis and infection with the bovine C-type leukemia virus (BLV) in cattle, *Vet. Microbiol.*, 1, 287, 1976.
35. Marshak, R. R. and Abt, D. A., The epidemiology of bovine leukosis, *Bibl. Haematol.*, 31, 166, 1968.
36. Abt, D. A., Marshak, R. R., Kulp, H. E., and Pollock, R. H., Studies on the relationship between lymphocytosis and bovine leukosis, *Bibl. Haematol.*, 36, 527, 1970.
37. Flensburg, J. C. and Stryffert, B., An evaluation of Danish leukosis control schemes, in *Bovine Leucosis: Various Methods of Molecular Virology*, Burny, A., Ed., Comm. European Commun., Luxembourg, 1977, 387.
38. Marshak, R. R., Hare, W. C. D., Abt, D. A., Crowshaw, J. E., Switzer, J. W., Ipsen, J., Dutcher, R. M., and Martin, J. E., Occurrence of lymphocytosis in dairy cattle herds with high incidence of lymphosarcoma, *Ann. N.Y. Acad. Sci.*, 108, 1284, 1963.
39. Kenyon, S. J. and Piper, C. E., Cellular basis of persistent lymphocytosis in cattle infected with bovine leukemia virus, *Infect. Immun.*, 16, 891, 1977.
40. Kenyon, S. J. and Peper, C. E., Properties of density gradient fractionated peripheral blood leukocytes from cattle infected with bovine leukemia virus, *Infect. Immun.*, 16, 898, 1977.
41. Bloom, J. C., Kenyon, S. J., and Gabuzda, T. G., Glucocorticoid effects on peripheral blood lymphoctyes in bovine leukemia virus-infected cows, *Blood*, 53, 899, 1979.
42. Paul, O. S., Pomeroy, K. A., Johnson, D. W., Muscoplat, C. C., Handwerger, B. S., Soper, F. F., and Sorensen, D. K., Evidence for the replication of bovine leukemia virus in the B lymphocytes, *Am. J. Vet. Res.*, 38, 873, 1977.
43. Ferrer, J. F., Bovine lymphosarcoma, in *Advances in Veterinary Science and Comparative Medicine*, Vol. 24, Brandly, C. A. and Cornelius, C., Eds., Academic Press, New York, 1980, 1.
44. Takashima, I. and Olson, C., Effect of mitogens and anti-bovine leukosis virus serums on DNA synthesis of lymphocytes from cattle, *Eur. J. Cancer*, 16, 639, 1980.
45. Thorn, R. M., Gupta, P., Kenyon, S. J., and Ferrer, J. F., Evidence that the spontaneous blastogenesis of lymphocytes from bovine leukemia virus-infected cattle is viral antigen specific, *Infect. Immun.*, 34, 84, 1981.
46. Ressang, A. A., Rumawas, W., Rondhuis, P. R., Haagsma, J., and Bercovich, Z., Studies on bovine leukosis. VII, *Zentralbl. Vet. Med. B*, 27, 576, 1980.
47. Matthaeus, W. and Straub, O. C., The immune response of normal and leukotic cattle to IBR-IPV-virus, *Zentralbl. Bakteriol. Hyg. A*, 240, 152, 1978.
48. Pierce, K. R., Young, M. F., McArthur, N. H., and Williams, J. D., Serum immunoglobulin concentrations of cattle in a herd with bovine leukosis, *Am. J. Vet. Res.*, 38, 771, 1977.

49. Kumar, S. P., Paul, P. S., Pomeroy, K. A., Johnson, D. W., Muscoplat, C. C., Van der Marten, J. J., Miller, J. M., and Sorensen, D. K., Frequency of lymphocytes bearing Fc receptors and surface membrane immunoglobulins in normal, persistent lymphocytotic and leukemic cows, *Am. J. Vet. Res.*, 39, 45, 1978.
50. Takashima, I., Olson, C., Driscoll, D. M., and Baumgartener, L. E., B-lymphocytes and T-lymphocytes in three types of bovine lymphosarcoma, *J. Natl. Cancer Inst.*, 59, 1205, 1977.
51. Wilkie, B. N., Caoili, F., and Jacobs, R. M., Bovine lymphocytes: erythrocyte rosettes in normal, lymphomatous and corticosteroid-treated cattle, *Can. J. Comp. Med.*, 43, 22, 1979.
52. Paul, P. S., Senogles, D. R., Muscoplat, C. C., and Johnson, D. W., Enumeration of T cells, B cells and monocytes in the peripheral blood of normal and lymphocytotic cattle, *Clin. Exp. Immunol.*, 35, 306, 1979.
53. Jacobs, R. M., Valli, W. E. D., and Wilkie, B. N., Response of cows with lymphoma to the intradermal injection of tumor cell antigens and phytohemagglutinin, *Can. J. Comp. Med.*, 45, 43, 1981.
54. Muscoplat, C. C., Alhaji, I., Johnson, D. W., Pomeroy, K. A., Olson, J. M., Larson, V. L., Stevens, J. B., and Sorensen, D. K., Characteristics of lymphocyte responses to phytomitogens: comparison of responses of lymphocytes from normal and lymphocytotic cows, *Am. J. Vet. Res.*, 35, 1053, 1974.
55. Muscoplat, C. C., Johnson, D. W., Pomeroy, K. A., Olson, J. M., Larson, V. L., Stevens, J. B., and Sorensen, D. K., Lymphocyte surface immunoglobulin: frequency in normal and lymphocytic cattle, *Am. J. Vet. Res.*, 35, 593, 1974.
56. Weiland, F. and Straub, O. C., Differences in the *in vitro* response of lymphocytes from leukotic and normal cattle to concanavalin A, *Res. Vet. Sci.*, 20, 340, 1976.
57. Jacobs, R. M., Valli, V. E. O., and Wilkie, B. N., Inhibition of lymphocyte blastogenesis by sera from cows with lymphoma, *Am. J. Vet. Res.*, 41, 372, 1980.
58. Trainin, Z. and Klopfer, U., Immunofluorescent studies of lymph nodes and spleens of leukotic cattle for cell producing IgM and IgG, *Cancer Res.*, 41, 1968, 1971.
59. Atluru, D., Johnson, D. W., Paul, P. S., and Muscoplat, C. C., B-lymphocyte differentiation, using pokeweed mitogen stimulation: *in vitro* studies in leukemic and normal cattle, *Am. J. Vet. Res.*, 40, 515, 1979.
60. Trainin, Z., Nobel, T. A., Klopfer, U., and Neumann, F., Absence of gamma macroglobulin (IgM) in the sera of leukotic cattle and the diagnosis of bovine leukosis, *Refu. Vet.*, 25, 185, 1968.
61. Jacobs, R. M., Valli, V. E. O., and Wilkie, B. N., Serum electrophoresis and immunoglobulin concentrations in cows with lymphoma, *Am. J. Vet. Res.*, 41, 1942, 1980.
62. Trainin, Z., Ungar-Waron, H., Meirom, L., Barnea, A., and Sela, M., IgG and IgM antibodies in normal and leukaemic cattle, *J. Comp. Pathol.*, 86, 571, 1976.
63. Ungar-Waron, H., Avraham, R., Gluckmann, A., and Trainin, Z., Reduced immunocompetence of antibodies produced in leukemia cattle, *Ann. Vet. Res.*, 9, 815, 1978.
64. Matthaeus, W. and Straub, O. C., Das serumglobulin bild bei leukose karnken rindern und schafen, *Zentralbl. Vet. Med. B*, 22, 758, 1975.
65. Matthaeus, W. and Straub, O. C., Studies on the distribution of γ-globulins and on abnormal globulins from serums of leukotic cattle and sheep, *Vet. Microbiol.*, 1, 363, 1976.
66. Matthaeus, W., Heterogenous properties of BLV-precipitating immunoglobulins in bovine leukemic serum, *Ann. Vet. Res.*, 9, 635, 1978.
67. Moulton, J. E. and Dungworth, D. L., Tumors of the lymphoid and hemopoietic tumors, in *Tumors in Domestic Animals*, 2nd ed., Moulton, J. E., Ed., University of California Press, Berkeley, 1978, 165.
68. Fujimoto, Y., Miller, J., and Olson, C., The fine structure of lymphosarcoma in cattle, *Pathol. Vet.*, 6, 15, 1969.
69. Sorensen, G. D. and Theilen, G. H., Electron microscopic observations of bovine lymphosarcoma, *Ann. N.Y. Acad. Sci.*, 3, 1231, 1963.
70. Dungworth, D. L., Theilen, G. H., and Ward, J. M., Early detection of the lesions of bovine lymphosarcoma, *Bibl. Haematol.*, 30, 206, 1968.
71. Smith, H. A., Jones, T. C., and Hunt, R. D., *Veterinary Pathology*, 4th ed., Lea & Febiger, Philadelphia, 1972, 208.
72. Jubb, K. V. and Kennedy, P. C., *Pathology of Domestic Animals 1*, 2nd ed., Academic Press, New York, 1970, 386.
73. Sonoda, M. and Marshak, R. R., Electron microscopic observations on the mononuclear cells in the peripheral blood of the clinically normal and lymphosarcoma cows, *Jpn. J. Vet. Res.*, 18, 9, 1970.
74. Olson, C., Miller, J. M., Miller, L. D., and Gillette, K. G., C-type virus and lymphocytic nuclear projections in bovine lymphosarcoma, *Am. J. Vet. Med. Assoc.*, 156, 1880, 1970.
75. Weber, A., Bingham, C., Pomeroy, K., and Dias, E., Enzootic bovine leukosis: prevalence of blood lymphocyte nuclear pockets in dairy bulls in the United States and foreign countries, *Am. J. Vet. Res.*, 41, 14, 1980.

76. Weber, A., Andrews, J., Dickinson, B., Larson, V., Hammer, R., Dirks, V., Sorensen, D., and Frommes, S., Occurrence of nuclear pockets in lymphocytes of normal, persistent lymphocytotic and leukemic adult cattle, *J. Natl. Cancer Inst.*, 43, 1307, 1969.
77. Valli, V. E. O., McSherry, B. J., Lumsden, J. H., Smart, M. E., Grenn, H. H., and Heath, B., Blood cell types associated with bovine lymphosarcoma, in *Proc. 2nd Int. Symp. Cancer Detection Prevention,* Intl. Congr. Series 322, Maltoni, C., Ed., Excerpta Medica, Amsterdam, 1973, 598.
78. Lukes, R. J. and Collins, R. D., A functional classification of malignant lymphomas, in *The Reticuloendothelial System,* Intl. Acad. Pathol. Monogr. No. 16, Rebuck, J. W., Berard, C. W., and Abell, M. R., Eds., Williams & Wilkins, Baltimore, 1975, 213.
79. Valli, V. E., McSherry, B. J., Dunham, B. M., Jacobs, R. M., and Lumsden, J. H., Histocytology of lymphoid tumors in the dog, cat and cow, *Vet. Pathol.*, 18, 494, 1981.
80. Gerard-Marchand, R., Hamlin, I., Lennert, K., Rilke, F., Stanfeld, A. G., and Van Unnik, J. A. M., Classification of non-Hodgkin's lymphomas, *Lancet,* 406, 1974.
81. Parodi, A. L., Mialot, M., Crespeau, F., Levy, D., Salmon, H., Nogues, G. and Gérard-Marchand, R., Attempt for a new cytological and cytoimmunological classification of bovine malignant lymphoma, in 4th Int. Symp. Bovine Leukosis, Straub, O. C., Ed., *Curr. Top. Vet. Med. Anim. Sci.*, 15, 1980. 561.
82. Dungworth, D. L., Theilen, G. H., and Lengyel, J., Bovine lymphosarcoma in California. II. The thymic form, *Pathol. Vet.*, 1, 323, 1964.
83. Muscoplat, C. C., Johnson, D. W., Pomeroy, K. A., Olson, J. M., Larson, V. L., Stevens, J. B., and Sorensen, D. K., Lymphocyte subpopulations and immunodeficiency in calves with acute lymphocytic leukemia, *Am. J. Vet. Res.*, 35, 1571, 1974.
84. Theilen, G. H. and Dungworth, D. L., Bovine lymphosarcoma in California: the calf form, *Am. J. Vet. Res.*, 26, 696, 1965.
85. Chander, S., Whitt, L. A., Greig, A. S., and Hare, W. C. D., Bovine lymphosarcoma in twin calves, *Can. J. Comp. Med.*, 41, 274, 1977.
86. Hugoson, G., Juvenile bovine leukosis. An epizootiological, clinical, patho-anatomical and experimental study, *Acta Vet. Scand. Suppl.*, 22, 5, 1967.
87. Richards, A. B., McArthur, N. H., and Young, M. F., Case studies of juvenile leukosis, *Cornell Vet.*, 71, 214, 1981.
88. Bundza, A., Greig, A. S., Chander, S., and Dukes, T. W., Sporadic bovine leukosis: a description of eight calves received at Animal Diseases Research Institute from 1974—1980, *Can. Vet. J.*, 21, 280, 1980.
89. Marshak, R. R., Hare, W. C. D., Dutcher, R. M., Schwartzman, R. M., Switzer, J. W., and Hubben, K., Observations on a heifer with cutaneous lymphoma, *Cancer,* 19, 724, 1966.
90. Clegg, F. W. and Moss, B., Skin lesions in a heifer — an unusual clinical history, *Vet. Rec.*, 77, 271, 1965.
91. Simoni, P. and Cinotti, S., Ultrastructural research on bovine cutaneous leukosis, in 4th Int. Symp. Bovine Leukosis, Straub, O. C., Ed., *Curr. Top. Vet. Med. Anim. Sci.*, 15, 1980, 583.
92. Stock, N. D. and Ferrer, J. F., Replicating C-type virus in phytohemagglutin-treated buffy-coat cultures of bovine origin, *J. Natl. Cancer Inst.*, 48, 985, 1972.
93. Driscoll, D. M., Baumgartener, L. E., and Olson, C., Concanavalin A and the production of bovine leukemia virus antigen in short-term lymphocyte cultures, *J. Natl. Cancer Inst.*, 58, 1513, 1977.
94. Van der Maaten, M. J., Miller, J. M., and Boothe, A. D., Replicating type-C virus particles in monolayer cell cultures of tissues from cattle with lymphosarcoma, *J. Natl. Cancer Inst.*, 52, 491, 1974.
95. Ferrer, J. F., Avila, L., and Stock, N. D., Serological detection of type-C viruses found in bovine cultures, *Cancer Res.*, 32, 1864, 1972.
96. Ressang, A. H., Mastenbroek, N., Quak, J., van Griensven, L. J. L. D., Calafat, J., Hilgers, J., Hageman, Ph.C., Soussi, T., and Swen, S., Studies on bovine leukemia. I. Establishment of type-C virus-producing cell lines, *Zentralbl. Vet. Med. B,* 21, 602, 1974.
97. Graves, D. C. and Ferrer, F., *In vitro* transmission and propagation of the bovine leukemia virus in monolayer cell culture, *Cancer Res.*, 6, 4152, 1976.
98. Diglio, C. A. and Ferrer, J. F., Induction of syncytia by the bovine C-type leukemia virus, *Cancer Res.*, 36, 1056, 1976.
99. Ferrer, J. F., Stock, N. D., and Lin, P. S., Detection of replicating C-type viruses in continuous cell cultures established from cows with leukemia: effect of culture medium, *J. Natl. Cancer Inst.*, 47, 613, 1971.
100. Van der Maaten, M. J. and Miller, J. M., Replication of bovine leukemia virus in monolayer cell cultures, *Bibl. Haematol.*, 43, 260, 1976.
101. Takashima, I. and Olson, C., Relation of bovine leukosis virus production on cell growth cycle, *Arch. Virol.*, 69, 141, 1981.
102. Van der Maaten, M. J., Communication to the Recipients of the FLK-BLV Cell Line, National Animal Disease Center, Ames, Iowa, July 11, 1977.

103. Hare, W. C. D., Lin, P., and Zachariasewycz, E., Studies on 12 cell lines established from mononuclear leukocytes of cattle with lymphosarcoma, *Cancer*, 22, 1074, 1968.
104. Mamoun, R. Z., Astier, T., and Guillemain, B., Establishment and propagation of a bovine leukemia virus-producing cell line derived from the leukocytes of a leukaemic cow, *J. Gen. Virol.*, 54, 357, 1981.
105. Kawakami, T. G., Moore, A. L., Theilen, G. H., and Munn, R. J., Comparisons of virus-like particles from leukotic cattle to feline leukosis virus, *Bibl. Haematol.*, 36, 471, 1970.
106. Dutta, S. K., Larson, V. L., Sorensen, D. K., Perman, A., Weber, A. F., Hammer, R. F., and Shope, R. E., Isolation of C-type virus particles from leukemia and lymphocytotic cattle, *Bibl. Haematol.*, 36, 548, 1970.
107. Calafat, J., Hageman, C., and Ressang, A. A., Structure of C-type virus particles in lymphocyte cultures of bovine origin, *J. Natl. Cancer Inst.*, 52, 1251, 1974.
108. Dekegel, D., Mammerickx, M., Burny, A., Portetelle, D., Cleuter, Y., Ghysdel, J., and Kettmann, R., Morphogenesis of bovine leukemia virus, in *Bovine Leucosis: Various Methods of Molecular Virology*, Burny, A., Ed., Comm. European Commun., Luxembourg, 1977, 31.
109. Calafat, J. and Ressang, A. A., Ultrastructure of bovine leukemia virus. Comparison with other RNA tumor viruses, in *Bovine Leucosis: Various Methods of Molecular Virology*, Burny, A., Ed., Comm. European Commun., Luxembourg, 1977, 13.
110. Dalton, A. J., Observations of the details of ultrastructure of a series of type-C viruses, *Cancer Res.*, 32, 1351, 1972.
111. Ferrer, J. F., Antigenic comparison of bovine type-C virus with murine and feline leukemia viruses, *Cancer Res.*, 32, 1871, 1972.
112. Calafat, J. and Ressang, A. A., Morphogenesis of bovine leukemia virus, *Virology*, 80, 42, 1977.
113. Mussgay, M., Dietzschold, B., Frenzel, B., Kaaden, O. R., Straub, O. C., and Weiland, F., The bovine leukosis virus, *Med. Micobiol. Immunol.*, 164, 131, 1977.
114. Weiland, F. and Übershär, S., Ultrastructural comparison of bovine leukemia virus (BLV) with C-type particles of other species, *Arch. Virol.*, 52, 187, 1976.
115. Graves, D. C., Diglio, C. A., and Ferrer, J. F., A reverse transcriptase assay for detection of the bovine leukemia virus, *Am. J. Vet. Res.*, 38, 1739, 1977.
116. Kettmann, R., Mammerickx, M., Dekegel, D., Ghyscael, J., Portetelle, D., and Burny, A., Biochemical approach to bovine leukemia, *Acta Haematol.*, 54, 201, 1975.
117. Fenner, F., The classification and nomenclature of viruses, *Intervirology*, 6, 1, 1976.
118. Pauli, G., Rohde, W., Ogura, H., Harms, E., and Bauer, H., Comparative immunological studies on bovine and ovine C-type particles, in *Bovine Leucosis: Various Methods of Molecular Virology*, Burny, A., Ed., Comm. European Commun., Luxembourg, 1977, 45.
119. Deshayes, L., Levy, D., Parodi, A. L., and Levy, J. P., Proteins of bovine leukemia virus. I. Characterization and reactions with natural antibodies, *J. Virol.*, 21, 1056, 1977.
120. Deshayes, L., Levy, D., and Parodi, A. L., Identification of structural polypeptides of bovine leukemia virus, in *Bovine Leucosis: Various Methods of Molecular Virology*, Burny, A., Ed., Comm. European Commun., Luxembourg, 1977, 57.
121. Kaaden, O. R., Frenzel, B., Weiland, F., Bruns, M., and Mussgay, M., Comparative electrophoretical analysis of the bovine leukemia virus polypeptides, in *Bovine Leucosis: Various Methods of Molecular Virology*, Burny, A., Ed., Comm. European Commun., Luxembourg, 1977, 83.
122. Onuma, M., Olson, C., Baumgartener, L. E., and Pearson, L. D., An ether-sensitive antigen associated with bovine leukemia virus infection, *J. Natl. Cancer Inst.*, 55, 1155, 1975.
123. Onuma, M., Olson, C., and Driscoll, D. M., Properties of two isolated antigens associated with bovine leukemia virus infection, *J. Natl. Cancer Inst.*, 57, 571, 1976.
124. Portetelle, D., Mammerickx, M., Bex, F., Burny, A., Cleuter, Y., Dekegel, D., Ghysdael, J., Kettmann, R., and Chantrenne, H., Purification of BLV gp70 and BLV p27. Detection by radioimmunoassay of antibodies directed against these antigens, in *Bovine Leucosis: Various Methods of Molecular Virology*, Burny, A., Ed., Comm. European Commun., Luxembourg, 1977, 131.
125. Miller, J. M., Van der Maaten, M. J., and Phillips, M., Studies of a glycoprotein associated with bovine leukemia virus, in *Bovine Leucosis: Various Methods of Molecular Virology*, Burny, A., Ed., Comm. European Commun., Luxembourg, 1977, 69.
126. Devare, S. G. and Stephenson, J. R., Biochemical and immunological chacterization of the major envelope glycoprotein of bovine leukemia virus, *J. Virol.*, 23, 443, 1977.
127. Rohde, W., Pauli, G., Paulsen, J., Harms, E., and Bauer, H., Bovine and ovine leukemia viruses. I. Characterization of viral antigens, *J. Virol.*, 26, 139, 1978.
128. Phillips, M., Miller, J. M., and Van der Maaten, M. J., Isolation of a precipitating glycoprotein antigen from cell cultures persistently infected with bovine leukemia virus, *J. Natl. Cancer Inst.*, 60, 213, 1978.
129. Devare, S. G., Stephenson, J. R., Sarma, P. S., Aaronson, S. A., and Chander, S., Bovine lymphosarcoma: development of a radioimmunologic technique for detection of the etiologic agent, *Science*, 194, 1428, 1976.

130. Devare, S. G., Oroszlan, S., and Stephenson, J. R., Application of radioimmunologic techniques to studies of the epidemiologic involvement of bovine leukemia virus in lymphosarcoma of domestic cattle, *Ann. Rech. Vet.*, 9, 689, 1978.
131. Burny, A., Bruck, C., Chantrenne, H., Cleuter, Y., Dekegel, D., Ghysdael, J., Kettman, R., Leclercq, M., Leunen, J., Mammerickx, M., and Portetelle, D., Bovine leukemia virus: molecular biology and epidemiology, in *Viral Oncology*, Klein, G., Ed., Raven Press, New York, 1980, 231.
132. Dietzschold, B., Kaaden, O. R., and Frenzel, B., Subunit and fine structure of the glycoprotein of bovine leukemia virus, *Ann. Rech. Vet.*, 9, 613, 1978.
133. Leamnson, R. N. and Halpern, M. S., Subunit structure of the glycoprotein complex of avian tumor virus, *J. Virol.*, 18, 956, 1976.
134. Ghysdael, J., Kettmann, R., and Burny, A., Translation of bovine leukemia virus genome information in heterologous protein synthesizing systems programmed with virion RNA and in cell-lines persistently infected by BLV, *Ann. Rech. Vet.*, 9, 627, 1978.
135. Sentsui, H., Thorn, R. M., Kono, Y., and Ferrer, J. F., Hemagglutination by bovine leukemia virus, *J. Gen. Virol.*, 59, 83, 1982.
136. Sentsui, H. and Kono, Y., Hemagglutination by equine infectious virus, *Infect. Immun.*, 14, 325, 1976.
137. Schäfer, W. and Szanto, J., Studies on mouse leukemia virus. II. Nachweiseines virus spezifischen hamagglutinins, *Z. Natursforsch.*, 24B, 1324, 1969.
138. Gilden, R. V., Long, C. W., Hanson, M., Toni, R., Charman, H. P., Oroszlan, S., Miller, J. M., and Van der Maaten, M. J., Characteristics of the major internal protein and RNA-dependent DNA polymerase of bovine leukemia virus, *J. Gen. Virol.*, 29, 305, 1975.
139. Miller, J. M. and Olson, C., Precipitating antibody to an internal antigen of the C-type virus associated with bovine lymphosarcoma, *J. Natl. Cancer Inst.*, 49, 1459, 1972.
140. Oroszlan, S., Copeland, T. D., Henderson, L. E., Stephenson, J. R., and Gilden, R. V., Amino-terminal sequence of bovine leukemia virus major internal protein: homology with mammalian type-C virus p30 structural proteins, *Proc. Natl. Acad. Sci. U.S.A.*, 76, 2996, 1979.
141. Oroszlan, S., Sarngadharan, M. G., Copeland, T. D., Kalyanaraman, V. S., Gilden, R. V., and Gallo, R. C., Primary structure analysis of the major internal protein p24 of human type-C T-cell leukemia virus, *Proc. Natl. Acad. Sci. U.S.A.*, 79, 1291, 1982.
142. McDonald, H. C., Graves, D. C., and Ferrer, J. F., Isolation and characterization of an antigen of the bovine C-type virus, *Cancer Res.*, 36, 1251, 1976.
143. Kettmann, R., Portetelle, D., Mammerickx, M., Cleuter, Y., Dekegel, D., Galoux, J., Ghysdael, J., Burny, A., and Chantrenne, H., Bovine leukemia virus: an exogenous RNA oncogenic virus, *Proc. Natl. Acad. Sci. U.S.A.*, 73, 1014, 1976.
144. Reitz, M. S., Poiesz, B. J., Ruscetti, F. W., and Gallo, R. C., Characterization and distribution of nucleic acid sequences of a novel type-C retrovirus isolated from neoplastic human T lymphocytes, *Proc. Natl. Acad. Sci. U.S.A.*, 78, 1887, 1981.
145. Ogura, H., Paulsen, J., and Bauer, H., Cross-neutralization of ovine and bovine C-type leukemic virus-induced syncytia formation, *Cancer Res.*, 37, 1486, 1977.
146. Paulsen, J., Rohde, W., Pauli, G., Harms, E., and Bauer, H., Comparative studies on ovine and bovine C-type particles, *Bibl. Haematol.*, 43, 190, 1976.
147. Kaaden, O. R., Frenzel, B., Dietzschold, B., Weiland, F., and Mussgay, M., Isolation of a p15 polypeptide from bovine leukemia virus and detection of specific antibodies in leukemic cattle, *Virology*, 77, 501, 1977.
148. Gupta, P. and Ferrer, J. F., Detection of a precursor of bovine leukemia virus structural proteins in purified virions, *Ann. Rech. Vet.*, 9, 619, 1978.
149. Burny, A., Bex, F., Chantrenne, H., Cleuter, Y., Dekegel, D., Ghysdael, J., Kettman, R., Leclercq, M., Leunen, J., Mammerickx, M., and Portetelle, D., Bovine leukemia virus involvement in enzootic bovine leukosis, *Adv. Cancer Res.*, 28, 251, 1978.
150. Ghysdael, J., Kettmann, R., and Burny, A., Translation of bovine leukemia virus virion RNAs in heterologous protein synthesizing systems, *J. Virol.*, 29, 1087, 1979.
151. Kaaden, O., Dietzschold, B., and Straub, O. C., Beitrag zur isolierung einer RNA-ab hängigen DNA-polymerase aus lymphozyten eines an lymphatischer leukose erkrankten rindes, *Zentralbl. Bakteriol. (Orig. A)*, 220, 101, 1972.
152. Dietzschold, B., Kaaden, O. R., Überschär, S., Weiland, F., and Straub, O. C., Suggestive evidence for an oncornavirus-specific DNA polymerase from C-type particles of bovine leukosis, *Z. Naturforsch. (C)*, 29, 72, 1974.
153. Kettman, R., Couez, D., and Burny, A., Restriction endonuclease mapping of linear unintegrated proviral DNA of bovine leukemia virus, *J. Virol.*, 38, 27, 1981.
154. Deschamps, J., Kettmann, R., and Burny, A., Experiments with cloned complete tumor-derived bovine leukemia information prove that the virus is totally exogenous to its target animal species, *J. Virol.*, 40, 605, 1981.

155. Shih, T. Y. and Scolnick, E. M., Molecular biology of mammalian sarcoma viruses, in *Viral Oncology*, Klein, G., Ed., Raven Press, New York, 1980, 135.
156. Onuma, M., Matsumoto, K., Moriguchi, R., Fujimoto, Y., Miyake, Y., Mikami, T., and Izawa, H., Transformed phenotypes in long-term cultures persistently infected with bovine leukemia virus, *Can. J. Comp. Med.*, 45, 154, 1981.
157. Onuma, M., Watarai, S., Suneya, M., Mikami, T., and Izawa, H., Induction of transformed phenotypes in sheep fibroblasts by culture fluids from cells persistently infected with bovine leukemia virus, *Microbiol. Immunol.* 25, 445, 1981.
158. Kettmann, R., Meunier-Rotival, M., Cortadas, J., Cuny, G., Ghysdael, J., Mammerickx, M., Burny, A., and Bernardi, G., Integration of bovine leukemia virus DNA in the bovine genome, *Proc. Natl. Acad. Sci. U.S.A.*, 76, 4822, 1979.
159. Couez, D., Mammerickx, M., Olson, C., Onuma, M., Parodi, A., Van der Maaten, M. J., Burny, A., and Kettmann, R., Base sequence variations in the bovine leukemia provirus, in 4th Int. Symp. Bovine Leukosis, Straub, O. C., Ed., *Curr. Top. Vet. Med. Anim. Sci.*, P.15, 1980.
160. Kettmann, R., Deschamps, J., Cleuter, Y., Couez, D., Burny, A., and Marbaix, G., Leukemogenesis by bovine leukemia virus: proviral DNA integration and lack of RNA expression of viral long terminal repeat and 3′ proximate cellular sequences, *Proc. Natl. Acad. Sci. U.S.A.*, 79, 2465, 1982.
161. Gallo, R. C. and Meyskens, F. L., Advances in the viral etiology of leukemia and lymphoma, *Semin. Hematol.*, 15, 379, 1978.
162. Kettmann, R., Cleuter, Y., Mammerickx, M., Meunier-Rotival, M., Bernardi, G., Burny, A., and Chantrenne, H., Genomic integration of bovine leukemia provirus: comparison of persistent lymphocytosis with lymph node tumor form of enzootic bovine leukosis, *Proc. Natl. Acad. Sci. U.S.A.*, 77, 2577, 1980.
163. Kettmann, R., Marbaix, G., Cleuter, Y., Portelle, D., Mammerickx, M., and Burny, A., Genomic integration of bovine leukemia provirus and lack of viral RNA expression in the target cells of cattle with different responses to BLV infection, *Leukemia Res.*, 4, 509, 1980.
164. Hare, W. C. D., Yang, T. J., and McFeely, R. A., A survey of chromosome findings in 47 cases of bovine lymphosarcoma (leukemia), *J. Natl. Cancer Inst.*, 38, 383, 1967.
165. Driscoll, D. M. and Olson, C., Bovine leukemia virus-associated antigens in lymphocyte cultures, *Am. J. Vet. Res.*, 38, 1897, 1977.
166. Baliga, V. and Ferrer, J. F., Expression of the bovine leukemia virus and its internal antigen in blood lymphocytes, *Proc. Soc. Exp. Biol. Med.*, 156, 388, 1977.
167. Ferrer, J. F., Cabradilla, C., and Gupta, P., Bovine leukemia: a model for viral carcinogenesis, in *Viruses in Naturally Occurring Cancers*, Vol. 7, Cold Spring Harbor Conf. Cell Proliferation, Cold Spring Harbor Laboratory, New York, 1980, 887.
168. Piper, C. E., Ferrer, J. F., Abt, D. D., and Marshala, R. A., Postnatal and prenatal transmission of the bovine leukemia virus under natural conditions, *J. Natl. Cancer Inst.*, 62, 165, 1979.
169. Kaaden, D. R., Lange, S., Romanowski, W., Marre, H., Pfeilsticker, J., and Roselius, R., Transient viraemia with bovine leukaemia virus in bulls, *Zentralbl. Vet. Med. B*, 29, 269, 1982.
170. Honma, T., Onuma, M., Suneya, M., Mikami, T., and Izawa, H., Cytotoxic antibody in cattle and sheep exposed to bovine leukemia virus, *Arch. Virol.*, 66, 293, 1980.
171. Driscoll, D. M., Onuma, M., and Olson, C., Inhibition of bovine leukemia virus release by antiviral antibodies, *Arch. Virol.*, 55, 139, 1977.
172. Watarai, S., Onuma, M., Mikami, T., and Izawa, H., Enhancement of antibody activity to bovine leukemia virus by complement, *Arch. Virol.*, 68, 135, 1981.
173. Gupta, P. and Ferrer, J. F., Expression of bovine leukemia virus genome is blocked by a non-Immunoglobulin protein in plasma from infected cattle, *Science*, 215, 405, 1982.
174. Hollinshead, A. and Valli, V. E., Preliminary findings of a tumor-associated antigen in bovine lymphosarcoma, *Bibl. Haematol.*, 43, 369, 1975.
175. Onuma, M. and Olson, C., Tumor-associated antigen in bovine and ovine lymphosarcoma, *Cancer Res.*, 37, 3249, 1977.
176. Jacobs, R. M., Valli, V. E. O., Wilkie, B. N., and Hollinshead, A. C., Partial purification of a common antigen in bovine lymphoma and its use in a lymphocyte blastogenesis assay, *Cancer Res.*, 41, 3000, 1981.
177. Jacobs, R. M., Valli, V. E. O., and Wilkie, B. N., Failure to detect soluble tumor-associated antigen by radioimmunoassay in bovine lymphoma, *Can. J. Comp. Med.*, 45, 82, 1981.
178. Onuma, M., Takashima, I., and Olson, C., Tumor-associated antigen and cell surface marker in cells of bovine lymphosarcoma, *Ann. Rech. Vet.*, 9, 825, 1978.
179. Onuma, M., Suneya, M., Mikami, T., and Izawa, H., Comparison of cell surface antigens expressed on BLV-transformed and bovine lymphosarcoma cells, in *10th Int. Symp. Comp. Res. Leukemia Related Dis.*, Yohn, D. S., Ed., 1981, 127.
180. Gillette, K. G., Olson, C., and Tekeli, S., Demonstration of abnormal antigen in bovine lymphosarcoma by immunofluorescence, *Am. J. Vet. Res.*, 30, 975, 1969.

181. Mussgay, M. and Kaaden, O., Progress in studies on the etiology and serologic diagnosis of enzootic bovine leukosis, *Curr. Top. Microbiol. Immunol.*, 79, 43, 1978.
182. House, C., House, J. A., and Glover, F. L., Antibodies to the glycoprotein antigen of bovine leukemia virus in the cattle population of five states, *Cornell Vet.*, 67, 510, 1977.
183. House, J. A. and House, C., Studies on bovine leukemia glycoprotein immunodiffusion test, in *Bovine Leucosis: Various Methods of Molecular Virology*, Burny, A., Ed., Comm. European Commun., Luxembourg, 1977, 237.
184. Ferrer, J., Piper, C., Abt, D. A., and Marshak, R. R., Diagnosis of bovine leukemia virus infections: evaluation of serologic and hematologic tests by a direct infectivity detection assay, *Am. J. Vet. Res.*, 38, 1977, 1977.
185. Paulsen, J. and Thies, E., Serological diagnosis of enzootic leukosis in cattle and sheep, in *Bovine Leucosis: Various Methods of Molecular Virology*, Burny, A., Ed., Comm. European Commun., Luxembourg, 1977, 223.
186. Miller, J. M. and Van der Maaten, M. J., Attempts to control spread of bovine leukemia virus infection in cattle by serologic surveillance with the glycoprotein agar gel immunodiffusion test, in *The Serological Diagnosis of Enzootic Bovine Leukosis*, Ressang, A. A., Ed., Comm. Euporean Commun., Luxembourg, 1978, 127.
187. Mammerickx, M., Cormann, A., Burny, A., Dekegel, D., and Portetelle, D., Eradication of enzootic bovine leukosis based on the detection of the disease by the GP immunodiffusion test, *Ann. Rech. Vet.*, 9, 885, 1978.
188. Miller, L. D., Export testing for enzootic bovine leukosis, *J. Am. Vet. Med. Assoc.*, 177, 620, 1980.
189. Gupta, P. and Ferrer, J. F., Comparison of various serological and direct methods for the diagnosis of BLV infection in cattle, *Int. J. Cancer*, 28, 179, 1981.
190. Schmerr, M. J. F., Miller, J. M., and Van der Maaten, M. J., Antigenic reactivity of a soluble glycoprotein associated with bovine leukemia virus, *Virology*, 109, 431, 1981.
191. Portetelle, D., Bruck, C., Mammerickx, M., and Burny, A., In animals infected by bovine leukemia virus (BLV) antibodies to envelope glycoprotein gp51 are directed against the carbohydrate moiety, *Virology*, 105, 223, 1980.
192. Phillips, M., Miller, J. M., and Van der Maaten, M. J., Isolation of a precipitating glycoprotein antigen from cell cultures persistently infected with bovine leukemia virus, *J. Natl. Cancer Inst.*, 60, 213, 1978.
193. Miller, J. M. and Van der Maaten, M. J., Use of glycoprotein antigen in the immunodiffusion test for bovine leukemia virus antibodies, *Eur. J. Cancer*, 13, 1369, 1977.
194. Miller, J. M. and Van der Maaten, M. J., Serologic detection of bovine leukemia virus, *Vet. Microbiol.*, 1, 195, 1976.
195. Deshayes, L., Levy, D., Parodi, A., and Levy, J. P., Spontaneous immune response of bovine leukemia-virus-infected cattle against five different viral proteins, *Int. J. Cancer*, 25, 503, 1980.
196. Burridge, M. J., Thurmond, M. C., Miller, J. M., Schmerr, M. J. F., and Van der Maaten, M. J., Fall in antibody titer to bovine leukemia virus in the periparturient period, *Can. J. Comp. Med.*, 46, 270, 1982.
197. Paulsen, J., Rudolph, R., and Miller, J. M., Antibodies to common ovine and bovine C-type virus-specific antigen in serum from sheep with spontaneous leukosis and from inoculated animals, *Med. Microbiol. Immunol.*, 159, 105, 1974.
198. Ferrer, J. F., Baliga, V., Diglio, C., Graves, D., Kenyon, S. J., McDonald, H., Piper, C., and Wau, K., Recent studies on the characterization of the bovine leukemia virus (BLV): development of new methods for the diagnosis of BLV infection, *Vet. Microbiol.*, 1, 159, 1976.
199. Frenzel, B., Kauden, O. R., Mussgon, M., Dietzscheld, B., Straub, O. C., and Weiland, F., Detection of bovine leukosis-associated antibodies by different tests, *Bibl. Haematol.*, 43, 366, 1976.
200. Ferrer, J. F., Bhatt, D., Abt, D. A., Marshak, R. R., and Baliga, V., Serological diagnosis of infection with the putative bovine leukemia virus, *Cornell Vet.*, 65, 527, 1975.
201. Trainin, Z., Meirom, R., and Gluckmann, A., Comparison between the immunodiffusion and the immunofluorescence tests in the diagnosis of bovine leukemia virus (BLV), *Ann. Rech. Vet.*, 9, 659, 1978.
202. Ressang, A. A., Ellens, O. J., Mastenbroek, N., Quak, J., Miller, J. M., and Van der Maaten, M. J., Studies on bovine leukemia. II. Haematological, serological, virological and electron microscopical diagnosis, *Zentralbl. Vet. Med. B*, 23, 566, 1976.
203. Mammerickx, M., Burny, A., Dekegel, D., Ghysdael, J., Kettmann, R., and Portetelle, D., Comparative study of four diagnostic methods of enzootic bovine leukemia, in *Bovine Leucosis: Various Methods of Molecular Virology*, Burny, A., Ed., Comm. European Commun., Luxembourg, 1977, 209.
204. Frenzel, B., Mussgay, M., Schneider, L. D., and Straub, O. C., Immunofluorescence test for bovine leukosis-associated complement-fixing antibodies, *Zentralbl. Vet. Med. B*, 22, 519, 1975.
205. Miller, J. M. and Van der Maaten, M. J., A complement fixation test for the bovine leukemia (type-C) virus, *J. Natl. Cancer Inst.*, 53, 1699, 1974.

206. Chander, S., Comparison between serological and hematological diagnosis of bovine leukosis, *Vet. Microbiol.*, 1, 239, 1976.
207. Hoff-Jorgensen, R. and Eskildsen, M., Demonstration of antibodies against bovine leukemia virus in bovine sera by complement fixation. Use of homologous complement component Clq and concentrated antigen in a direct test, in *Bovine Leucosis: Various Methods of Molecular Virology*, Burny, A., Ed., Comm. European Commun., Luxembourg, 1977, 179.
208. Mammerickx, M., Burny, A., Dekegel, D., Ghysdael, J., Kettman, R., and Portetelle, D., Studies on the diagnosis of enzootic bovine leukosis by complement fixation, *Zentralbl. Vet. Med. B*, 24, 349, 1977.
209. Levy, D., Deshayes, L., Guillemain, B., and Parodi, A. L., Bovine leukemia virus-specific antibodies among French cattle. I. Comparison of complement fixation and hematological tests, *Int. J. Cancer*, 19, 822, 1977.
210. Devare, S. G., Chander, S., Samagh, B. S., and Stephenson, J. R., Evaluation of radioimmunoprecipitation for the detection of bovine leukemia virus infection in domestic cattle, *J. Immunol.*, 119, 277, 1977.
211. Tabel, H., Chander, S., Van der Maaten, M. J., and Miller, J. M., Bovine leukosis. V. Epidemiological study of bovine C-type virus by the use of the complement fixation test, *Can. J. Comp. Med.*, 40, 350, 1976.
212. Ressang, A. A., Gielbeus, A. L. J., Quak, S., Mastenbroek, N., Tuppert, C., and de Castro, A., Studies on bovine leukosis. VI. Enzyme linked immunosorbent assay for detection of antibodies to bovine leukosis virus, *Ann. Rech. Vet.*, 9, 663, 1978.
213. Bianchfiori, F. and Cenci, G., ELISA test for detection of antibodies to enzootic bovine leukosis, in 4th Int. Symp. Bovine Leukosis, Straub, O. C., Ed., *Curr. Top. Vet. Med. Anim. Sci.*, 15, 1980. 167.
214. Behrens, F., Ziegelmaiser, R., Forschner, E., Manz, D., and Wiegand, D., Comparative studies on the enzyme-linked immunosorbent assay and immunodiffusion test in the diagnosis of enzootic bovine leukosis, in 4th Int. Symp. Bovine Leukosis, Straub, O. C., Ed., *Curr. Top. Vet. Med. Anim. Sci.*, 15, 1980. 173.
215. Poli, G., Balsari, A., Toniolo, A., Ponti, W., and Vacirca, G., Microplate enzyme-linked immunosorbent assay for bovine leukemia virus antibody, in 4th Int. Symp. Bovine Leukosis, Straub, O. C., Ed., *Curr. Top. Vet. Med. Anim. Sci.*, 15, 1980. 211.
216. Todd, D., Adair, B. M., and Wibberley, G., An enzyme-linked immunosorbent assay for enzootic bovine leukosis virus antibodies, *Vet. Rec.*, 107, 124, 1980.
217. Graves, D. C., McQuade, M., and Weibel, K., Comparison of the enzyme-linked immunosorbent assay with an early polykaryocytosis inhibition assay and the agar-gel immunodiffusion test for the detection of antibodies to bovine leukemia virus, *Am. J. Vet. Res.*, 43, 960, 1982.
218. Collet-Cassart, D., Magnusson, C. G. M., Jay, R. F., Reicher, V., and Holy, H. W., Automated methods for the determination of bovine leukosis, in 4th Int. Symp. Bovine Leukosis, Straub, O. C., Ed., *Curr. Top. Vet. Med. Anim. Sci.*, 15, 1980, 185.
219. Miller, J. M., Schmerr, M. J. F., and Van der Maaten, M. J., Comparison of four serologic tests for the detection of antibodies to bovine leukemia virus, *Am. J. Vet. Res.*, 42, 5, 1981.
220. McDonald, C. and Ferrer, F., Detection quantitation and characterization of the major internal virion antigen of the bovine leukemia virus by radioimmunoassay, *J. Natl. Cancer Inst.*, 57, 875, 1976.
221. Chander, S., Devare, S. G., and Stephenson, J. R., Comparison of radioimmunoassay and Bendixen's hematological evaluation in adult and calf forms of bovine lymphosarcoma, in *Bovine Leucosis: Various Methods of Molecular Virology*, Burny, A., Ed., Comm. European Commun., Luxembourg, 1977, 247.
222. Bex, F., Bruck, C., Mammerickx, M., Portetelle, D., Ghysdael, J., Cleuter, Y., Leclercq, M., Dekegel, D., and Burny, A., Humoral antibody response to bovine leukemia virus infection in cattle and sheep, *Cancer Res.*, 39, 1118, 1979.
223. Levy, D., Deshayes, L., Parodi, A. L., Levy, J. P., Stephenson, J. R., Devare, S. G., and Gilden, R. V., Bovine leukemia virus-specific antibodies among French cattle. II. Radioimmunoassay with the major structural protein (BLV p24), *Int. J. Cancer*, 20, 543, 1977.
224. Schmerr, M. F. and Goodwin, K. R., Optimum conditions for the radioimmunoprecipitation assay for the major internal protein of bovine leukemia virus, *Vet. Immunol. Immunopathol.*, 2, 291, 1981.
225. Devare, S. G. and Stephensen, J. R., Radioimmunoassay for the major internal antigen (p24)and envelope glycoprotein (gp51) of bovine leukemia virus, in *The Serological Diagnosis of Enzootic Bovine Leukosis*, Ressang, A. A., Ed., Comm. European Commun., Luxembourg, 1978, 43.
226. Gupta, P. and Ferrer, J. F., A critical comparison of the virus neutralization radioimmunoprecipitation and immunodiffusion tests for the serological diagnosis of BLV infection, *Ann. Rech. Vet.*, 9, 683, 1978.

227. Levy, D., Deshayes, L., Parodi, A. L., and Levy, J. P., Bovine leukemia virus-specific antibodies in French cattle. III. Prevalence of the BLV-gp51 radioimmunoassay for the detection of BLV-infected animals, *Int. J. Cancer*, 25, 147, 1980.
228. Poli, G., Pozza, O., Ponti, W., Balsari, A., and Vacirca, G., Application of counter-immunoelectrophoresis for a rapid serodiagnosis of enzootic bovine leukosis, *Br. Vet. J.*, 136, 251, 1980.
229. Wuu, K. D., Graves, D. C., and Ferrer, J. F., Inhibition of the reverse transcriptase of bovine leukemia virus by antibody in sera from leukemic cattle and immunological characterization of the enzyme, *Cancer Res.*, 37, 1438, 1977.
230. Ferrer, J. F. and Cabradilla, C. D., The phenomenon of polykaryocytosis induced by BLV in mixed cultures: specificity, mechanisms and application to the diagnosis of BLV infection in cattle, *Ann. Rech. Vet.*, 9, 721, 1978.
231. Ferrer, J. F. and Diglio, C., Development of an *in vitro* infectivity assay for the C-type bovine leukemia virus, *Cancer Res.*, 36, 1068, 1976.
232. Greig, A. S., Chander, S., Samagh, B., and Bouillant, A. M. P., A simple, rapid, syncytial-inhibition test for antibodies to bovine leukemia virus, *Can. J. Comp. Med.*, 42, 446, 1978.
233. Guillemain, B., Mamoun, R., Astier, T., Duplan, J. F., and Parodi, A. L., Early polykaryocytosis inhibition test: evaluation of its performance in a seroepidemiological survey of bovine leukemia virus-induced antibody in cattle, *Ann. Rech. Vet.*, 9, 1978, 709.
234. Ferrer, J. F., Piper, C. E., and Baliga, V., Diagnosis of BLV infection in cattle of various ages, in *Bovine Leucosis: Various Methods of Molecular Virology*, Burny, A., Ed., Comm. European Commun., Luxembourg, 1977, 323.
235. Olson, C., Baumgartener, L. E., Miller, J. M., and Van der Maaten, M. J., A comparison of tests on reference serums for BLV antibody, *Vet. Microbiol.*, 1, 275, 1976.
236. Guillemain, B., Mamoun, R., Levy, R., Astier, T., Irgens, K., and Parodi, A. A., Early polykaryocytosis inhibition: a simple quantitative *in vitro* assay for the detection of bovine leukemia virus infection in cattle, *Eur. J. Cancer*, 14, 811, 1978.
237. Astier, T., Mamoun, R., Guillemain, B., Parodi, A. L., and Duplan, J. F., Immunological nature of the inhibition of BLV-induced early polykaryocytosis by bovine sera, *Int. J. Cancer*, 22, 47, 1978.
238. Zavada, J., Cerny, L., Zavadova, Z., Bozonova, J., and Altstein, A. D., A rapid neutralization test for antibodies to bovine leukemia virus with the use of rhabdovirus pseudotypes, *J. Natl. Cancer Inst.*, 62, 95, 1979.
239. Zajac, V., Altaner, C., Zavada, J., and Cerny, L., Comparison of radioimmunoassay for internal protein of bovine leukemia virus with neutralization test employing VSV-BLV pseudotype, *Neoplasma*, 27, 517, 1980.
240. Jerabek, L., Gupta, P., and Ferrer, J. F., An infectivity assay for bovine leukemia virus using the immunoperoxidase technique, *Cancer Res.*, 39, 3952, 1979.
241. Onuma, M., Olson, C., and Baumgartener, L. E., Inhibition of bovine leukemia virus release, *J. Natl. Cancer Inst.*, 54, 1199, 1975.
242. Schmerr, M. V., Van der Maaten, M. J., and Miller, M. J., Application of radioimmunoassay for detection of the major internal antigen (p24) of bovine leukemia virus from cultured lymphocytes of cattle, *Comp. Immunol. Microbiol. Infect. Dis.*, 3, 327, 1980.
243. Van der Maaten, M. J. and Miller, J. M., Use of a continuous feline cell line for virologic and serologic investigations of bovine leukemia virus infections, *Am. J. Vet. Res.*, 41, 1789, 1980.
244. Paul, P. S., Castro, A. E., Pomeroy, K. A., Johnson, D. W., and Muscoplat, C. C., A microculture syncytia assay for bovine leukemia virus, *Vet. Microbiol.*, 3, 77, 1978.
245. Diglio, C. A., Piper, C. E., and Ferrer, J. F., An improved syncytia infectivity assay for the bovine leukemia virus, *In Vitro*, 14, 502, 1978.
246. Paul, P. S., Pomeroy, K. A., Castro, A. E., Johnson, D. W., Muscoplat, C. C., and Sorensen, D. K., Detection of bovine leukemia virus in B-lymphocytes by the syncytia induction assay, *J. Natl. Cancer Inst.*, 59, 1269, 1977.
247. Guillemain, B., Mamoun, R., Astier, T., and Duplan, J. F., Mechanisms of early and late polykaryocytosis induced by the bovine leukemia virus, *J. Gen. Virol.*, 57, 227, 1981.
248. Ferrer, J. F., Cabradilla, C., and Gupta, P., Use of a feline cell line in the syncytia infectivity assay for the detection of bovine leukemia virus infection in cattle, *Am. J. Vet. Res.*, 42, 9, 1981.
249. Irgens, K., Pinelli, C., Guillemain, B., Levy, D., and Parodi, A. L., Early syncytium formation induced by bovine leukemia virus in mixed cultures, *Biomedicine*, 27, 49, 1977.
250. Benton, C. V., Soria, A. E., and Gilden, R. V., Direct syncytial assay for the quantitation of bovine leukemia virus, *Infect. Immun.*, 20, 307. 1978.
251. Baumgartener, L. E., Olson, C., Miller, J. M., and Van der Maaten, M. J., Survey for antibodies to leukemia (C-type) virus in cattle, *J. Am. Vet. Med. Assoc.*, 166, 249, 1975.
252. Burridge, M. J., Puhr, D. M., and Hennemann, J. M., Prevalence of bovine leukemia virus infection in Florida, *J. Am. Vet. Med. Assoc.*, 179, 704, 1981.
253. Olson, C., Hoss, H. E., Miller, J. M., and Baumgartener, L. E., Evidence of bovine C-type (leukemia) virus in dairy cattle, *J. Am. Vet. Med. Assoc.*, 163, 355, 1973.

254. Piper, C., Abt, D. A., Ferrer, J., and Marshak, R. R., Seroepidemiological evidence for horizontal transmission of bovine C-type virus, *Cancer Res.*, 35, 2714, 1975.
255. Burridge, M. J., Wilcox, C. J., and Hennemann, J. M., Influence of genetic factors on the susceptibility of cattle to bovine leukemia virus infection, *Eur. J. Cancer*, 15, 1395, 1979.
256. Huber, N. L., DiGiacomo, R. F., Evermann, J. F., and Studer, E., Bovine leukemia virus infection in a large Holstein herd: cohort analysis of the prevalence of antibody-positive cows, *Am. J. Vet. Res.*, 42, 1474, 1981.
257. Van der Maaten, M. J. and Miller, J. M., Appraisal of control measures for bovine leukosis, *J. Am. Vet. Med. Assoc.*, 175, 1287, 1979.
258. Langston, A., Ferdinand, G. A. A., Ruppanner, R., Theilen, G. H., Drlica, S., and Behymer, D., Comparison of production variables of bovine leukemia virus antibody-negative and antibody-positive cows in two California dairy herds, *Am. J. Vet. Res.*, 39, 1093, 1978.
259. Ferrer, J., Piper, C., Abt, D. A., Marshak, R. R., and Bhatt, D., Natural mode of transmission of the bovine C-type leukemia virus (BLV), *Bibl. Haematol.*, 43, 235, 1976.
260. Olson, C., Kaja, R., Stauffacher, R., and Zehner, C. E., Development of bovine leukosis virus infection in cattle, *Ann. Rech. Vet.*, 9, 845, 1978.
261. Van der Maaten, M. J., Miller, J. M., and Schmerr, M. J. F., *In utero* transmission of bovine leukemia virus, *Am. J. Vet. Res.*, 42, 1052, 1981.
262. Straub, O. C. and Lorenz, R. J., The influence of colostrum and milk on the development of lymphocytosis in the bovine, *Vet. Microbiol.*, 1, 327, 1976.
263. Chander, S., Samagh, B. S., and Greig, A. S., BLV-antibodies in serial sampling over five years in a bovine leukosis herd, *Ann. Rech. Vet.*, 9, 797, 1978.
264. Levy, D., Deshayes, L., Guillemain, B., Irgens, K., and Parodi, A. L., Studies on bovine leukemia virus infection. A serological survey in French cattle, in *Bovine Leucosis: Various Methods of Molecular Virology*, Burny, A., Ed., Comm. European Commun., Luxembourg, 1977, 199.
265. Bech-Neilsen, S., Piper, C. E., and Ferrer, J. F., Natural mode of transmission of the bovine leukemia virus: role of blood-sucking insects, *Am. J. Vet. Res.*, 39, 1089, 1978.
266. Onuma, M., Watarai, S., Ichijo, S., Ishihara, K., Ohtanis, T., Sonoda, M., Mikami, T., Izawa, H., and Konishi, T., Natural transmission of bovine leukemia virus among cattle, *Microbiol. Immunol.*, 24, 1121, 1980.
267. Oshima, K., Okada, K., Numakunai, S., Yoneyama, Y., Sato, S., and Takahashi, K., Evidence on horizontal transmission of bovine leukemia virus due to blood-sucking Tabonid flies, *Jpn. J. Vet. Sci.*, 43, 79, 1981.
268. Buxton, B., Schultz, R., and Collins, W. E., Role of insects in the transmission of bovine leukosis virus: potential for transmission by mosquitoes, *Am. J. Vet. Res.*, 43, 1458, 1982.
269. Van der Maaten, M. J. and Miller, J. M., Susceptibility of cattle to bovine leukemia virus infection by various routes of exposure, in *Adv. Comp. Leukemia Res.*, Beutvelzen, P., Hilgers, J., and Yohn, D. S., Eds., Elsevier/North Holland, Amsterdam, 1978, 29.
270. Wilesmith, J. W., Straub, O. D., and Lorenz, R. J., Some observations on the epidemiology of bovine leucosis virus infection in a large dairy herd, *Res. Vet. Sci.*, 28, 10, 1980.
271. Thurmond, M. C. and Burridge, M. J., A study of the natural transmission of bovine leukemia virus: preliminary results, in 4th Int. Symp. Bovine Leukosis, Straub, O. C., Ed., *Curr. Top. Vet. Med. Anim. Sci.*, 15, 1980, 244.
272. Enke, K. H., Vermutlicher übertragungsweg der leukose vom rind auf das schaf, *Monatsh. Veterinaermed. Sonderhft.*, 19, 45, 1964.
273. Wilesmith, J. W., Straub, O. C., and Lorenz, R. J., Untersuchungen zur iatrogenen übertragung des virus der rinderleukose, *Tieraerztl. Umschau.*, 33, 519, 1978.
274. Roberts, D. H., Lucas, M. H., Wibberley, G., and Chasey, D., Investigation of the possible role of the tuberculin intradermal test in the spread of enzootic bovine leukosis, *Vet. Res. Commun.*, 4, 301, 1981.
275. Miller, J. M. and Van der Maaten, M. J., Infectivity tests of secretions and excretions from cattle infected with bovine leukemia virus, *J. Natl. Cancer Inst.*, 62, 425, 1979.
276. Ferrer, J. F., Kenyon, S. J., and Gupta, P., Milk of dairy cows frequently contains a leukemogenic virus, *Science*, 213, 1014, 1981.
277. Ferrer, J. F. and Piper, C. E., Role of colostrum and milk in the natural transmission of the bovine leukemia virus, *Cancer Res.*, 41, 4906, 1981.
278. Ferrer, J. F., Bovine leukosis: natural transmission and principles of control, *J. Am. Vet. Med. Assoc.*, 175, 1281, 1979.
279. Van der Maaten, M. J., Miller, J. M., and Schmerr, M. J. F., Factors affecting the transmission of bovine leukemia virus from cows to their offspring, in 4th Int. Symp. Bovine Leukosis, Straub, O. C., Ed., *Curr. Top. Vet. Med. Anim. Sci.*, 15, 1980. 225.
280. Baumgartener, L., Olson, C., and Onuma, M., Effect of pasteurization and heat treatment on bovine leukemia virus, *J. Am. Vet. Med. Assoc.* 169, 1189, 1976.

281. Ferrer, J. F., Eradication of bovine leukemia virus infection from a high prevalence herd using radioimmunoassay for identification of infected animals, *J. Am. Vet. Med. Assoc.*, 180, 890, 1982.
282. Lucas, M. H., Dawson, M., Chasey, D., Wibberley, G., and Roberts, D. H., Enzootic bovine leucosis virus in semen, *Vet. Rec.*, 106, 128, 1980.
283. Ressang, A. A., Mastenbroek, N., and Quak, J., Studies on bovine leucosis. IX. Excretion of bovine leucosis virus, *Zentralbl. Vet. Med. B*, 29, 137, 1982.
284. Gupta, P. and Ferrer, J. F., Detection of bovine leukemia virus antigen in urine from naturally infected cattle, *Int. J. Cancer*, 25, 663, 1980.
285. Baumgartener, L. E., Crowley, J., Entine, S., Olson, C., Hugoson, G., Hansen, H. J., and Dreher, W. H., Influence of sire on BLV infection in progeny, *Zentralbl. Vet. Med. B*, 25, 202, 1978.
286. Ressang, A. A., Mastenbroek, N., and Quak, J., Transmission and immune response in bovine enzootic leukosis, *Tijdsch. Diergeneesk.*, 105, 657, 1980.
287. Roberts, D. H., Lucas, M. H., Wibberley, G., and Chasey, D., An investigation into the susceptibility of cattle to bovine leukosis virus following inoculation by various routes, *Vet. Rec.*, 110, 222, 1982.
288. Marín, C., de López, N., de Alvaraz, L., Castaños, H., España, W., León, A., and Bello, A., Humoral spontaneous response to bovine leukemia virus infection in zebu, sheep, buffalo, and capybara, in 4th Int. Symp. Bovine Leukosis, Straub, O. C., Ed., *Curr. Top. Vet. Med. Anim. Sci.*, 15, 1980, 310.
289. Olson, C. and Baumgartener, L. E., Epizootiology of natural infection with bovine leukemia virus, *Bibl. Haematol.*, 43, 255, 1976.
290. Kenyon, S. J., Ferrer, J. F., McFeely, R. A., and Graves, D. C., Induction of lymphosarcoma in sheep by bovine leukemia virus, *J. Natl. Cancer Inst.*, 67, 1157, 1981.
291. Onuma, M., Baumgartener, L. E., Olson, C., and Pearson, L. D., Fetal infection with bovine leukemia virus in sheep, *Cancer Res.*, 37, 4075, 1977.
292. Mammerickx, M., Portetelle, D., and Burny, A., Experimental cross-transmissions of bovine leukemia virus (BLV) between several animal species, *Zentralbl. Vet. Med. B*, 28, 69, 1981.
293. Van der Maaten, M. J. and Miller, J. M., Serological evidence of transmission of bovine leukemia virus to chimpanzees, *Vet. Microbiol.*, 1, 351, 1976.
294. Baumgartener, L. E. and Olson, C., Host range of bovine leukosis virus: preliminary report, in 4th Int. Symp. Bovine Leukosis, Straub, O. C., Ed., *Curr. Top. Vet. Med. Anim. Sci.*, 15, 338, 1980.
295. Hoss, H. E. and Olson, C., Infectivity of bovine C-type (leukemia) virus for sheep and goats, *Am. J. Vet. Res.*, 35, 633, 1974.
296. Barthold, S. S., Baumgartener, L. E. and Olson, C., Lack of infectivity of bovine leukemia (C-type) virus to rats, *J. Natl. Cancer Inst.*, 56, 643, 1976.
297. Olson, C., Miller, L. D., and Miller, J. M., Role of C-type virus in bovine lymphosarcoma, *Bibl. Haematol.*, 39, 198, 1973.
298. Van der Maaten, M. J. and Miller, J. M., Sites of in vivo replication of bovine leukemia virus in experimentally infected cattle, *Ann. Rech. Vet.*, 9, 831, 1978.
299. Olson, C., Miller, L. D., Miller, J. M., and Hoss, H. E., Brief communication: transmission of lymphosarcoma from cattle to sheep, *J. Natl. Cancer Inst.*, 49, 1463, 1972.
300. Mammerickx, M., Dekegel, D., Burny, A., and Portetelle, D., Study of the oral transmission of bovine leukosis to sheep, *Vet. Microbiol.*, 1, 347, 1976.
301. Van der Maaten, M. J., Miller, J. M., Marshak, R. R., and Bhatt, D. M., Induction of lymphoid tumors in sheep with cell-free preparations of bovine leukemia virus, *Bibl. Haematol.*, 43, 377, 1976.
302. Olson, C. and Baumgartener, L. E., Pathology of lymphosarcoma in sheep induced with bovine leukemia virus, *Cancer Res.*, 36, 2365, 1976.
303. Olson, C., Kettmann, R., Burny, A., and Kaja, R., Goat lymphosarcoma from bovine leukemia virus, *J. Natl. Cancer Inst.*, 67, 671, 1981.
304. McClure, H. M., Keeling, M. E., Custer, R. P., Marshak, R. R., Abt, D. A., and Ferrer, J. F., Erythroleukemia in two infant chimpanzees fed milk from cows naturally infected with the bovine C-type virus, *Cancer Res.*, 34, 2745, 1974.
305. Irvin, A. D., Brown, C. G. D., Kanhai, G. K., and Stagg, D. A., Transplantation of bovine lymphosarcoma cells to athymic (nude) mice, *Res. Vet. Sci.*, 22, 53, 1977.
306. Onuma, M., Matsumoto, K., Kodama, H., Fujimoto, Y., Moriguchi, R., Mikami, T., and Izawa, H., Heterotransplantation of natural bovine lymphosarcoma to nude mice, *Eur. J. Cancer*, 17, 61, 1981.
307. Donawick, W. J., Johnstone, C., Martens, J. G., Dodd, D. C., Martin, J. E., and Marshak, R. R., Successful transplantation of lymphosarcoma in calves treated with anti-lymphocyte serum, *J. Natl. Cancer Inst.*, 44, 467, 1970.
308. Marshak, R. R., Hare, W. C. D., Dodd, D. C., McFeely, R. A., Martin, J. E., and Dutcher, R. M., Transplantation of lymphosarcoma in calves, *Cancer Res.*, 27, 498, 1967.
309. Miller, J. M. and Van der Maaten, M. J., Evaluation of an inactivated bovine leukemia virus preparation as an immunogen in cattle, *Ann. Rech. Vet.*, 9, 871, 1978.

310. Miller, J. M., Van der Maaten, M. J., and Schmerr, M. J. F., Vaccination of cattle with binary ethylenimine-treated bovine leukemia virus, *Am. J. Vet. Res.*, 44, 64, 1983.
311. Portetelle, D., Bruck, C., Burny, A., Dekegel, D., Mammerickx, M., and Urbain, J., Detection of complement-dependent lytic antibodies in sera from bovine leukemia virus-infected animals, *Ann. Rech. Vet.*, 9, 607, 1978.
312. Theilen, G. H., Miller, J. M., Higgins, J., Ruppanner, R. N., and Garrett, W., Vaccination against bovine leukemia virus infection, in 4th Int. Symp. Bovine Leukosis, Straub, O. C., Ed., *Curr. Top. Vet. Med. Anim. Sci.*, 15, 1980,547.
313. Bause, I., Maas-Inderwiesen, F., and Schmidt, F. W., Results of an epidemiologic survey of enzootic bovine leukosis in the northern part of Lower Saxony and a preliminary communication of an examination into relationship between BLV-antibody development and calving, *Ann. Rech. Vet.*, 9, 765, 1978.
314. Ferdinand, G. A. A., Langston, A., Ruppanner, A., Drlica, S., Theilen, G. H., and Behymer, D. E., Antibodies to bovine leukemia virus in a leukosis dairy herd and suggestions for control of infection, *Can. J. Comp. Med.*, 43, 173, 1979.
315. Bürki, F., Experiences gained and progress achieved with BLV elimination from Austrian livestock, in 4th Int. Symp. Bovine Leukosis, Straub, O. C., Ed., *Curr. Top. Vet. Med. Anim. Sci.*, 15, 1980, 516.
316. Vries, G. de., Progress and problems in the eradication of EABL in the Netherlands, in 4th Int. Symp. Bovine Leukosis, Straub, O. C., Ed., *Curr. Top. Vet. Med. Anim. Sci.*, 15, 1980, 510.
317. Flensburg, J. C., Serological evidence of infection with bovine leukemia virus (BLV) in Danish cattle: an account of the enzootic bovine leukosis (EBL) control program 1979/80, in *4th Int. Symp. Bovine Leukosis*, Straub, O. C., Ed., *Curr. Top. Vet. Med. Anim. Sci.*, 15, 1980, 500.
318. Schmidt, F. W., Results and observations of an EBL eradication programme based on AGIDT diagnosis and culling of reactors, in 4th Int. Symp. Bovine Leukosis, Straub, O. C., Ed., *Curr. Top. Vet. Med. Anim. Sci.*, 15, 1980, 491.
319. Thurmond, M. C., Carter, R. L., Puhr, D. M., and Burridge, M. J., Decay of colostrol antibodies to bovine leukemia virus with application to detection of calf hood infection, *Am. J. Vet. Res.*, 43, 1152, 1982.
320. Burridge, M. J., Review of epidemiologic studies investigating possible association between bovine leukosis and human disease, in *Proc. Bovine Leukosis Symp.*, May 22, U.S. Department of Agriculture, Washington, D.C., 1979, 135.
321. Trainin, Z., Meirom, R., Barnea, A., and Ungar-Waron, H., Common reactivity of bovine and human sera towards bovine lymphoid tumor cells, *Bibl. Haematol.*, 43, 232, 1976.
322. Donham, K. H., Berg, J. W., and Sawin, R. S., Epidemiologic relationships of the bovine population and human leukemia in Iowa, *Am. J. Epidemiol.*, 112, 80, 1980.
323. Donham, K. H., Van der Maaten, M. J., Miller, J. M., Kruse, B. C., and Rubino, M. J., Seroepidemiologic studies on the possible relationships of human and bovine leukemia: brief communication, *J. Natl. Cancer Inst.*, 59, 851, 1977.
324. Olson, C. and Driscoll, D. M., Bovine leukosis: investigation of risk for man, *J. Am. Vet. Med. Assoc.*, 173, 1470, 1978.
325. Burridge, M. J., The zoonotic potential of bovine leukemia virus, *Vet. Res. Commun.*, 5, 117, 1981.
326. Caldwell, G. G., Bovine leukemia virus, public health serologic studies, in *Proc. Bovine Leukosis Symp.*, May 22, U.S. Department of Agriculture, Washington, D.C., 1979, 143.
327. Blair, A. and Hayes, H. M., Cancer and other causes of death among U.S. veterinarians, 1966—1976. *Int. J. Cancer*, 25, 181, 1980.

Chapter 3

JAAGSIEKTE: AN INFECTIOUS PULMONARY ADENOMATOSIS OF SHEEP

D. W. Verwoerd, R. C. Tustin, and A. Payne

TABLE OF CONTENTS

I. INTRODUCTION

Jaagsiekte (JS), or ovine pulmonary adenomatosis, has been known as a disease entity for more than a century. Nevertheless, it remained an enigmatic condition until recently when its etiology was elucidated. The discovery that the causal agent is a retrovirus answered some of the questions, but also raised others, and thus, for various reasons, stimulated a wider interest in the disease.

First of all, JS is a convenient model of a transmissible mammalian adenocarcinoma of viral origin. It can be transmitted experimentally with high efficiency and with a relatively short incubation period to outbred, newborn lambs. Apart from murine mammary tumor virus, the jaagsiekte retrovirus (JSRV) is to date the only retrovirus shown to be involved in the neoplastic transformation of mammalian epithelial cells. As such, it can play a vital role in establishing whether the oncogene-mediated mechanism of transformation, which was discovered for the classical retroviruses involved in tumors of the blood cells and connective tissue, also applies to carcinomas.

Of equal interest is the fact that JS is widely regarded as a "slow viral disease" and has often been confused with maedi, another lung condition of sheep that is the archetype of this group of diseases. Maedi is also caused by a retrovirus (MVV), which is the type species of the Lentivirinae subgroup of retroviruses. Other members of this subgroup include the etiological agents of ovine progressive pneumonia (PPV) and caprine arthritis-encephalitis (CAEV). These viruses usually cause chronic inflammatory lesions rather than neoplastic transformation and are, therefore, biologically distinct from JSRV. It would be of considerable importance to determine whether any relationship exists between these two groups of retroviruses that sometimes coexist in one host. Whether they are related or not, a comparative study of the molecular mechanisms involved in their pathogenesis should yield some information useful for the elucidation of viral carcinogenesis.

Finally, it is a disease that is widely distributed throughout the world and is of considerable economic importance in those countries that have large sheep populations.

II. HISTORY

The earliest reports concerning JS that have been traced emanated from the Cape of Good Hope (now the Cape Province of the Republic of South Africa) during the early part of the 19th century.[1] It can be assumed that the condition alluded to was indeed JS because Hutcheon in 1891 defined it as a distinct disease entity and described its symptomatology and macroscopic pathology in great detail, as well as attempts to control it.[2] The occurrence of JS in England was first described in 1894 (albeit unwittingly, for McFadyean, who reported it at the time, considered the lesion to be caused by lungworms).[3] It was first recorded in Germany in 1899,[4] in France in 1926,[5] and has subsequently been diagnosed in many other countries of the world.

The disease has been confused in the past with several other conditions affecting the ovine lung, the most noteworthy of which are ovine progressive pneumonia, maedi, "Laikipia lung disease", "Graaff-Reinet disease", and "epithelialization" of the lung. Ovine progressive pneumonia has been described in several countries, under various names including Montana progressive pneumonia in the U.S., zwoegerziekte in the Netherlands, la bouhite in France, and Maedi in Iceland.[1,6,7] "Laikipia lung disease" is a term coined in Kenya to describe a complex of lung conditions that includes bacterial, lungworm, and possibly chlamydial pneumonias, ovine progressive pneumonia, and JS,[6] while "Graaff-Reinet disease" is a now defunct name that was used to describe a chronic form of pneumonia of undetermined etiology. It occurred in the Graaff-Reinet district of the Republic of South Africa, but has not been diagnosed in that area for more than half a century. The term "epithelialization" implies a metaplasia of the lining epithelium of the pulmonary alveoli to a basically cuboidal type (in most instances probably due to a Type II pneumocyte proliferation) and is a nonspecific change caused by a variety of etiologic factors. It serves no useful purpose to compound and perpetuate this confusion by documenting in detail these past mistakes and inaccuracies.

The probable infectious nature of JS was realized at a very early stage by the scientists who were the first to be confronted with field outbreaks and with the problem of controlling it. In fact, Hutcheon emphasized that there was strong evidence that JS was contagious in character and that, once it commenced in a flock, it was apt to increase.[2] Although he did not know what the cause was, but considered it "something special", he advocated immediate slaughter of every animal as soon as it was observed to be affected in an attempt to arrest a further spread of the disease in the flock.

De Kock[8,9] was probably the first to transmit the disease experimentally. In 1929 he reported that small numbers of healthy sheep that had been maintained in close contact with diseased animals developed lesions, which were small, localized and asymptomatic. Because he had not been able to reproduce what he regarded as the typical disease, he was rather hesitant to claim that the disease had, in fact, been transmitted. Nine years later Dungal[10] in Iceland showed unequivocally that JS could be transmitted by cohabitation of healthy with diseased sheep. He housed eight uninfected animals with diseased animals. One of these sheep died of the disease after 7 months, and typical lesions were present in six of the others when slaughtered after 10 months. He also demonstrated by means of transmission experiments that the exhaled respiratory air and bronchial secretions of affected sheep contained the etiologic agent.[11]

III. EPIZOOTIOLOGY

Jaagsiekte is present or has been diagnosed in many countries in four continents of the world (Table 1). It has not been recorded from Australia or New Zealand despite the large sheep populations in those two countries. The disease was introduced into Iceland in 1933, and after a widespread epizootic, it was finally eradicated in that country in 1952 after an intensive slaughter campaign.[12] It is endemic in most European and some African countries with significant sheep populations, while sporadic cases have been found elsewhere.

To date, JS has been confirmed only in the domestic sheep. A number of case reports in goats were probably based on incorrect diagnoses, as all attempts by various workers to transmit the disease to goats and other experimental animals were unsuccessful. The only exception was the successful transplantation of JS tumor cells into nude mice.[15]

All sheep breeds seem to be susceptible, but to different degrees. In Iceland, up to 90% of the animals of the Gottorp breed were affected, compared with only 10% in the Adalbol breed.[10] The fact that in Great Britain JS is found mainly in parts of Scotland, and only sporadically in England, suggests differences in susceptibility among different breeds.[16] In South Africa there are some indications of a lower incidence in English breeds than in the Merino and Karakul and their cross-breeds, which make up the bulk of the sheep population.

The incidence of JS in South African flocks can vary from less than 1% to about 20%, depending on breed susceptibility, flock management, and other factors. In a recent survey in Scotland, 20% of all mature sheep necropsied at the Veterinary Investigation Centres were found to have JS lesions in their lungs.[17]

IV. CLINICAL SIGNS

The duration of the natural incubation period is not known precisely, but, by ex-

Table 1
COUNTRIES IN WHICH JAAGSIEKTE HAS BEEN RECORDED[1,13]

Country	Incidence
America	
North America	Sporadic[13]
U.S.	
Canada	Sporadic[14]
South America	Sporadic
Chile	
Peru	Endemic — enzootic
Asia	
India	Sporadic
Europe	
U.K.	Endemic
France	Sporadic
E. Germany	Endemic
W. Germany	Endemic
Bulgaria	Endemic
Yugoslavia	Sporadic
S.E. Rusia	Endemic
European Russia	Endemic
Greece	Endemic
Italy	Endemic
Spain	Endemic
Portugal	Endemic
Czechoslovakia	Sporadic
Rumania	Endemic
Israel	Endemic
Turkey	Sporadic
Iceland	Eradicated in 1952
Africa	
Republic of South Africa	Endemic
South West Africa/Namibia	Endemic
Tanzania	Sporadic
Kenya	Endemic

trapolation of results obtained experimentally, it probably varies from about 9 months to 2 to 3 years. However, after inoculation of lambs with high concentrations of the etiologic agent, very small lesions have been observed in the lungs within as short a period as 21 days.[19,20] It must also be borne in mind that in a small percentage of infected animals, the lung lesions do not progress to the stage where they will induce clinical manifestations. Because of the long incubation period, clinical signs in natural cases are very rarely encountered in animals under the age of 9 months.

Clinical signs are manifested only when lung lesions are relatively advanced. The onset is insidious, and the first sign is that the respiration rate is more rapid than normal after an affected animal has been driven. At this stage the animal may be in good condition, but later on it loses weight rapidly. As the condition progresses, the animal lags behind the flock when driven. Marked respiratory distress is evident on exercise, the respiratory movements being short and jerky. Dyspnea is progressive. Moist rales are heard on auscultation of the chest. Spasmodic bouts of coughing occur. There is a great increase in the amount of secretion from the lungs, and if a severely affected sheep is upended with the head down, the fluid streams out of the nostrils. This is regarded as a pathognomonic sign of JS. There is, characteristically, no fever, but this may occur as a result of secondary bacterial infections, which are a common complication of JS. Appetite becomes impaired as the disease progresses.

The duration of JS in its clinical form varies considerably. A sheep may survive for a few weeks or months, and cases are on record where affected animals have survived for longer than a year, if well cared for. Once symptoms are seen, the disease is irreversible, however. When newborn lambs are experimentally infected, the disease is often acute, and animals die within a few days of first showing clinical signs.[19]

V. TRANSMISSION

An epithelial cell culture, (15.4), established from a lung lesion of a field case of JS, was shown to possess a male karyotype even though the donor was female.[21] This suggested transplantation of tumor cells as a possible mode of natural transmission. By intratracheal inoculation of the same and other tumor cell lines, it was confirmed experimentally that the disease can be transmitted by means of the transplantation of viable tumor cells, which were also found to be present in considerable numbers in lung exudate. In the same series of experiments, however, inoculation of male cells into female animals in some cases produced female tumors, an indication of the transformation of the recipients' cells. Transformation by means of a subcellular fraction was proved by injecting a homogenate of male cells into a female animal and producing a female tumor.[21] It is clear, therefore, that two modes of transmission coexist in nature: transplantation and transformation. The relative importance of these two mechanisms is not known, but in both cases droplet infection via the respiratory system is indicated as the natural route. there is no evidence so far for either vertical transmission or for intrauterine infection.

Experimental transmission of the disease has been achieved by means of parenteral inoculation of JS-lung suspensions by the intrapleural, intrapulmonary, intratracheal, and subcutaneous routes as well as by exposing sheep to aerosol sprays of cell-free lung extracts.[1,6] During transmission studies with intratracheal injections of semipurified lung rinse material, the success rate of experimental transmission was strongly dependent on the age of the recipient animal, however. Newborn lambs were found to be highly susceptible, with the efficiency of transmission decreasing to almost zero at the age of 6 months. It can be expected, therefore, that natural infection also occurs predominantly in young animals. Newborn lambs inoculated with concentrated lung rinse material developed an acute form of the disease, showing clinical signs as soon as 3 to 5 weeks after infection.[19,30]

VI. ETIOLOGY

Even though JS was known to be an infectious disease long before its neoplastic nature was established in the 1920s, many years of painstaking research in various countries failed to reveal the etiological agent. Transmission of the disease by means of inoculations of cell-free filtrates obviously suggested a viral etiology, though at one stage *Mycoplasma ovipneumonia* was also though to be a candidate etiological agent.

Herpesvirus ovis has been demonstrated in or isolated from adenomatous lungs, or cell cultures derived from it, by various workers.[22-25] All attempts to transmit the disease with herpesvirus isolates failed, however, and a serological survey indicated that the virus is widely distributed in sheep populations. Furthermore, molecular hybridization studies failed to reveal any correlation between the presence of herpesvirus DNA sequences in the cell genomes and the occurrence of the disease.[26] The virus is therefore regarded, at present, as a passenger with no direct role in the etiology of JS.

The presence of Type C retrovirus particles in electron micrographs of adenomatous tissue was first described by workers in Israel, who also obtained biochemical evidence for the presence of reverse transcriptase activity in affected lungs.[27,28] Morphologically typical retroviruses were also demonstrated in cell cultures derived from material orig-

Table 2
SERIAL TRANSMISSIONS OF JAAGSIEKTE TO NEW-BORN LAMBS LEADS TO AN INCREASE IN VIRAL CONCENTRATION (RDP ACTIVITY) IN THE LUNGS ACCOMPANIED BY A DECREASE IN THE INCUBATION PERIOD (LAG)[19]

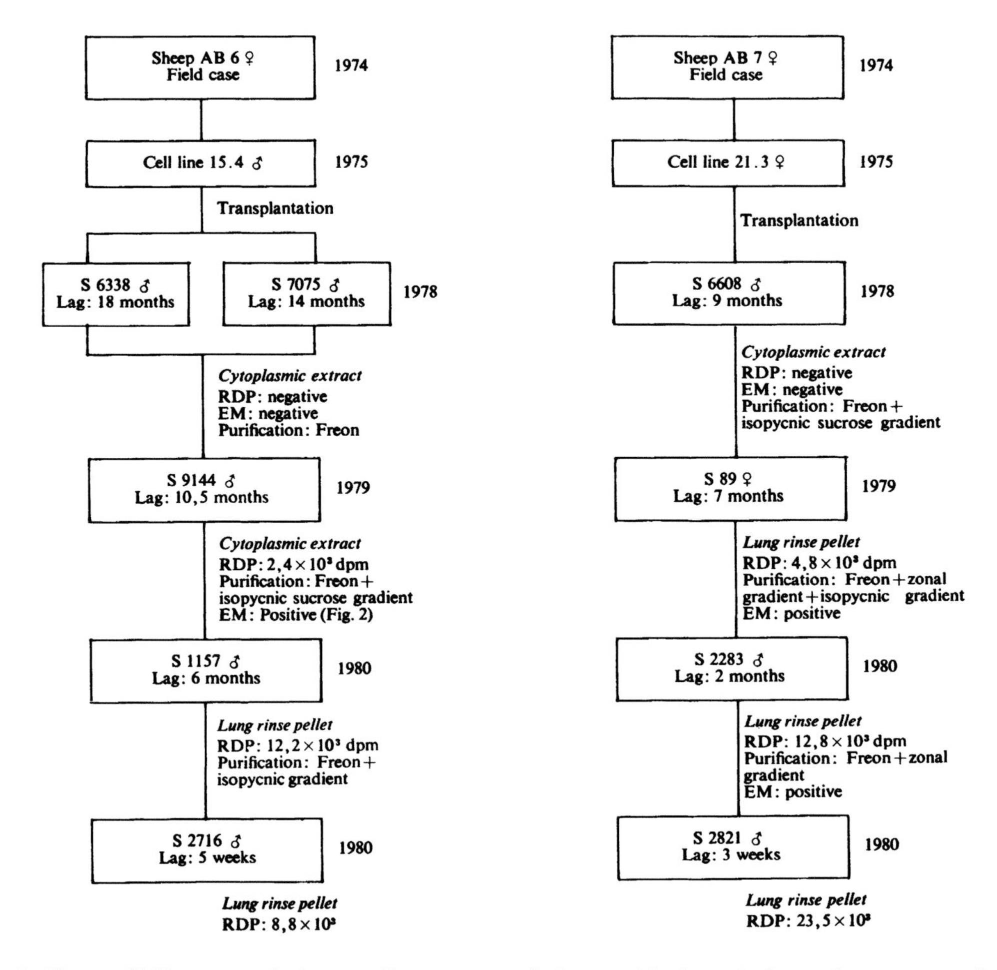

in Kenya.[23] No transmission studies were carried out with these isolates, however, and no attempt was made to distinguish the particles observed from those of maedi-visna virus, a retrovirus of sheep present in both countries.

Transmission of JS with lung exudate possessing reverse transcriptase activity was the first real evidence for the involvement of a retrovirus in the etiology of the disease.[7,19,29] Early attempts to isolate and cultivate the putative JSRV from lung exudate all failed, however. It also proved possible to transmit the disease by means of cytoplasmic extracts from tumor cell lines established from lung lesions, but again no virus could be isolated from these fractions. Serial transmissions to newborn lambs over a number of years resulted in a gradual increase of virus concentration, as measured in terms of reverse transcriptase (RDP) activity in both cytoplasmic extracts and lung exudate (Table 2). Eventually the point was reached in which concentration and purification of lung rinse material yielded virions of sufficient purity to demonstrate that they possess the size, density, and structure typical of retroviruses. These preparations were highly efficient in producing acute cases of JS and this, together with an inverse relationship between the RDP activity of the inoculum and the lag phase before symp-

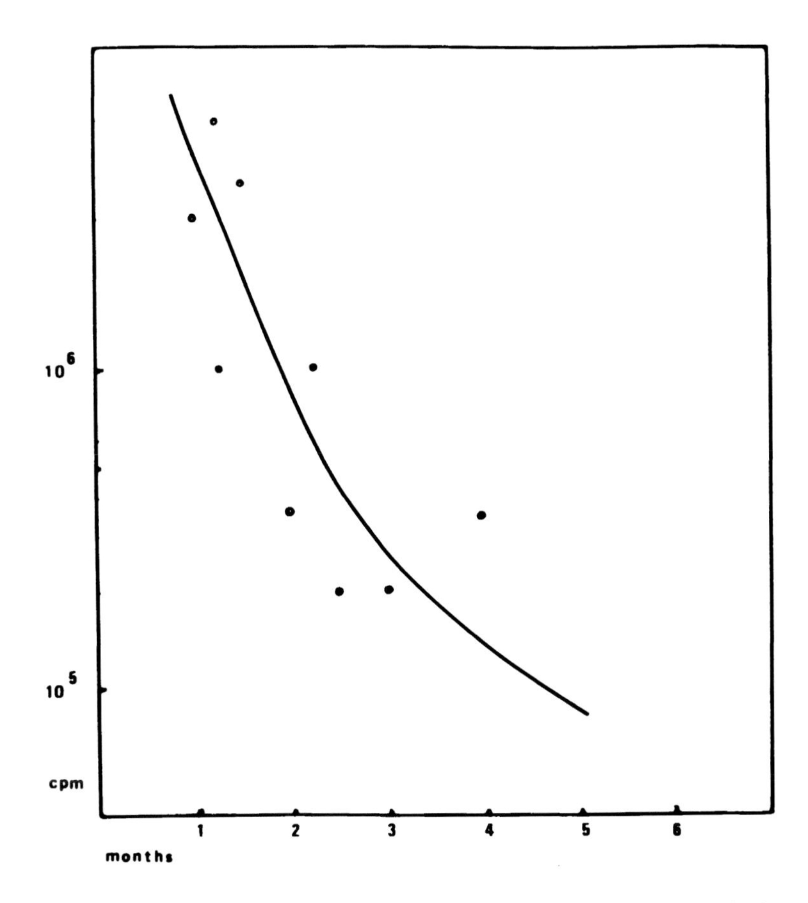

FIGURE 1. Inverse relationship between virus concentration (RDP activity in cpm) and incubation period (months) in a group of newborn lambs inoculated intratracheally with concentrated and semipurified lung rinse fluid from jaagsiekte lungs.

toms appeared (Figure 1), constituted conclusive evidence that a retrovirus is the etiological agent of JS.[19,30]

VII. CHARACTERISTICS OF THE JAAGSIEKTE RETROVIRUS (JSRV)

A. Isolation and Purification

Until recently, when short-term RDP activity was also detected in primary cultures established from adenomatous tissue,[30] the only source of virus had been the fluid obtained from adenomatous lungs, and all published reports on the characteristics of the virus are based on material derived from this source.[19,20] A procedure for the isolation and partial purification of virus from lung rinse material has recently been published.[31]

The production of virus consisted of inoculating newborn lambs intratracheally with semipurified virus. Advanced lesions usually developed within 2 to 3 months, at which stage the lambs were slaughtered and the lungs rinsed thoroughly by pouring 1 to 2 *l* of buffer into the trachea and massaging the lungs well before collecting the rinse fluid. After the removal of cells and cell debris by low speed centrifugation, crude viral pellets were obtained by high speed centrifugation. These pellets contained large amounts of unidentified contaminants, and it was only possible to attempt further purification after the surprising discovery that most of this material can be removed by extraction

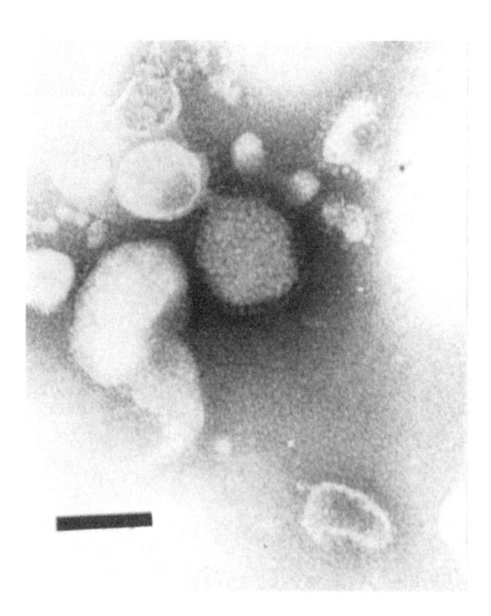

FIGURE 2. Negatively stained, fixed, JSRV particle. Bar = 100 nm.

with fluorocarbons without loss of reverse transcriptase activity. Similar treatment completely degrades other retroviruses. Further purification consisted of zonal or isopyknic centrifugation in sucrose gradients and chromatography on Sephacryl 1000 columns. Virus purified in this way was shown to consist of immune complexes, mainly with IgA, which probably explained both its resistance to fluorocarbons and its noninfectivity for cell cultures. The immunoglobulin can also be expected to have some influence on some of the physical characteristics, such as density, reported for the virus.

B. Physicochemical Properties

Gauged from electron microscopy,[32] mature virions have a diameter of 104 to 120 nm. Their density has been reported to be between 1.17 and 1.18 g/mℓ,[19,20] but values of 1.16 to 1.19 have been found, depending on the degree of purification and of viral disruption in various preparations. It should be borne in mind that RDP activity was used in all these experiments to indicate viral concentration due to the lack of a practical assay for biological activity. This can be very misleading, as the relationship between viral integrity and enzyme activity is unknown.

The viral genome consists of single-stranded RNA with a sedimentation constant of 60 to 70S.[31] This is the typical genome structure of retroviruses. As in the case of the other retroviruses, the JSRV-RNA is contained in a core particle with a density of 1.21 g/mℓ, which forms part of a complex capsid structure.[20,31] Preliminary electrophoretic studies indicate that the capsid consists of at least seven polypeptides, the main components varying in size between 25,000 and 84,000 daltons.

C. Morphology and Morphogenesis

The morphology of JSRV has been studied electron microscopically, using both purified virus preparations and thin sections of affected lungs derived from experimentally produced cases.[19,30,32] Negatively stained, fixed preparations revealed particles with an average diameter, which includes an envelope with spikes on the surface, of 107 nm (Figure 2).

The spikes are spaced at intervals of 11 to 13 nm, are 10 to 12 nm long and consist of narrow spines with knob-like extremities. The general structure of negatively stained JSRV is thus very similar to that of other retroviruses with surface spikes, such as

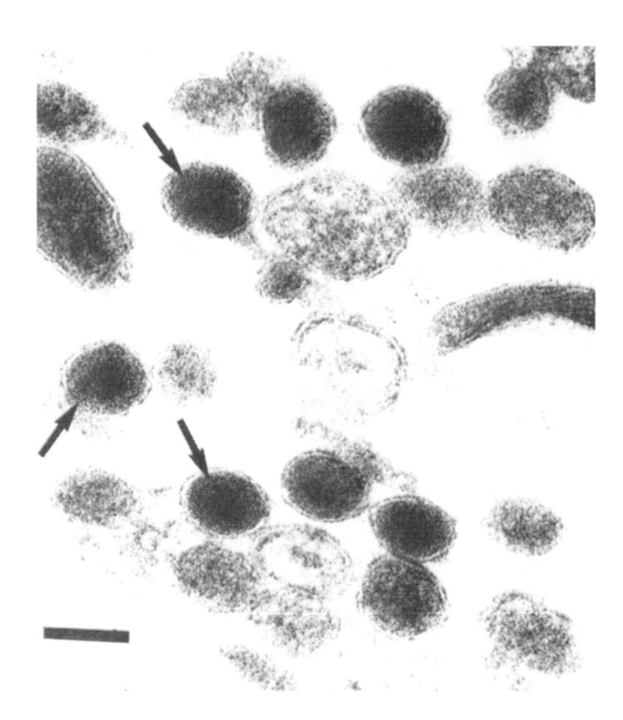

FIGURE 3. JSRV particles (arrows) in a pellet purified from a lung wash. Bar = 100 nm.

mouse mammary tumor virus, but the length and spacing of the spikes can serve as a distinguishing feature.

Positively stained virions of JSRV are also similar to certain retroviruses in some respects, but distinct in others. The particles vary in size from about 104 nm in the case of purified viruses to 121 nm in thin sections, and possess a slightly eccentric spherical nucleoid surrounded by a unit membrane. The most distinguishing feature of the South African virus is its electron-dense appearance, due to the electron density of its perinucleoidal space (Figure 3).[32] However, the virus observed in experimentally induced JS lungs in Scotland does not have an electron-dense perinucleoidal space.[30]

Immature intracytoplasmic particles are found in some JS tumor cells.[32] These particles are on average 74 nm in diameter and are usually found singly or in small clusters of two or three. They usually have a small electron-lucent center surrounded by a dense inner and less electron-dense outer shell, which sometimes carries projections (Figure 4 A). The particles are often associated with centrioles, suggesting a possible mechanism of transport to the cell surface. A more condensed form of intracytoplasmic particle, found adjacent to the plasma membrane (Figure 4B), suggests some form of maturation before it buds from the tips of microvilli into the intercellular spaces (Figure 4C) or alveolar lumen.[32] Immature extracellular virions are rarely seen, so that maturation is apparently completed before or during the budding process.

The morphogenesis of JSRV is therefore quite distinct from either that of typical Type C retroviruses or of maedi-visna virus, in which the virions are assembled at the plasma membrane in a typical crescent shape before budding as an immature particle. These virions mature only after completion of the budding process. Types B and D retroviruses bud with a complete core, as does JSRV, but their intracytoplasmic particles differ morphologically from those of JSRV.

To summarize, JSRV has a structure typical of retroviruses in general, but it can be distinguished from other members of the family in certain aspects of its morphology and morphogenesis.

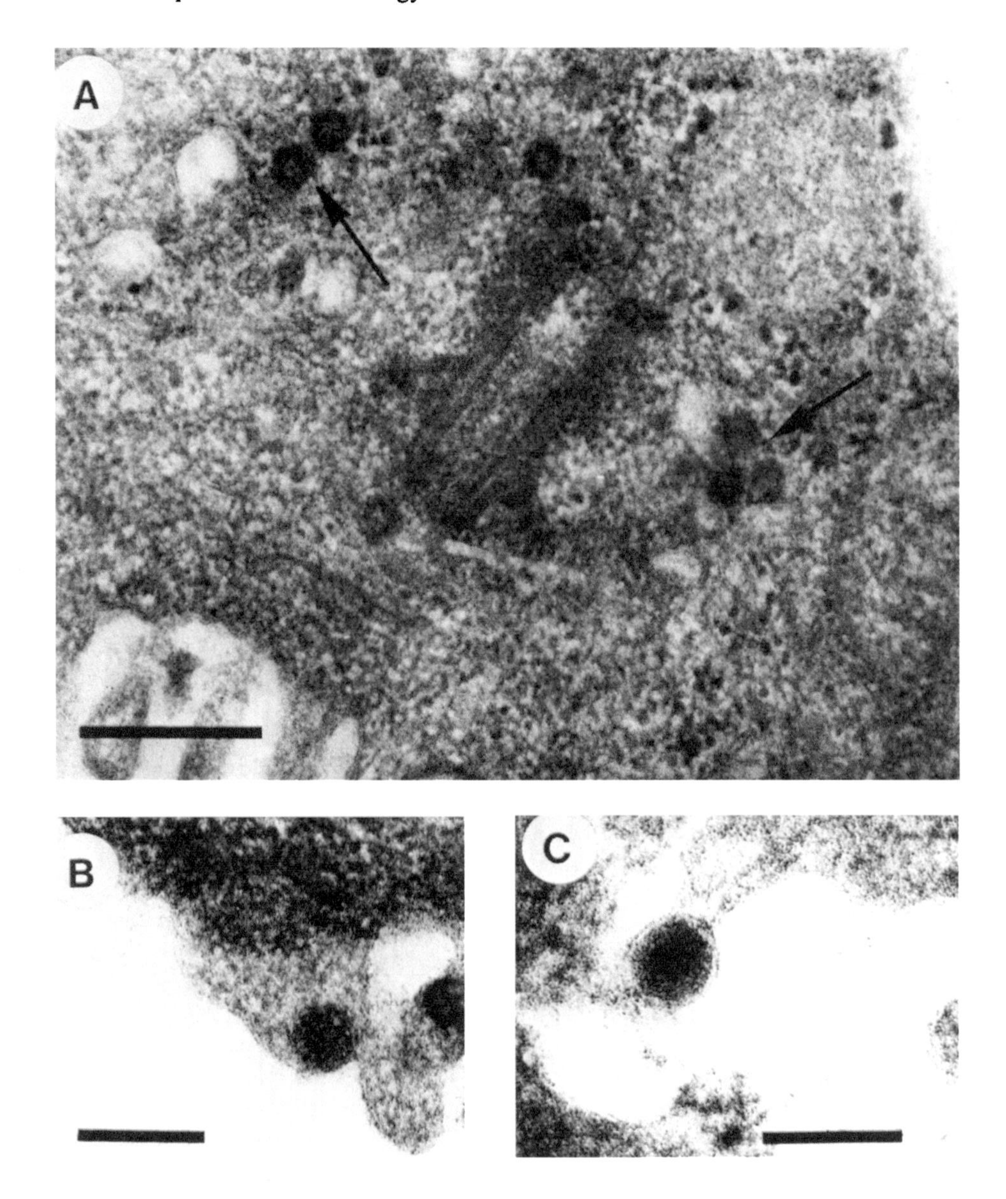

FIGURE 4. (A) Intracytoplasmic JSRV (arrow) associated with a centriole in a jaagsiekte tumor cell. Barr = 300 nm. (B) JSRV particle at the surface of a tumor cell. Bar = 150 nm. (C) Final stage of budding of a JSRV particle from the tip of a tumor cell microvillus. Bar = 150 nm.

D. Relationship to Other Retroviruses

It is not uncommon for one species to harbor several retroviruses at once. Murine cells, for example, can simultaneously contain and even produce MMTV, a Type B retrovirus, and several Type C viruses. It is, therefore, not surprising that the sheep should be host to two groups of retroviruses, both causing "slow viral disease", but biologically quite distinct. The lentiviruses (MVV, PPV, CAEV) are involved in the etiology of disease conditions characterized by chronic inflammatory lesions in the lungs, central nervous system (CNS) and joints. JSRV, on the other hand, elicits a neoplastic disease by transforming epithelial cells in the lungs. The morphology and morphogenesis of the two groups, as well as their biochemical chacteristics, discussed in the previous sections, also differ significantly. The lentiviruses resemble Type C viruses, whereas JSRV, in its density (1.175 g/mℓ) and its preference for Mg^{++} as divalent ion for its reverse transcriptase, as well as in its morphology and morphogenesis, more closely resembles Type B and Type D retroviruses.

Molecular hybridization experiments, using cloned MMTV (Type B) and BLV proviruses as probes, failed to reveal any significant homology of the genomes of these viruses with JSRV, however.[33] Antisera neutralizing the reverse transcriptases of Types B, C, and D as well as BLV failed to neutralize the JSRV enzyme. A more sensitive ELISA test, using disrupted whole virus as antigen, did not reveal any relationship between JSRV and these viruses, or the lentiviruses either.[31]

On the other hand, using a sensitive immunoblotting technique, antisera against the group-specific P27 antigen of Type D Mason-Pfizer monkey virus and against whole MMTV cross-reacted with a P25 antigen in material derived from JS lungs, but not in MVV.[34] Matters are further complicated by the recent isolation, in cell cultures, of viruses from JS lung material that cross-reacted with MVV serologically and in molecular hybridization studies.[33,35] These isolates could represent new viruses or recombinants of the previously known ones.

It is difficult at present to interpret these conflicting results. Much more work is needed in this field, as it is essential to be able to distinguish between the different viruses for diagnostic and control purposes.

VIII. PATHOGENESIS

The target cell for JSRV appears to be the Type II alveolar epithelial cell of the sheep lung.[30,32] Viral replication was observed in these cells, which also constitute the main population of transformed tumor cells of the typical adenomatous lesion of JS. The other cell type that may be involved is the nonciliated bronchial epithelial cell, also known as the Clara cell. No immature intracytoplasmic particles or other evidence for replication was found in the fibroblastic elements or in the macrophages present in large numbers in adenomatous lungs, or in other organs, in our studies. In Israel, Type C particles were observed budding from fibroblast-like cells and plasma-like cells in JS tumor tissue.[36,37] Antibodies to the virus have been found in the lung exudate, mostly in the form of immune complexes of both IgA and IgG with the virions. Circulating antibodies and viral antigens have not yet been demonstrated in the serum of affected animals. These observations, as well as the absence of a fever reaction, even in acute experimental cases, suggest the absence of a viremia and the localization of the infection to the respiratory system.

The primary lesion is the transformation of the Type II epithelial cell, causing the cell to proliferate and eventually obliterate the alveolus. Normal respiration is thereby progressively affected and the animal ultimately dies from asphyxia. A contributory factor to this sequence of events is the copious amount of exudate produced in typical adenomatous lungs. The transformed cells retain their secretory function and produce the excess of frothy, surfactant-containing, clear viscous fluid that accumulates in the air passages, giving the lungs their edematous appearance and aggravating the animal's respiratory distress. The resulting coughing gives rise to aerosol formation that in turn can lead to droplet infection, by inhalation, of other animals.

The pathogenesis of natural JS cases is usually complicated by secondary infections. The most common of these are pasteurellosis and mycoplasmosis, respectively causing acute and chronic pneumonic symptoms. Very often the immediate cause of death in animals suffering from JS is pneumonia.

Very little is known about the role of the macrophages in the pathogenesis of jaagsiekte. They may form part of a cellular immune response to the viral infection or to the transformed cells. No evidence for any cytotoxic effect towards tumor cells could be demonstrated, however. In normal lungs alveolar macrophages play a major role in the clearance of surfactant material. It is conceivable, therefore, that the characteristic

FIGURE 5. Macroscopic appearance of the lungs of an uncomplicated experimentally produced case of jaagsiekte.

accumulation of macrophages in jaagsiekte lungs may be due to the increased production of surfactant material.

In the case of maedi-visna virus, it has been shown that the macrophage is the primary target cell for the virus, and that macrophages are essential for viral expression in fibroblasts, which can undergo a nonproductive infection. No evidence has so far been reported for such a role in JS, except for our observation that suppression of macrophages with antimacrophage serum decreased the efficiency of experimental infection.[21]

IX. PATHOLOGY

A. Macroscopic Pathology

In an advanced case of the disease, the lungs do not collapse normally when the chest is opened, and a chronic adhesive pleuritis, due mainly to secondary bacterial infection, and some fluid in the thorax may be present. The lungs are much heavier than normal. Both lungs are usually involved, but not necessarily to the same degree. The more cranial parts of the lungs are usually more severely affected, but any part may show lesions (Figure 5). Large areas, if not the entire part of one or more lobes, may be involved and patches and nodules of diseased tissue, the smallest being less than a millimeter in diameter, are scattered in varying numbers throughout the more normal lung tissue. The larger lesions have a solid tumor-like appearance, are grayish-white, and have a tough consistency. In chronic cases some evidence of fibroplasia may be present. Small lesions are grayish-white and semitransparent. Lesions induced by experimental intratracheal inoculation are clearly multifocal in origin and, therefore, grow into large tumors at a much greater rate.

The primary lesion or lesions grow by expansion and by intrapulmonary metastasis of the neoplastic cells or by spread of infection or both, resulting in the dissemination of the condition. Growth and coalescence of new foci with neighboring ones eventually result in large parts of the lobe becoming affected. Secondary bacterial pneumonia with or without abscess formation frequently complicates the picture.

Intra- and extrathoracic metastasis of the neoplasm has very rarely been seen in South Africa, but is reported to be common in Israel. The most common site is the broncheal and mediastinal lymph nodes. Spread to the visceral and pariental pleura has been observed in two cases. Extrathoracic localization has been in such distant organs and tissues as the pelvic peritoneal serosa, skeletal musculature, liver, spleen, heart, and mesenteric lymph nodes.[1,6,38]

B. Histopathology

In the International Histological Classification of Tumors of Domestic Animals of the World Health Organization,[39] the JS lesion is classified as a bronchiolo-alveolar adenocarcinoma and, as the name implies, very early lesions may be observed to involve the epithelium of either the terminal and respiratory bronchioles or the alveoli. The latter is, however, the most frequent initial site. The earliest observable lesion with the light microscope consists of a small number of neoplastic cuboidal to columnar epithelial cells grouped together on the basement lamina and adjacent to normal cells (Figure 6A). These proliferate, almost fill the lumen of the affected alveolus or bronchiolus and then infiltrate and expand into the surrounding lung tissue forming acinar-like structures. Multicentric lesions may coalesce.

The acini of the neoplasm are lined by well-differentiated cuboidal to columnar epithelial cells that are usually arranged in a single layer. The acini themselves vary in shape and size, and, in some, papillary infoldings into the lumen are present (Figure 6B). Multilayered acini are much less common and the cells in them tend to be more flattened. The neoplastic cells have large nuclei with several nucleoli and the cytoplasm in many is vacuolated. The cytoplasm of the vacuolated cells stains intensely with the PAS reagent before, but not after, treatment with diastase, an indication that glycogen is present.[40]

In experimentally produced cases, the lesions are clearly multifocal in origin, with small acini, more or less equal in size, distributed throughout the lung (Figure 7).

The connective tissue stroma separating acini is generally scant and delicate, but in large and apparently relative mature lesions this tissue may be quite extensive and dense (Figure 8A) and, in a small number of cases, even myxomatous in parts (Figure 8B).

Alveolar macrophage response within the tumor, but particularly in the surrounding nonneoplastic lung tissue, is variable in extent, but in older lesions it may be very marked (Figure 9A). In some cases, peribronchial, peribronchiolar, and perivascular lymphoid hyperplasia of the lung tissue in and around lesions is prominent (Figure 9B). A purulent or fibrinous inflammatory reaction, caused by secondary invaders, is generally present in and around extensive lesions. In many terminal cases this tends to obscure the underlying neoplasm.

The appearance of extrapulmonary metastases resembles that of the primary lung neoplasm.

C. Ultrastructure of Jaagsiekte Lesions

Jaagsiekte tumor cells are covered with abundant microvilli (Figure 10A), which distinguish them from normal sheep granular pneumocytes in scanning electron microscopy studies. Jaagsiekte lesions vary from a few tumor cells at one point of an alveolus through continuous layers of cells lining the alveolus (Figure 10B) to large bunches of cells that have proliferated to fill the alveolus (Figure 11) and have spread to adjacent alveoli.[41]

In early lesions tumor cells are cuboidal or columnar. However, in more advanced lesions, the shape of the cells varies according to the amount of packing[41,42] (Figure 12). The nuclei in cuboidal cells tend to be large and central, while in columnar cells they are located towards the base of the cells. In some cells the nuclei are convoluted, but this is not characteristic of all tumor cells. Margination of the nucleoli is usually observed with peripherally located chromatin.[41]

A characteristic of the tumor cells on the periphery of the lesions is the presence of secretory granules in the apical portion of the cells. These granules resemble the secretory granules responsible for surfactant production in granular pneumocytes. In some JS lesions the granules are small and few in number, whereas in others the granules are very prominent. There is a great deal of variation in the appearance of the secretory

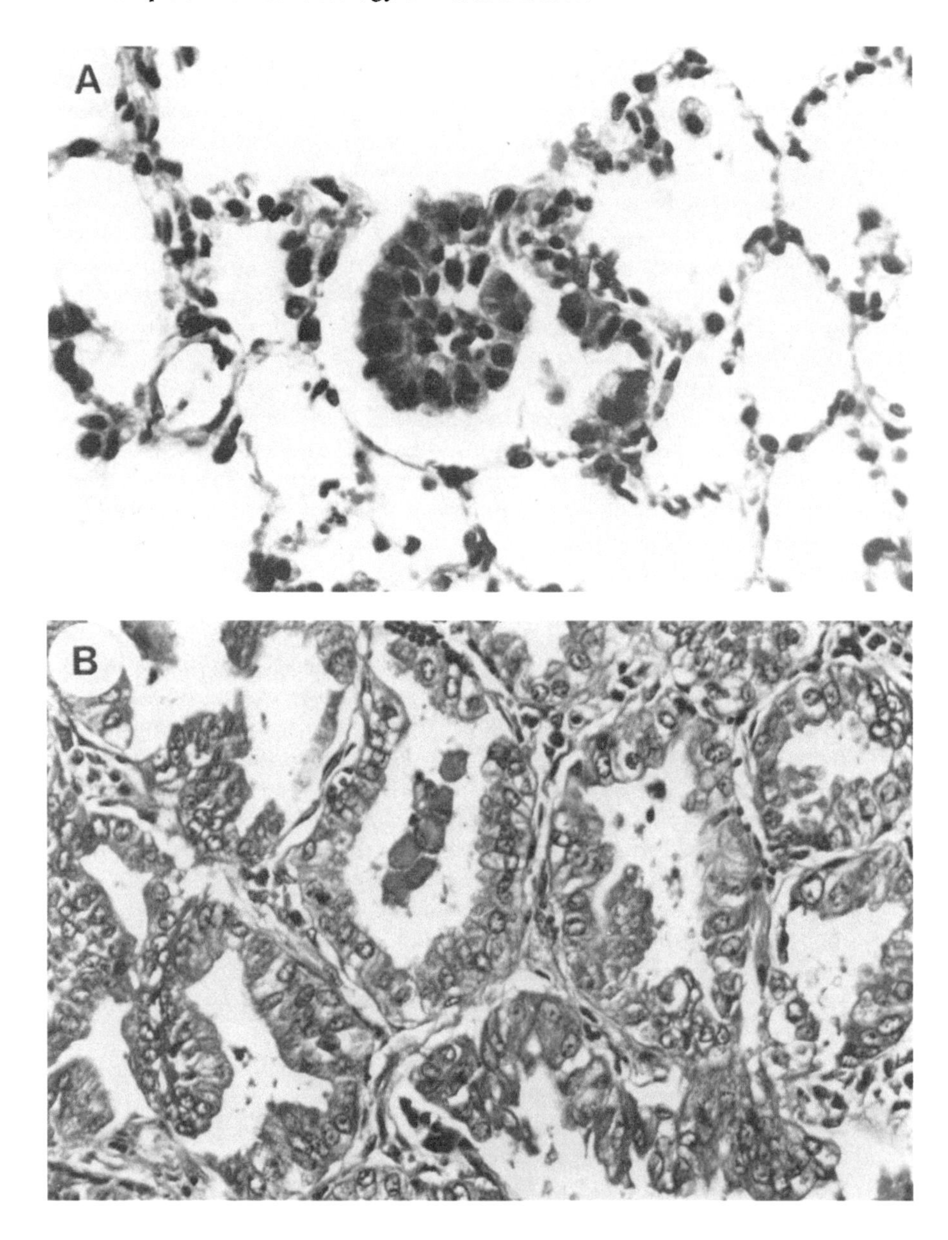

FIGURE 6. (A) An early jaagsiekte lesion comprising a group of neoplastic cuboidal to columnar epithelial cells replacing the normal alveolar epithelial cells. (H.E. stain; magnification × 480.) (B) A typical jaagsiekte lesion. The tumor cells have formed acini supported by small amounts of connective tissue. The neoplastic cells are well-differentiated and, in most parts, form a single layer. In some acini papillary folds project into the lumens. (H.E. stain; magnification × 300.)

granules. Some contain almost no electron-dense material, whereas others are uniformly electron-dense or contain electron-dense myelinoid whorls (Figure 13).[41] Some granules have been observed releasing their contents into the alveolar lumen. The content of the secretory granules is thought to be related to the rate of surfactant production, which again relates to the increased amount of lung exudate found in JS. The myelin whorls were observed only in tumor cell secretory granules and represent intracellular surfactant reserves.[43] Extracellular surfactant reserves take the form of tubular myelin[43] and are present in large quantities in the alveolar lumen of JS lungs (Figure

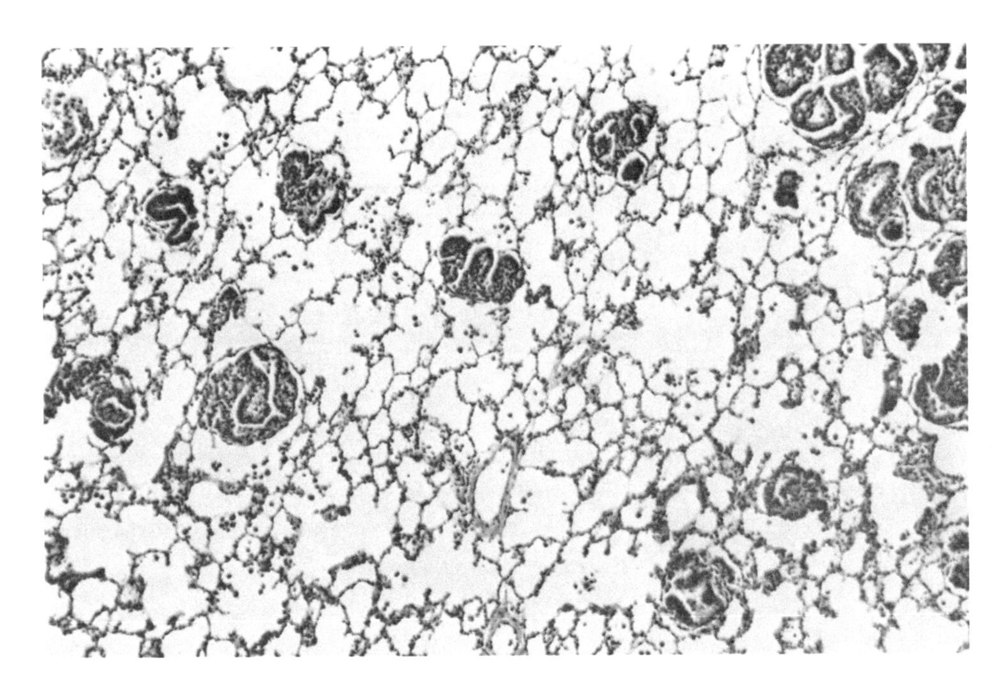

FIGURE 7. Multiple foci of tumor development in an experimentally induced case of jaagsiekte. Each focus in this case probably represents a primary site of virus infection and multiplication with consequent tumor induction, rather than a site of intrapulmonary metastasis of the neoplastic cells. Infection was induced by the intratracheal inoculation of 4 mℓ of a suspension consisting of a Freon® extracted lung rinse pellet from a previous case. The lamb was inoculated 4 days after birth and killed 6 weeks later, showing acute symptoms of jaagsiekte. (H.E. stain; magnification × 75.)

14). In these lungs the surfactant reserves are much higher than in normal sheep lungs, which tend to have electron-lucent secretory granules and very little tubular myelin. It is interesting to note that granular pneumocytes of the lungs of fetal lambs contain more electron-dense material than mature granular pneumocytes.[44] This implies that the tumor cells resemble the embryonal stages of the granular pneumocyte, which is an indication of malignancy.

Most tumor cells are in a state of rapid protein synthesis as the high incidence of polysomes and nucleolar margination indicates (Figure 12). As in other malignant cells, the mitochondria are often abnormal and fragile. The Golgi apparatus is usually well-developed and associated with membrane bound vesicles. Glycogen granules are seldom seen in acute cases of experimentally induced JS.[41] However, in chronic natural cases, clusters of glycogen granules have been reported to be characteristic of tumor cells.[37,13] Most tumor cells have large numbers of filaments running through the cytoplasm (Figure 15).

JSRV particles are comparatively rare in JS lungs, and not all tumor cells appear to be actively producing virus. Electron microscopic studies of experimentally induced cases of JS reveal intracytoplasmic particles in some tumor cells that bud with a condensed nucleoid to form mature extracellular particles[41] (Figure 4). These particles are generally found in close proximity to the tumor cells in the alveolar lumen, but have also been observed in the intertumor cell spaces as well as in tumor cell vacuoles.

Macrophages are usually found in large clusters in the alveoli (Figure 16), but are also present as individual cells attached to normal or tumor cells. The large size of the macrophages, the large number of lysosomes and phagolysosomes in the cytoplasm, and the ruffled surface of these cells indicate that they are in a highly activated state.

Plasma cells are found in the interstitial spaces of JS lesions, often in close associa-

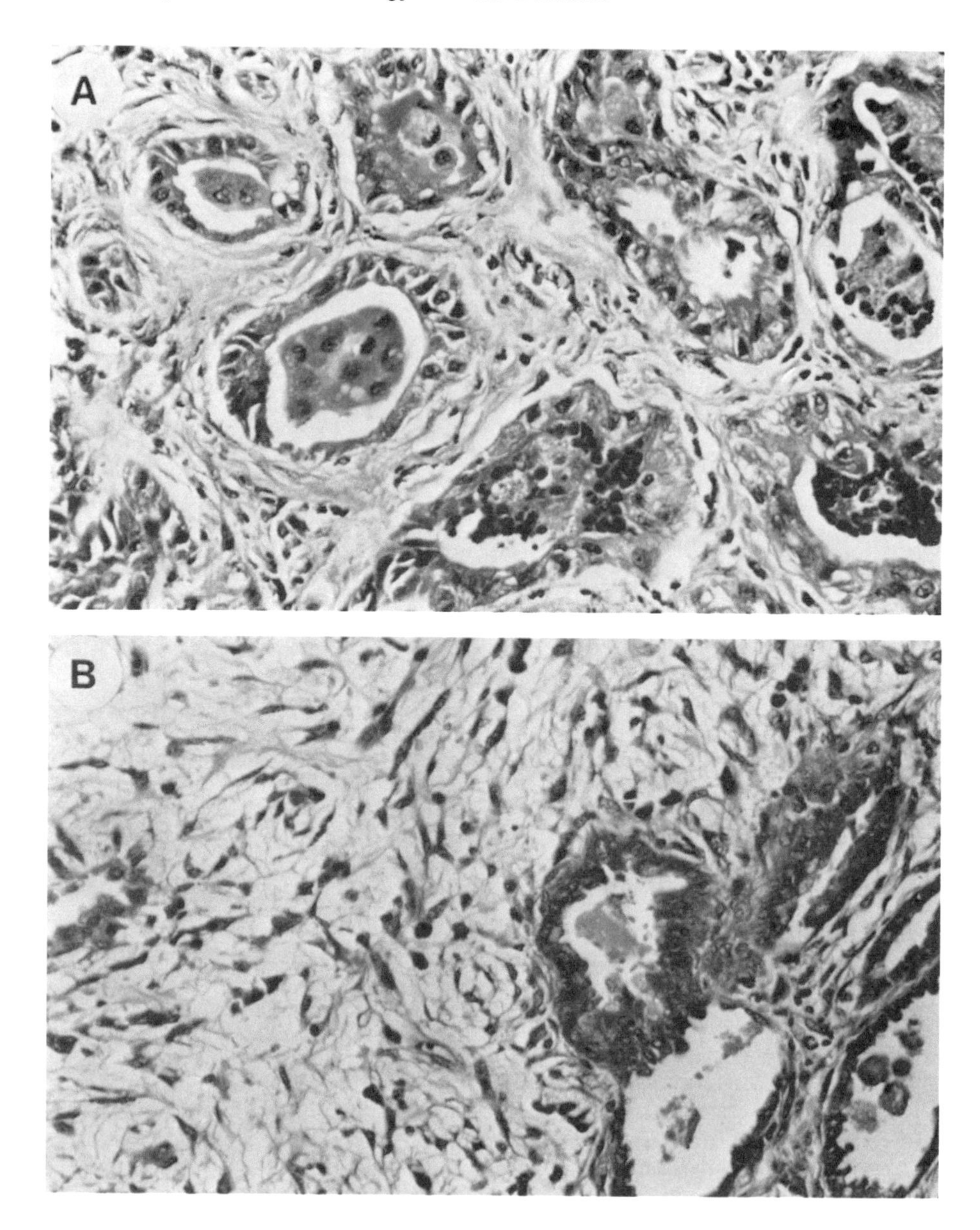

FIGURE 8. (A) Part of a lesion containing dense mature collagenous connective tissue stroma in which are a few infiltrating round cells. (H.E. stain; magnification × 300.) (B) In some of the older lesions parts of the stroma of the neoplasm are myxomatous in nature. (H.E. stain; magnification × 300.)

tion with lymphocytes.[41] The plasma cells probably play an important role in the production of local antibody against JSRV. JSRV is often found in the form of immune complexes in the lung, but the site of immunoglobulin production has not yet been elucidated. In cases where there is an inflammatory reaction, neutrophils are seen in the alveolar lumen, between tumor cells, and in the interstitial space. Although monocytes are observed in the lesions, they are comparatively rare compared with the other types of immune cells.

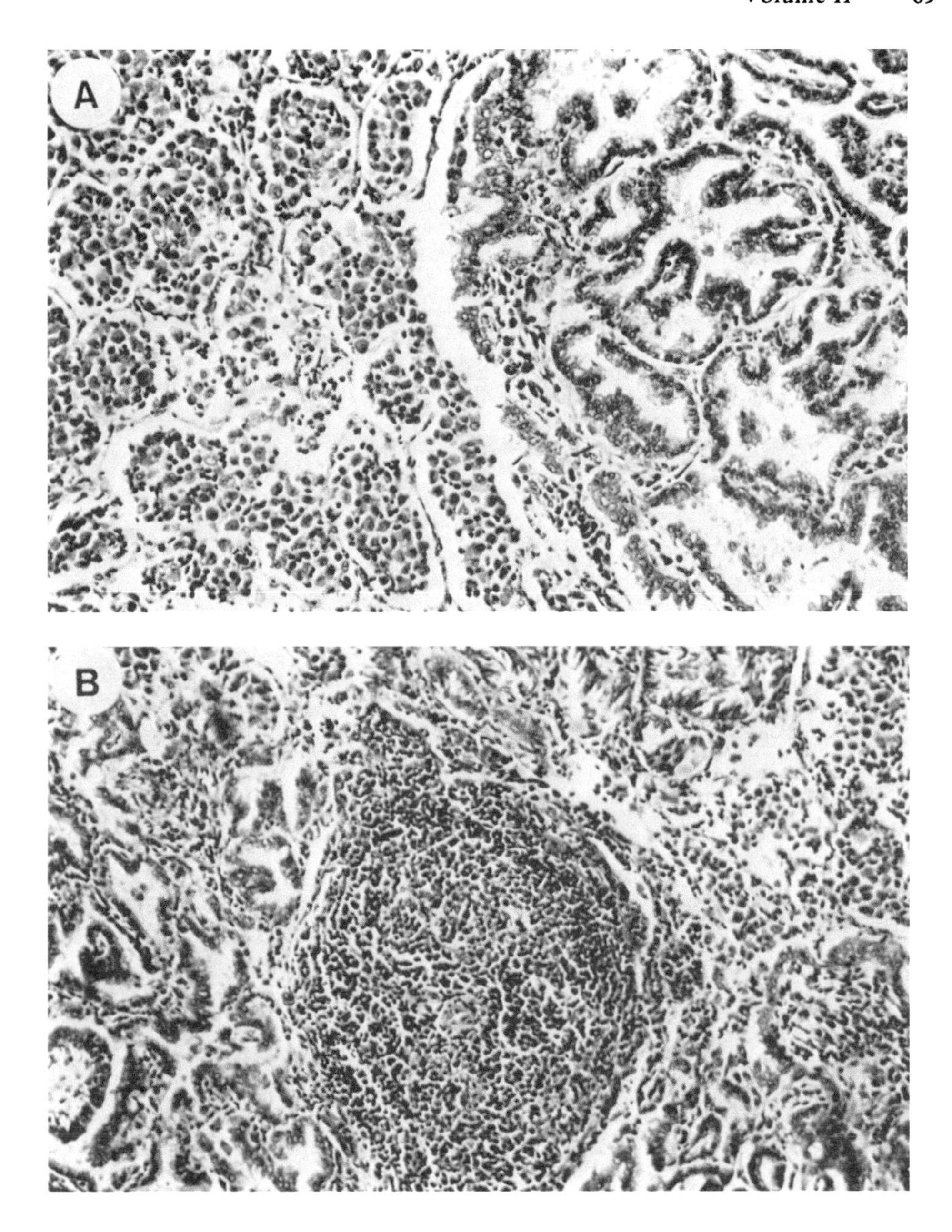

FIGURE 9. (A) Alveoli of the lung adjacent to a jaagsiekte focus containing numerous infiltrating macrophages as well as a few lymphocytes. This response is considered to be secondary, as is that of the lymphoreticular portrayed in Figure 9B. (H.E. stain; magnification × 150.) (B) Conspicuous perivascular lymphoreticular cuffing in a jaagsiekte lesion. Macrophage and lymphocyte infiltration is also present in parts. (H.E. stain; magnification × 120.)

X. IMMUNOBIOLOGY

There is abundant evidence, on a clinical level, for the existence of acquired resistance to infection with the JS retrovirus. It is well illustrated by the experience in Iceland when flocks, previously free of the disease, first became infected. In the first few years the incidence of the disease was as high as 50 to 60%. Thereafter it decreased rapidly to 5 to 10%, more or less on a par with countries where the disease is endemic.[12] A

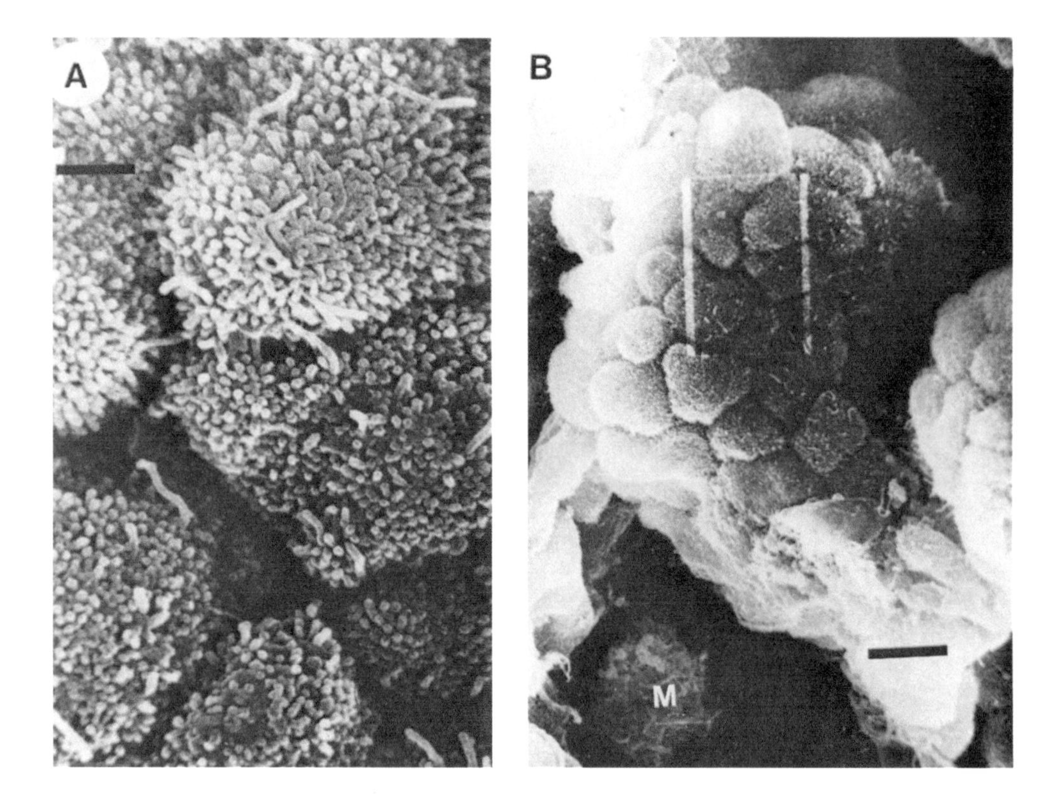

FIGURE 10. (A) Microvilli on the surface of jaagsiekte tumor cells. Bar = 1.5 μm. (B) Tumor cells lining an alveolus. Note the macrophage (M) in the adjacent alveolus. Bar = 7.5 μm.

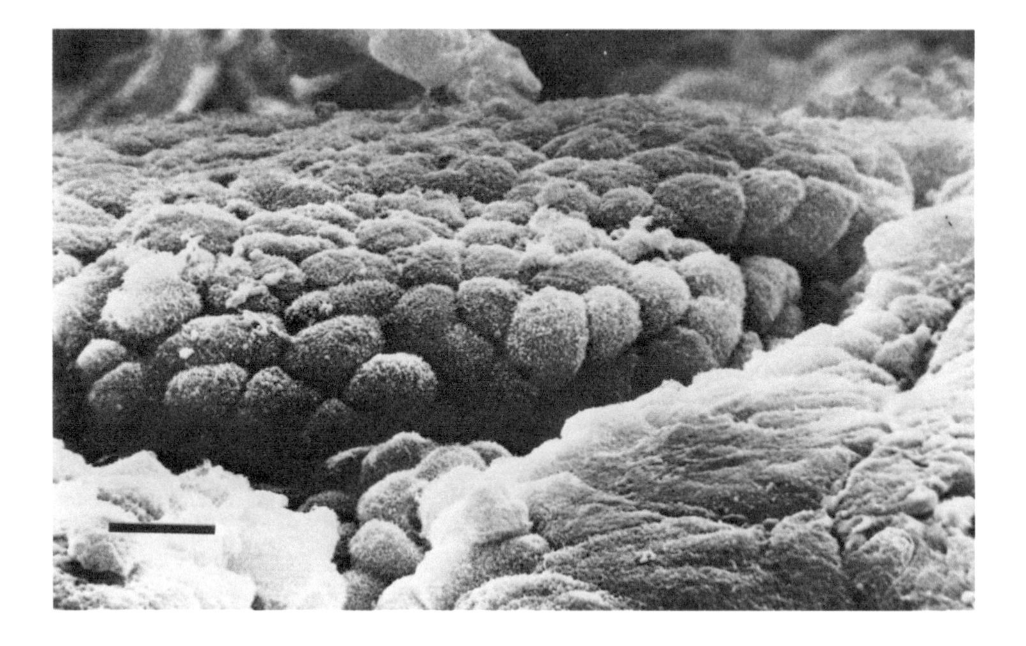

FIGURE 11. An alveolus filled with tumor cells. Bar = 6 μm.

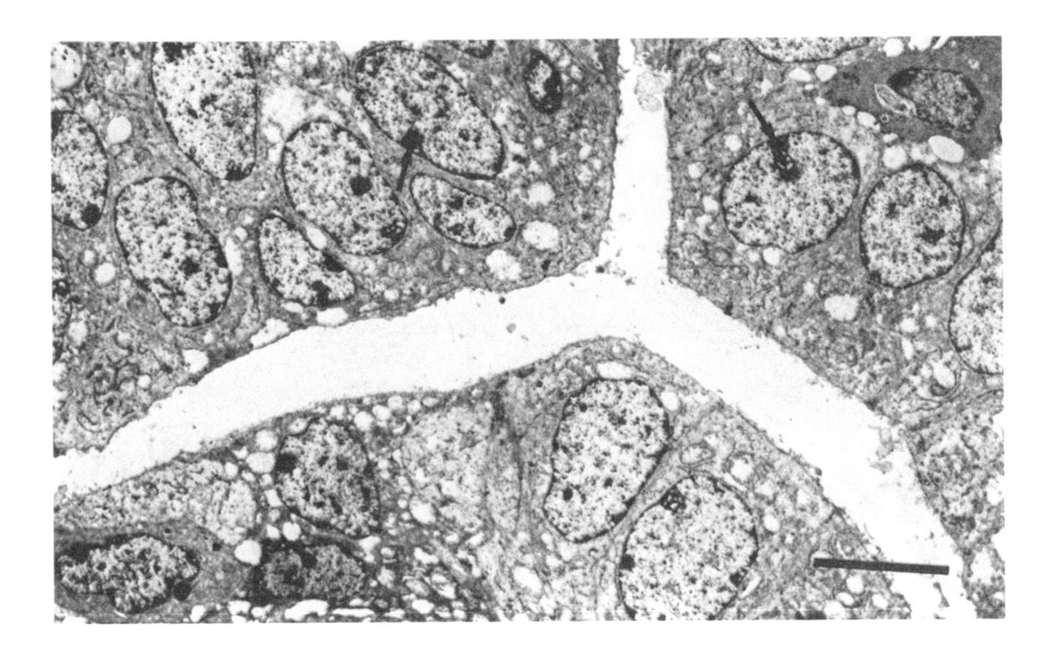

FIGURE 12. Tumor cells lining the alveolar lumen. Note the marginated nuclei (arrow). Bar = 5 μm.

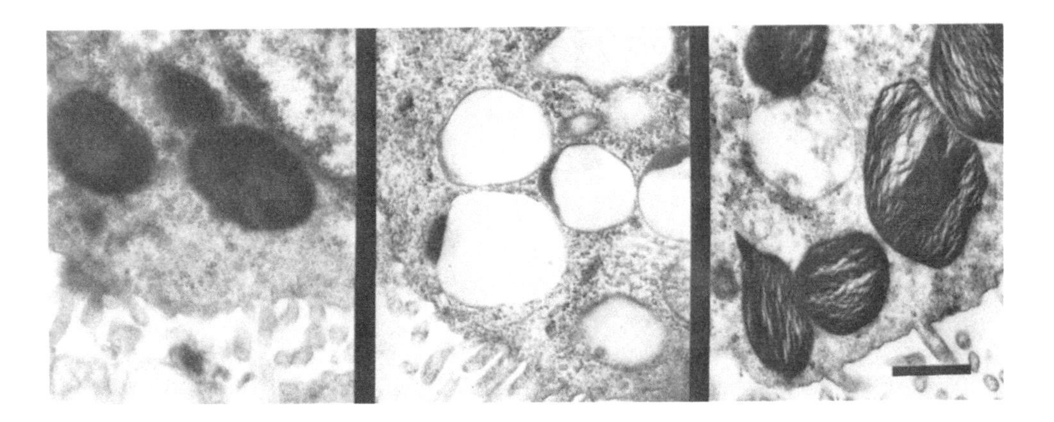

FIGURE 13. A comparison of the different forms of secretory granules found in jaagsiekte tumor cells. Bar = 1 μm.

similar experience has often been reported by farmers who introduce an infected animal into a clean flock. The number of JS cases usually increases for a number of years, then decreases. Although suggestive, this is no proof of an immunological phenomenon. In fact, circulating antibodies to the virus have not yet been conclusively demonstrated, owing largely to the lack of purified virus or viral antigen. In a preliminary survey of 100 sera from JS cases, we found only a small number containing antibodies to the virus. However, this result could be misleading as it was later found that the semipurified virus used as antigen consisted largely of immune complexes.

In contrast to the serum, lung fluid from adenomatous lungs contains high concentrations of both antiviral IgA and IgG, and virus purified from the fluid was shown to be complexed, mainly with IgA.[31] This suggests that the immune response to JSRV may depend mainly on a localized IgA mediated reaction. It is interesting to note that

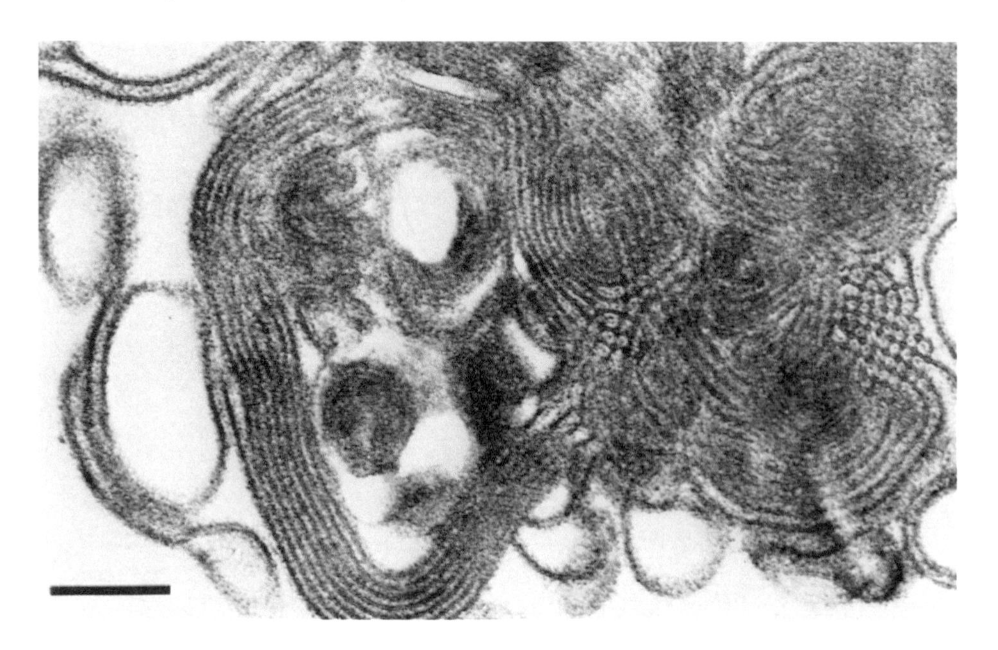

FIGURE 14. Tubular myelin in the alveolar lumen of a jaagsiekte lung. Bar = 2.2 μm.

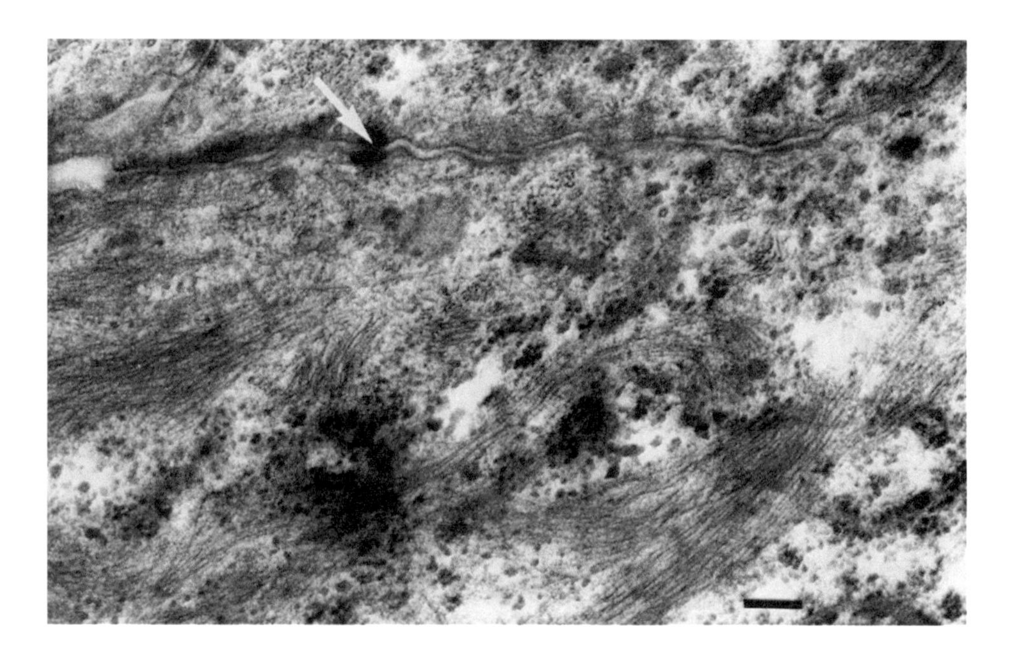

FIGURE 15. Filaments and bundles of filaments which are characteristic of jaagsiekte tumor cells. Note the desmosome (arrow). Bar = 200 nm.

IgA is generally regarded as a secretion of the secretory epithelial cells of mucous membranes. If, in the case of the lung alveolus, the Type II virus producing tumor cells play this role, it is possible that all virions budding from these cells are simultaneously coated with IgA. This would explain both the low infectivity of JSRV, its stability towards fluorocarbons and, possibly, the absence of circulating antibodies.

No evidence has so far been reported for antibodies to tumor-specific antigens or for

FIGURE 16. A cluster of macrophages (M) in a jaagsiekte lung. Bar = 5 μm.

cellular immunity to either virus or tumor cells. An accumulation of macrophages is sometimes regarded as pathognomonic for JS. Attempts to demonstrate a cytotoxic effect of these cells against cultured tumor cells were unsuccessful, however. We regard it as likely that the macrophages have some other, nonimmunological role to play.

XI. DIAGNOSIS

A. Clinical Diagnosis

The symptoms of JS have been described under Section IV. Clinical Signs. Of special diagnostic significance are symptoms associated with respiratory distress, such as dyspnea, accelerated and forced respiratory movements and coughing, in the absence of fever. A copious, clear, viscous lung exudate, often running out of the nostrils, is pathognomonic for the disease. Chasing the animals to accentuate the dyspnoea and upending them to detect the lung exudate are common diagnostic aids.

B. Histological Diagnosis

Although a definite clinical diagnosis can sometimes be made, it is not always possible. Field cases of the disease are more often than not complicated by secondary infections that can obscure the clinical picture. Histological confirmation of all diagnoses is, therefore, always advisable.

The histopathology of JS has already been discussed fairly comprehensively. There is no difficulty in distinguishing the disease from maedi, chronic pneumonia, or lungworm infestation, the main conditions from which it has to be differentiated. Very early JS lesions, however, can be difficult to distinguish from epithelialization of the alveolar epithelium.

C. Detection of JSRV and Viral Antigens

Viral particles are difficult to find in sections of adenomatous lungs by electron microscopy, and even in high-speed pellets of lung exudate and lung rinse fluid they are difficult to demonstrate, especially in field cases. The detection of viral reverse transcriptase activity in lung rinse pellets is a more feasible way of demonstrating the presence of a retrovirus in affected lungs, but the test is not specific for JSRV.

D. Serology

An ELISA test has been used successfully to detect both viral antigens and immune complexes of JSRV and IgA or IgG antiviral antibodies in lung rinse fluid. This test has a potential for diagnosis, but its general application will be hampered by the practical problem of sample collection. A serum-based assay would obviously be the solution, but attempts to detect circulating antibodies have so far failed (see Section X. Immunobiology).

XII. PROPHYLAXIS AND CONTROL

There is no effective vaccine against JS at present. The discovery that a retrovirus is the etiological agent of the disease has not increased the prospect of the development of a vaccine. Retrovirus infections have been shown to persist in the presence of high antibody titers and, apart from feline leukemia, successful immunization has not been achieved in diseases associated with horizontal transmission of retroviruses, such as enzootic bovine leukosis. The fact that many retroviruses can be transmitted vertically, and that suppression of viral replication after the initial transforming event has taken place cannot be expected to alter the course of a retrovirus-induced tumor, are complicating factors to be kept in mind.

Nevertheless, claims for successful immunization with vaccines prepared from affected lungs have been made. It is possible that antitumor and not antiviral antibodies may have been involved in the immunization with these crude preparations, but the recent demonstration of the local immune response to JSRV infection adds to the doubts raised about these "vaccines".[1] However, once the virus can be cultured in vitro, and when more information on the immunobiology of the disease becomes available, a vaccine could possibly be developed to stimulate local immunity to viral infection or even to enhance rejection of the transformed cells.

In the absence of effective immunization, the only prophylactic measure available at present is to prevent the introduction of the disease into a clean flock by new animals. Once the disease has been diagnosed, its incidence can be limited by a policy of strict isolation and elimination of positive cases.

Because of the long incubation period between infection of young animals and the appearance of symptoms, the progeny of affected ewes should also be eliminated. Complete eradication of the disease has been achieved only in Iceland and that by means of a severe slaughter-out policy.

A more rational control of the disease will be possible once a sensitive test for the detection of early cases or virus carriers has been developed. The infectivity of the virus is low, and it is relatively unstable, even in the form of an immune complex. It should, therefore, be possible to eradicate the virus, provided it is not endogenous, i.e., integrated into the cell genome.

XIII. CONCLUSION

Progress has been slow in the efforts to elucidate the pathobiology of JS. Since the discovery that a retrovirus causes the disease, attention has mainly been focused on the characterization of the virus and the elucidation of its relationship to other retroviruses. These studies have been severely hampered by the lack of an in vitro system for the cultivation of the virus.

The pathology and pathogenesis of the disease have been studied extensively and present few difficulties in interpretation on a cellular level. As far as the molecular mechanisms of cell transformation by the virus are concerned, however, very little information is available. We do not know, for example, whether JSRV is endogenous. We do not know whether the viral genome contains a transforming oncogene and, if it does, whether it is related to the other known viral or cellular oncogenes. Much further work is necessary to elucidate the molecular biology of the virus and, ultimately, the way in which it transforms its host cell.

Furthermore, little is known about the host animal's immune response. A localized, secretory IgA response has been demonstrated, but it has not yet been shown conclusively whether this is the only immunological defense mechanism. In order to develop control measures against the disease, either by vaccination or by eradication, it is essential that sensitive techniques be developed for detecting circulating antibodies or viral antigens. Much more information is also needed about the possible role of genetic factors in the acquired resistance to JSRV infection.

However, the foundation has now been laid, and if the technical problems of in vitro cultivation and purification of the virus can be solved, progress towards the goal of understanding viral transformation and controlling the disease should be greatly expedited.

REFERENCES

1. Tustin, R. C., Ovine jaagsiekte, *J. S. Afr. Vet. Med. Assoc.*, 40, 3, 1969.
2. Hutcheon, D., Reply to query No. 191 about jagziekte or chronic catarrhal pneumonia, *Agric. J. Cape Good Hope*, 4, 87, 1891.
3. McFadyean, J., Verminous pneumonia in the sheep, *J. Comp. Pathol. Ther.*, 7, 31, 1894.
4. Eber, A., Uber multiple adenombildung in den Lungen der Schafe, *Z. Tiermed.*, 3, 161, 1899.
5. Aynaud, M., Origine vermineuse du cancer pulmonaire de la brebis, *C. R. Seances Soc. Biol., Paris*, 95, 1540, 1926.
6. Wandera, J. G., Sheep pulmonary adenomatosis (jaagsiekte), *Adv. Vet. Sci. Comp. Med.*, 15, 251, 1971.
7. Perk, K., Slow virus infections of ovine lung, *Adv. Vet. Sci. Comp. Med.*, 26, 267, 1982.
8. De Kock, G., Further observations on the etiology of jaagsiekte in sheep, *Rep. Vet. Res. Union S. Afr.*, 15, 1169, 1929.
9. De Kock, G., The transformation of the lining of the pulmonary alveoli with special reference to adenomatosis in the lungs (jagziekte) of sheep, *Am. J. Vet. Res.*, 19, 261, 1958.
10. Dungal, N., Epizootic adenomatosis of the lungs of sheep in relation to verminous pneumonia and jaagsiekte, *Proc. R. Soc. Med.*, 31, 497, 1928.
11. Dungal, N., Experiments with jaagsiekte, *Am. J. Pathol.*, 22, 737, 1946.
12. Sigurdsson, B., Adenomatosis of sheep's lungs. Experimental transmission, *Arch. Gesamte Virusforsch.*, 8, 51, 1958.
13. Cutlip, R. C. and Young, S., Sheep pulmonary adenomatosis (Jaagsiekte) in the United States, *Am. J. Vet. Res.*, 43, 2108, 1982.
14. Stevenson, R. G. and Rehmtulla, A. J., Pulmonary adenomatosis (jaagsiekte) in sheep in Canada, *Can. Vet. J.*, 21, 267, 1980.
15. Verwoerd, D. W., Meyer-Scharrer, E., and Du Plessis, J. L., Transplantation of cultured jaagsiekte (sheep pulmonary adenomatosis) cells into athymic nude mice, *Onderstepoort J. Vet. Res.*, 44, 271, 1977.
16. Sharp, J. M., personal communication, 1983.
17. Hunter, A. and Munro, R., The diagnosis, occurrence and distribution of sheep pulmonary adenomatosis in Scotland 1975—1981, *Br. Vet. J.*, 139, 153, 1983.

18. Verwoerd, D. W., unpublished observations, 1981.
19. Verwoerd, D. W., Williamson, Anna-Lise, and De Villiers, E.-M., Aetiology of jaagsiekte: transmission by means of subcellular fractions and evidence for the involvement of a retrovirus, *Onderstepoort J. Vet. Res.*, 47, 275, 1980.
20. Herring, A. J., Sharp, J. M., Scott, F. M. M., and Angus, K. W., Further evidence for a retrovirus as the aetiological agent of sheep pulmonary adenomatosis (jaagsiekte), *Vet. Microbiol.*, 8, 237, 1983.
21. Verwoerd, D. W., De Villiers, Ethel-Michele, and Tustin, R. C., Aetiology of jaagsiekte: experimental transmission to lambs by means of cultured cells and cell homogenates, *Onderstepoort J. Vet. Res.*, 47, 13, 1980.
22. Mackay, J. M. K., Tissue culture studies of sheep pulmonary adenomatosis (jaagsiekte). II. Transmission of cytopathic effects to normal cultures, *J. Comp. Pathol.*, 79, 147, 1969.
23. Malmquist, W. A., Krauss, H. H., Moulton, J. E., and Wandera, J. G., Morphologic study of virus-infected lung cell cultures from sheep pulmonary adenomatosis (jaagsiekte), *Lab. Invest.*, 26, 528, 1972.
24. Cvetanovic, V., Forsek, Z., Nevjestic, A., and Rukanvia, Lj., Isolation of the sheep pulmonary adenomatosis (SPA) virus, *Veterinaria*, 21, 493, 1972.
25. De Villiers, E.-M., Els, H. J., and Verwoerd, D. W., Characteristics of an ovine herpes virus associated with pulmonary adenomatosis (jaagsiekte) in sheep. *S. Afr. J. Med. Sci.*, 40, 165, 1975.
26. De Villiers, E.-M., and Verwoerd, D. W., Presence of *Herpesvirus ovis* DNA sequences in cellular DNA from sheep lungs affected with jaagsiekte (pulmonary adenomatosis), *Onderstepoort J. Vet. Res.*, 47, 109, 1980.
27. Perk, K., Hod, I., and Nobel, T. A., Pulmonary adenomatosis of sheep (jaagsiekte). I. Ultrastructure of the tumor, *J. Natl. Cancer Inst.*, 46, 525, 1971.
28. Perk, K., Michalides, R., Spiegelman, S., and Schlom, J., Biochemical and morphological evidence for the presence of an RNA tumor virus in pulmonary carcinoma of sheep (jaagsiekte), *J. Natl. Cancer Inst.*, 53, 131, 1974.
29. Martin, W. B., Scott, M. M., Sharp, G. M., Angus, K. W., and Norval, M., Experimental production of sheep pulmonary adenomatosis (jaagsiekte), *Nature (London)*, 264, 1976.
30. Sharp, J. M., Angus, K. W., Gray, E. W., and Scott, F. M. M., Rapid transmission of sheep pulmonary adenomatosis (jaagsiekte) in young lambs, *Arch. Virol.*, 78, 89, 1983.
31. Verwoerd, D. W., Payne, A., York, D. F., and Myer, M. S., Isolation and preliminary characterization of the jaagsiekte retrovirus (JSRV), *Onderstepoort J. Vet. Res.*, 50, 309, 1983.
32. Payne, A., Verwoerd, D. W., and Garnett, H. M., The morphology and morphogenesis of jaagsiekte retrovirus (JSRV), *Onderstepoort J. Vet. Res.*, 50, 317, 1983.
33. De Villiers, E.-M., Verwoerd, D. W., and Payne, A., Comparative studies on ovine retroviruses, presented at CEC Symp. Slow Virus Diseases Sheep Goats, September 13, Edinburgh, 1983.
34. Sharp, J. M. and Herring, A. J., Sheep pulmonary adenomatosis: demonstration of a protein which cross-reacts with the major core proteins of Mason-Pfizer monkey virus and mouse mammary tumour virus, *J. Gen. Virol.*, 64, 2323, 1983.
35. Perk, K., personal communication, 1983.
36. Hod, I., Perk, K., Nobel, T. A., and Klopfer, V., Lung carcinoma of sheep (jaagsiekte). III. Lymph node, blood and immunoglobulin, *J. Natl. Cancer Inst.*, 48, 487, 1972.
37. Hod, I., Herz, A., and Zimber, A., Pulmonary carcinoma (jaagsiekte) of sheep. Ultrastructural study of early and advanced tumor lesions, *Am. J. Pathol.*, 86, 545, 1977.
38. Nobel, T. A., Neumann, F., and Klopfer, U., Histological patterns of the metastasis in pulmonary adenomatosis of sheep (jaagsiekte), *J. Comp. Pathol.*, 79, 537, 1969.
39. Stünz, H., Head, K. W., and Nielsen, S. W., Tumours of the lung, in International Histological Classification of Tumours of Domestic Animals, *Bull. W. H. O.*, 50, 9, 1974.
40. Markson, L. M. and Terlecki, S., The experimental transmission of ovine pulmonary adenomatosis, *Pathologia Vet.*, 1, 269, 1964.
41. Payne, A. and Verwoerd, D. W., A scanning and transmission electronmicroscopy study of jaagsiekte lesions, *Onderstepoort J. Vet. Res.*, 50, 1, 1984.
42. Nisbet, D. I., Mackay, J. M. K., Smith, W., and Gray, E. W., Ultrastructure of sheep pulmonary adenomatosis (jaagsiekte), *J. Pathol.*, 103, 157, 1971.
43. Massaro, P., The cellular and molecular basis of pulmonary alveolar stability, *J. Lab. Clin. Med.*, 98, 155, 1981.
44. Kikkawa, Y., Motoyama, E. K., and Cook, C. P., The ultrastructure of the lungs of lambs, *Am. J. Pathol.*, 47, 877, 1965.

Chapter 4

CANINE PARAINFLUENZA VIRUS

Wolfgang Baümgartner

TABLE OF CONTENTS

I. INTRODUCTION AND HISTORY

Canine parainfluenza (CPI) virus, an enveloped RNA virus of the paramyxovirus group, is an upper respiratory viral pathogen of dogs. It is closely related to simian virus (SV-5), which was first isolated from rhesus and cynomolgous kidney cell cultures in 1956.[1] Between 1967 and 1970, numerous instances of CPI virus isolations (synonyms: SV-5-like virus, parainfluenza-2 virus) from dogs with respiratory disease were reported.[2-4] Clinically, affected dogs exhibited signs of sudden onset of fever, malaise, coughing, and occurrence of copious amounts of nasal discharge. This virus has been shown to be one of several etiologic agents of kennel-cough, an important clinicopathologic entity of dogs. Recently, Evermann et al.[5] isolated a CPI virus from the cerebrospinal fluid (CSF) of a dog with neurological dysfunction. This viral isolate was subsequently found capable of inducing acute and chronic central nervous system (CNS) lesions under experimental conditions.[6,7]

II. CHARACTERIZATION OF CANINE PARAINFLUENZA (CPI) VIRUS

Canine parainfluenza virus is a typical member of the genus *Paramyxovirus*. The paramyxovirus virion consists of a membrane envelope, covered with surface projections that enclose a helical ribonucleoprotein nucleocapsid. Most of the spherical virions are 150 to 200 nm in diameter; however, large particles of 500 to 700 nm also occur.[8,9] The virion is covered with spikes 8 to 13 nm in length. The nucleocapsids are flexible helical structures 17 to 18 nm in diameter with a central hole about 5 nm in diameter. The nucleocapsid is single-stranded with a length of approximately 1 μm, and contains 4 to 5% RNA. Paramyxoviruses have six to seven proteins with a molecular weight between 74,000 and 38,000 daltons. The major proteins are the envelope associated matrix protein (M-protein), two glycoproteins, a larger and a smaller one, a nucleocapsid protein (NC-protein), and the P-protein. Hemagglutinin and neuraminidase activity are associated with the larger glycoprotein (HN-protein). The smaller glycoprotein (F-protein) is responsible for hemolysis, cell fusion and virus penetration.[10,11]

III. PATHOGENESIS OF CPI VIRUS INFECTION IN VITRO

The virus replicates readily in primary cultures of canine kidney cells and in many different continuous cell lines (Vero cells, Madin Darby Canine Kidney, BHK-21, etc.). Forty-eight hours after infection, multinucleated syncytial giant cell formation begins. With time, the syncytial cells enlarge in size and number. Associated with this progressing cytopathic effect (cpe), individual cells become vacuolated and rounded before detachment from the culture vessel. By 6 days postinfection, approximately 95% of the cells detach from the monolayer. In addition, small syncytia and/or no cpe (persistent) types of SV-5 and CPI virus infection in tissue culture had been reported.[12-14]

Virus-infected cells show hemadsorption (Had) at 4°C with guinea pig, chicken, dog, and human (type O) erythrocytes. Furthermore, eosinophilic intracytoplasmic inclusion bodies were observed (Figure 1); occasionally intranuclear inclusion bodies are found in the late stage of infection with the CNS-origin CPI virus.[9] Vero cells, persistently infected with CPI, are readily established from cells surviving a lytic infection. Between 85 and 100% of cells persistently infected with CPI contain viral antigen as determined by immunofluorescence (Figure 2). Roughly 10 to 15% of these persistently infected cells are Had-positive.

IV. SEROLOGY

Humoral antibodies induced by CPI infection are determined by virus-neutraliza-

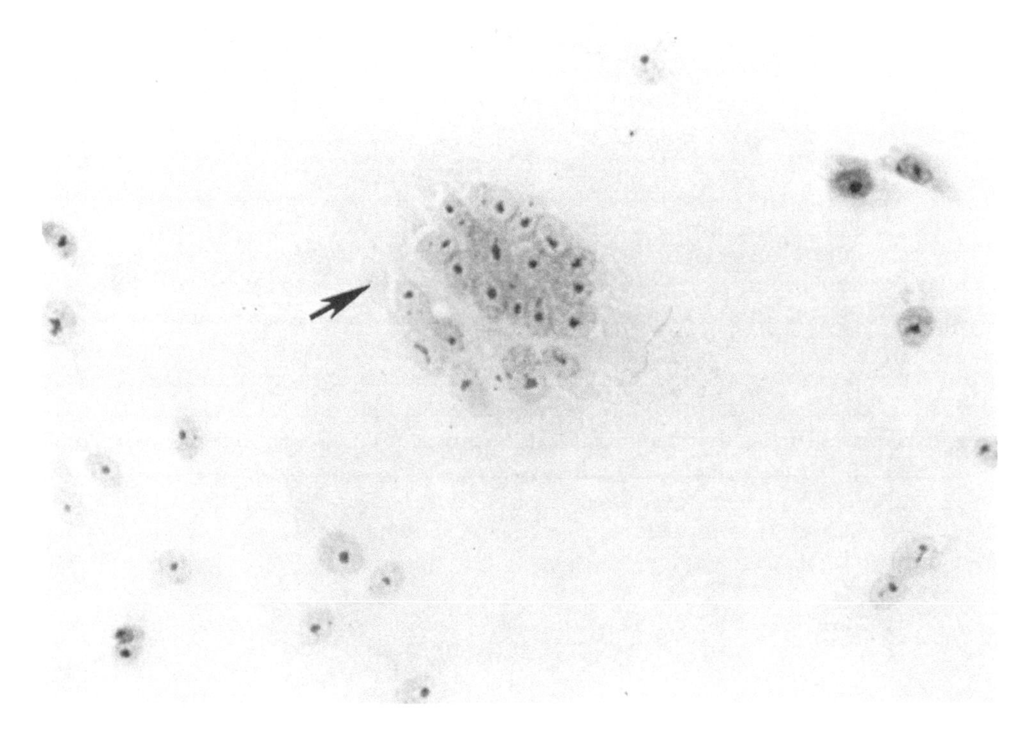

FIGURE 1. Multinucleated syncytial giant cell with intracytoplasmic viral inclusion bodies (arrow) in a Vero cell monolayer infected 36 hr previously with CPI virus.

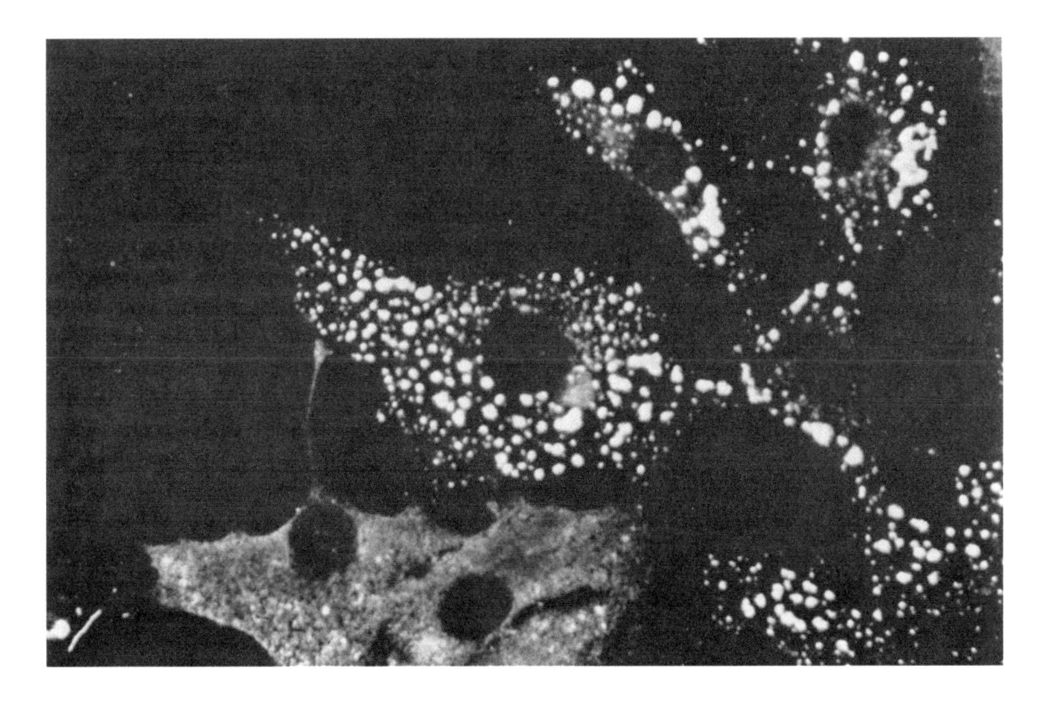

FIGURE 2. Vero cell monolayers with persistent CPI virus infection. Viral antigen is demonstrated by indirect immunofluorescence methods.

tion, hemagglutination inhibition (HAI), or CPI-specific ELISA assays. To enhance hemagglutination (HA) titers of viral preparations, the virus is treated with Tween® 80 and ether.[15] The HAI tests are performed at 4°C to reduce neuraminidase activity. Canine parainfluenza virus cross-reacts with SV-5 virus and, to a lesser extent, with human mumps virus; the latter is a one-way reaction that does not occur when CPI virus is tested with antisera against mumps virus.[16] Therefore, reports[17] of mumps virus infection in dogs based upon serology only should be cautiously interpreted.

Several studies have evaluated the epizootiology of CPI infection in various canine populations. A survey in eastern Washington showed that 19% of the dogs had antibodies against CPI virus.[18] Others found 14% of the dogs were seropositive in various geographic areas of the U.S.[19] It has been shown that CPI virus is highly communicable within a closely confined kennel population as demonstrated by increase in the number of seropositive dogs from 3 to 72% within 3 weeks after a CPI virus infection outbreak.[20] Thirty percent of 456 canine sera, including 320 sera from kennels with respiratory disease problems, had antibodies against CPI, as shown by a study conducted in West Germany.[21] Under experimental conditions, seroconversion occurred between 7 and 11 days postinfection (PID), reached peak values 4 to 7 weeks postinfection and then declined thereafter, though still detectable 6 months later.[6,16]

V. CLINICAL AND PATHOLOGICAL FINDINGS IN DOGS WITH CPI VIRUS INFECTION

Two recognized clinicopathologic forms of disease (upper respiratory and CNS) associated with CPI virus infection are known to occur. The respiratory form occurs under natural conditions and within the canine population is by far the most common and important manifestation of disease. Two to three days after aerosol exposure, 100% of infected dogs without preexisting antibodies develop a rise in temperature (0.5 to 1.0°C), which lasts for 2 days. This is accompanied by subsequent slight nasal discharge and a tracheobronchitis characterized by spontaneous cough of 2 to 12 days duration. Histopathologic lesions include catarrhal rhinitis, interstitial pneumonia, bronchopneumonia, bronchitis, and bronchiolitis[2,16] (Figure 3). With the fluorescent antibody test, viral antigen was found in the epithelium of the nasal mucosa, trachea, lungs, tonsils, and pharynx. Under natural conditions, CPI virus infections occur in combination with other agents, e.g., *Bordetella bronchiseptica,* canine adenovirus-2, canine distemper virus, and canine herpesvirus.[3,21] Furthermore, the clinical signs of CPI-induced respiratory infection are unspecific and not distinguishable from other causes of acute respiratory infections in dogs. Once acquired by vaccination or natural infection, circulating antibodies prevent reinfection and multiplication of virus.[2,16,23]

In the experimental and clinical cases described above, it is important to note that an overtly viremic phase of infection was not apparent. However, reports of successful isolation of CPI from cerebrospinal fluid,[5] spleen, kidney, and liver[3] imply that the virus, even under natural conditions, will spread beyond the upper respiratory tract.

The second or neurological form of CPI-related disease is inadequately defined at this time. A CPI virus was isolated from the CSF of a dog with a history of incoordination and posterior paresis of 2 weeks duration.[5] Subsequently, locomotor activity returned to normal.[5] Intracerebral inoculation of dogs with this neuropathogenic CPI virus resulted in the development of a nonsuppurative meningoencephalitis with focal necrosis 7 to 10 PID in half of the infected animals (Figure 4). The diagnosis of CPI virus infection was confirmed by seroconversion by PID 7, by virus reisolation from brain tissue, and by demonstration of virus in CNS tissues by immunofluorescence and ultrastructural examination.[6,7] Surviving animals developed internal hydrocephalus (Figure 5) with and without aqueductal stenosis.[6] Since the virus was shown to replicate

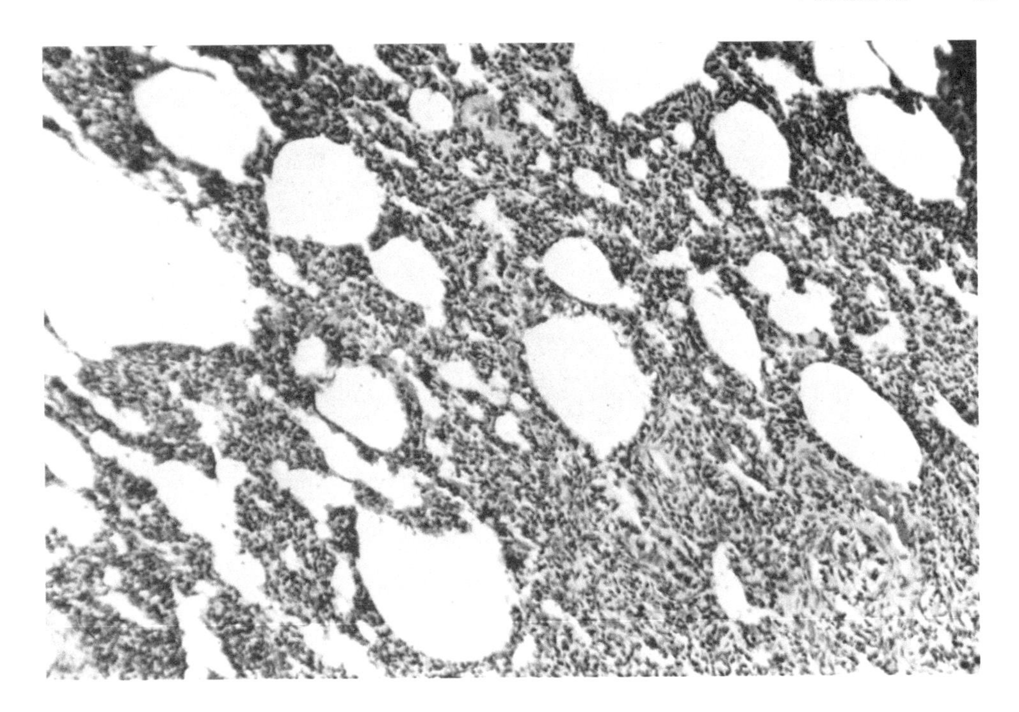

FIGURE 3. Interstitial pneumonia and atelectasia (PID 10) induced by CPI virus infection in a gnotobiotic puppy inoculated intracerebrally at 7 days of age.

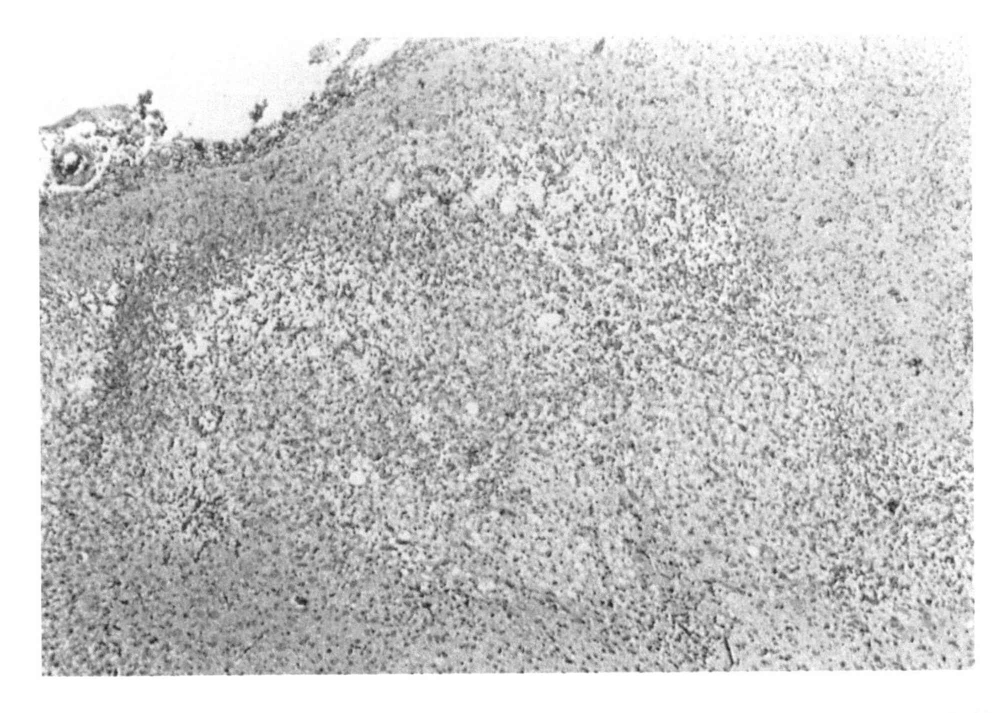

FIGURE 4. Acute encephalomalacia (arrows) and meningoencephalitis induced in a dog inoculated with CPI virus (PID 10).

in ependymal lining cells it is likely that hydrocephalus developed subsequently to this effect.

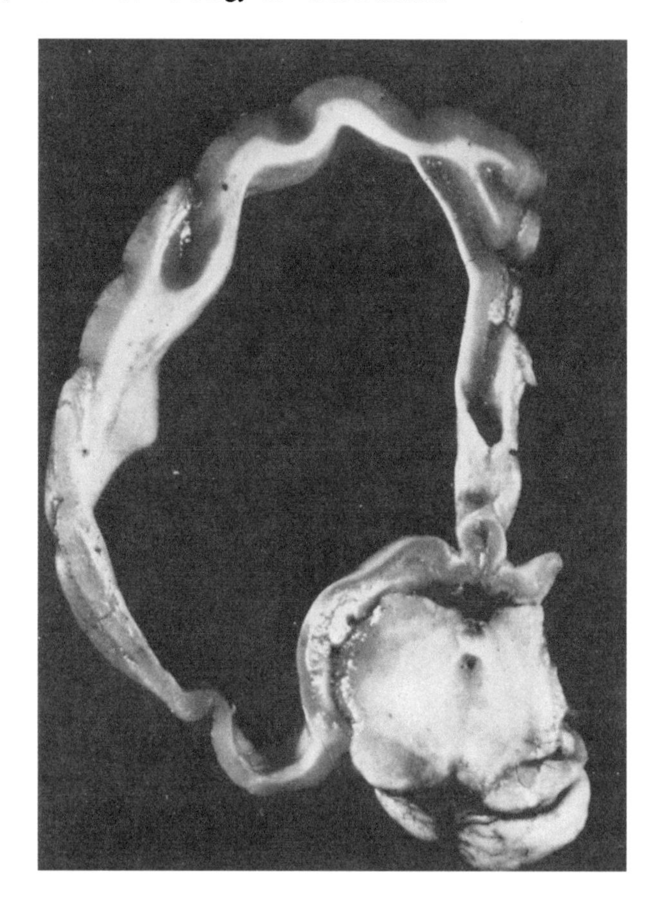

FIGURE 5. Severe internal hydrocephalus that developed 6 months after intracerebral infection with CPI virus. (From Baumgärtner, W. K., Krakowka, S., Koestner, A., and Evermann, J., *Vet. Pathol.*, 19, 79, 1982. With permission.)

VI. TREATMENT AND PREVENTION

Generally, the respiratory form has a mild to moderate course and thus does not necessitate treatment. More severe cases are complicated by secondary bacterial infections that will respond to proper antibiotic and symptomatic therapeutic regimens. Occasional fatalities attributed solely to CPI respiratory infection have been reported.[24]

To prevent CPI virus infections and to avoid complicated secondary infections, a modified-live canine parainfluenza virus vaccine has been developed. Dogs vaccinated i.m. developed a significantly greater immune response than dogs given the same dose subcutaneously.[25] Attenuated CPI virus is frequently included in the multiple agent vaccine products currently used by veterinarians.

REFERENCES

1. Hull, R. N., Minner, J. R., and Smith, J. W., New viral agents recovered from tissue cultures of monkey kidney cells, *Am. J. Hyg.*, 63, 204, 1956.

2. Appel, M. J. G. and Percy, D. H., SW-5-like parainfluenza virus in dogs, *J. Am. Vet. Med. Assoc.*, 156, 1778, 1970.
3. Binn, L. N., Eddy, A., Lazar, E. C., Helms, J., and Murnane, T., Viruses recovered from laboratory dogs with respiratory disease, *Proc. Soc. Exp. Biol. Med.*, 126, 140, 1967.
4. Carandell, R. A., Brumlow, W. B., and Davison, V. E., Isolation of parainfluenza virus from sentry dogs with upper respiratory disease, *Am. J. Vet. Res.*, 29, 2141, 1968.
5. Evermann, J. F., Lincoln, J. D., and McKeirnan, A. J., Isolation of a paramyxovirus from the cerebrospinal fluid of a dog with posterior paresis, *J. Am. Vet. Med. Assoc.*, 177, 1132, 1980.
6. Baümgartner, W. K., Krakowka, S., Koestner, A., and Evermann, J., Acute encephalitis and hydrocephalus in dogs caused by canine parainfluenza virus, *Vet. Pathol.*, 19, 79, 1982.
7. Baümgartner, W. K., Krakowka, S., Koestner, A., and Evermann, J., Ultrastructural evaluation of acute encephalitis and hydrocephalus in dogs caused by canine parainfluenza virus, *Vet. Pathol.*, 19, 205, 1982.
8. Choppin, P. W. and Stoeckenius, W., The morphology of SV-5 virus, *Virology*, 23, 195, 1964.
9. Baümgartner, W. K., Metzler, A., Krakowka, S., and Koestner, A., *In vitro* identification and characterization of a virus isolated from a dog with neurological dysfunction, *Infect. Immunol.*, 31, 1177, 1981.
10. Choppin, P. W., Compans, R. W., Wagner, R. R., and Bader, J. P., in *Comprehensive Virology*, Vol. 4, Fraenkel-Conrat, H. and Wanger, W. W., Eds., Plenum Publishing, New York, 1975, 95.
11. Rott, R., Molecular basis of infectivity and pathogenicity of myxovirus, *Arch. Virol.*, 285, 1978.
12. Baümgartner, W. K., unpublished data, 1984.
13. Rhim, J. S. and Schell, K., Cytopathic effects of the parainfluenza virus SV-5 in Vero cells, *Nature (London)*, 216, 271, 1967.
14. Compans, R. W., Holmes, K. V., Dales, S., and Choppin, P. W., An electron microscopic study of moderate and virulent cell interaction of the parainfluenza virus SV-5, *Virology*, 30, 411, 1964.
15. John, T. J. and Fulginiti, V. A., Parainfluenza 2 virus: increase in hemagglutinin titer on treatment with Tween 80 and ether, *Proc. Soc. Exp. Biol. Med.*, 121, 109, 1966.
16. Rosenberg, F. J., Lief, F. S., Todd, J. D., and Reif, J. S., Studies of canine respiratory viruses. I. Experimental infection of dogs with an SV-5-like canine parainfluenza agent, *Am. J. Epidemiol.*, 94, 147, 1971.
17. Binn, L. N., A review of viruses recovered from dogs, *J. Am. Vet. Med. Assoc.*, 156, 1672, 1970.
18. Fulton, R. W., Ott, R. L., Duenwald, J. L., and Gorham, J. R., Serum antibodies against canine respiratory viruses: prevalence among dogs of eastern Washington, *Am. J. Vet. Res.*, 35, 853, 1974.
19. Bittle, J. L. and Emery, J. B., The epizootiology of canine parainfluenza, *J. Am. Vet. Med. Assoc.*, 156, 1771, 1970.
20. Binn, L. N. and Lazar, E. C., Comments on epizootiology of parainfluenza SV-5 in dogs, *J. Am. Vet. Med. Assoc.*, 156, 1774, 1970.
21. Bibrack, B. and Benary, F., Seroepizootologische Untersuchungen über die Bedeutung von parainfluenza-2-infectionen beim Zwingerhusten in Deutschland, *Zentralbl. Vet. Med. B*, 22, 610, 1975.
22. Cornwell, H. I. C., McCandlish, I. A. P., Thompson, H., Laird, H. M., and Wright, N. G., Isolation of parainfluenza virus SV-5 from dogs with respiratory disease, *Vet. Rec.*, 98, 301, 1976.
23. Packard, M. E. and Grafton-Packard, B., The effect of canine parainfluenza vaccine on the spread of tracheobronchial coughs in a boarding kennel, *Am. Anim. Hosp. Assoc.*, 15, 241, 1979.
24. Ajiki, M., Takamura, K., Hiramatsu, K., Nakai, M., and Susaki, N., Isolation and characterization of parainfluenza 5 virus from a dog, *Jpn. J. Vet. Sci.*, 44(4), 607, 1982.
25. Emery, J. B., House, J. A., Bittle, J. L., and Spotts, A. M., A canine parainfluenza viral vaccine: immunogenicity and safety, *Am. J. Vet. Res.*, 37, 1323, 1976.

Chapter 5

THE PATHOBIOLOGY OF FELINE CALICIVIRUSES

R. Charles Povey

TABLE OF CONTENTS

I. INTRODUCTION AND HISTORY

A. Caliciviruses in General

Caliciviruses are so named because of the characteristic morphology of cup-like (from Latin calyx, cup) depressions on the virion surface. These viruses were formerly classified as a genus of the family Picornaviridae. They are now recognized as a separate family of single-stranded RNA viruses, the Caliciviridae.[1]

Caliciviruses have been found to infect and cause disease in pigs, cats, and pinnipeds. The pinniped hosts are sea lions (*Zalophus californianus*), northern fur seals (*Callorhinus ursinus*), and northern elephant seals (*Mirounga angustirostris*). The predominant manifestations of caliciviral disease in these species are vesicular and erosive lesions of the skin and oral cavity and pneumonia. A calicivirus neutralized by antiserum to a feline strain has also been isolated from a dog with glossitis.[2]

Calicivirus serotypes antigenically similar to the isolates made from pinnipeds have been recovered off the California coast from the opaleye fish (*Girella nigricans*),[3] and from a lung fluke, genus *Zalophatrema*, in a sea lion dying of verminous pneumonia.[4] One of the opaleye isolates was pathogenic for pigs.[3] This provides a link between saltwater fish and the pig that may explain the origins of vesicular exanthema of swine, a disease which first occurred in pigs in California in 1932.[5]

Virus particles with the characteristic calicivirus morphology have been observed in human,[6] bovine,[7] and porcine diarrhoeic feces,[8] and in the intestinal contents of chicks[9] or guinea fowl (*Numida meleagris*).[10] As yet these are only putative members of the caliciviruses as they have not been cultured in vitro or been characterized. Their pathogenicity has not been definitely established although the human calici-like viruses have been strongly implicated in outbreaks of acute diarrhea in infants and old people.[11-16] The isolation of a possible calicivirus from tissues of mink with penumonia has recently been reported.[17]

B. Caliciviruses of Cats

The first isolations of what are now referred to as feline caliciviruses (FCV) were made during attempts to identify the causal agent of feline panleukopenia.[18-19] In fact Bolin initially claimed his isolate was the cause of panleukopenia, and his strain of FCV is still referred to as "FPL". The association of this virus with panleukopenia was subsequently disproven.[20-21] Fastier designated his isolate "kidney cell degenerating" virus (KCD). These two isolates, FPL and KCD, produced a rapid cytopathic effect in feline cell cultures. This showed as rounding and increased refractility of cells without inclusion bodies or syncytia formation. The viruses were distinct from the herpes virus causing feline viral rhinotracheitis (FVR) described by Crandell and Maurer,[22] which produced both inclusions and polykaryons.

During 1958 several investigators in the U.S. reported isolations of viruses similar to FPL and KCD from the upper respiratory tract, conjunctiva, and mouth of kittens with respiratory disease, conjunctivitis and stomatitis, respectively.[23] Crandell and Madin[24] compared their California feline isolate (CFI) with KCD and Bolin's FPL viruses. The viruses produced the same cytopathic effect, but were serologically distinct in cross-neutralization tests. Crandell et al.[25] were able to classify 33 similar isolates from cats, including strains CFI and FPL, into five distinct antigenic groups by serum neutralization. However, complement fixation tests demonstrated cross-reactions between the isolates.[26]

The first recoveries of similar feline viruses in Britain were designated as feline influenza isolates,[27] but these included some identified later as FVR virus. Feline caliciviruses have subsequently been reported from domestic cats in many other parts of the world including Western and Eastern Europe, Scandinavia, Australasia, and Japan. Beisdes the domestic cat, there are confirmed isolations from cheetahs,[28] and all members of the Felidae are probably susceptible. The isolation of an apparent feline strain from a dog with glossitis[2] is the only report of a feline calicivirus from a nonfeline host.

C. Taxonomy of Caliciviruses

Burki[29] drew attention to the similarities between the rapidly cytopathic, ether and chloroform resistant, small RNA viruses and the picornaviruses of other species. He designated them "feline picornaviruses", and classified them as having properties intermediate between enteroviruses and rhinoviruses. However, the electron-microscopic structure of the feline viruses[30,31] was distinct from that of other picornaviruses, except for the virus of vesicular exanthema of swine (VESV), which the feline virus isolates resembled closely. VESV and the feline isolates were also similar in biochemical and biophysical characteristics.[32] Wawrzkiewicz et al.[32] suggested that VESV and the "feline picornaviruses" might be members of a new subgroup of the picornaviruses. Further biochemical and biophysical evidence for the resemblance of VESV to, and yet distinction from, other picornaviruses was presented by Oglesby et al.[33]

Taking account of this evidence, the International Committee on Nomenclature of Viruses, renamed the International Committee for the Taxonomy of Viruses (ICTV), proposed that VESV type A48 be the type virus of a new genus of the family Picornaviridae. Feline picornaviruses were included in this new genus. The name Calicivirus was proposed for this genus because of the unique cup-like depressions seen by electron microscopy in negative-stained preparations.[34]

In 1974 Burroughs and Brown[35] argued that the differences in physicochemical properties between the caliciviruses and other picornaviruses were sufficient to consider the caliciviruses as a new virus family. In addition to the previous evidence of distinct morphology, larger size, higher sedimentation coefficient, and lower RNA content, Burroughs and Brown found that VESV and FCV had only one major polypeptide. This was in contrast to the four major polypeptides found in other picornaviruses.

Studdert and O'Shea[36] described the ultrastructural cytopathology of FCV infection. They pointed out that the accumulation of calicivirus particles in single or multiple linear arrays associated with microfibrils in the cytoplasm, and the condensation of nuclear chromatin usually in a single rounded central mass, were unlike the cell changes associated with replication of picornaviruses.

At the molecular level of calicivirus replication, there are similarities between caliciviruses and picornaviruses. The genome similarities are lack of a methylated cap at the 5′ end of the RNA and the presence of a genome-associated virion protein (VPg), covalently linked to the RNA.[37-40] There are, however, major differences in genome strategy. The VPg of caliciviruses is required for infectivity.[40] Also a subgenomic messenger RNA coding for calicivirus capsid protein, with a genome-sized RNA probably acting as a messenger for nonstructural proteins, contrasts with the synthesis of structural and nonstructural picornavirus proteins, which are formed by cleavage of a polyprotein that is the translation product of genome-sized mRNA.[37,41-44]

Recognizing the differences between caliciviruses and picornaviruses, ICTV established a Caliciviridae Study Group to examine the question of separate family status, and this group supported recognition of a family, Caliciviridae with one genus, *Calicivirus*.[45]

II. PHYSICOCHEMICAL PROPERTIES OF CALICIVIRUSES AND FELINE CALICIVIRUSES IN PARTICULAR

A. Morphology

Calicivirus virions are nonenveloped icosahedrons, 35 to 40 nm in diameter. They have distinctive surface morphology that appear as 32 cup-shaped depressions or

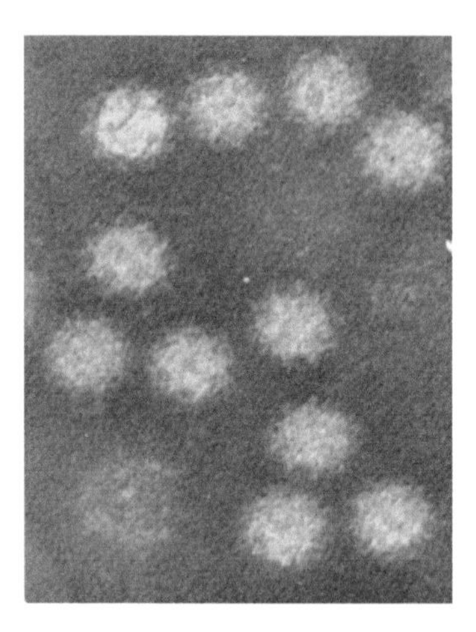

FIGURE 1. Electron micrograph of feline calicivirus showing typical calicivirus morphology of cup-like surface depressions. (Courtesy of D. Cubitt.)

"holes" in negatively stained electron-microscope preparations (Figure 1). The various explanations made for this appearance have been a double-shell structure;[30] a reverse ("negamere") arrangement of 32 morphological units in pentamer/hexamer icosahedral formation;[31] 20 identical morphological units consisting of a total of 60 structural units;[46] 60 dimeric subunits arranged to give 32 "holes";[47] and 60 units made up of polypeptide trimers with 32 "holes".[48] The last seems the most supportable hypothesis.

B. Chemical Composition

The virion is composed of a single positive-strand, nonsegmented, linear RNA with a covalent linked protein (VPg) that is necessary for infection.[39,49-52] The RNA has a sedimentation coefficient of 36 to 38S,[42,53] a molecular weight of 2.5×10^6 to 2.8×10^6 daltons,[33] and accounts for approximately 18% of the virion mass.[48]

The remaining 82% of the virion consists of proteins. There is a single major structural polypeptide of 60×10^3 to 70×10^3 daltons and a minor polypeptide of 15×10^3 to 19×10^3 daltons.[35,50]

Other nonstructural proteins, some of which may be precursors of structural polypeptides, have been identified.[41,43,44] No lipid or carbohydrate has been detected in caliciviruses, and no viral enzymes have been identified.

C. Virion Mass

Buoyant density measurements of caliciviruses usually range from 1.36 to 1.39 g/m*ℓ* in cesium chloride, but vary with conditions including centrifugation, time, pH, and storage.[32-35,54-57] Sedimentation coefficients of caliciviruses are most frequently calculated to be 183S,[50,56,58] although Burroughs and Brown[35] found a strain of FCV sedimenting at 170S. The estimated virion mass is 1.4×10^7 to 1.6×10^7 daltons.[53]

D. Stability of Infectivity

Caliciviruses are typically unstable at pH 3, variably stable at pH 3 to pH 5, and stable between pH 5 and pH 10. However, one strain of FCV, 17-FRV, was found to be moderately stable at pH 2.5.[59]

At room and incubator temperatures FCV are stable for several days and survive without loss of titer for at least 18 months at −20°C or below.[24] There is inactivation within 1 hr at 50°C, and the presence of 1 to 2 *M* Ca^{++} or Mg^{++} ions does not protect against and may enhance inactivation.[29,60] High concentrations (1 to 2 *M*) of Na^+ may be partially protective against heat inactivation.[59]

FCV are resistant to organic solvents such as 20% ether at 4°C for 18 hr,[29,60] 5% chloroform at room temperature for 10 min,[29,61] and 0.2% sodium deoxycholate.[60,62] Burki and Pichler[62] found that two of the seven FCV strains that they tested were partially inactivated by 0.5% trypsin.

Of the common disinfectants, phenolic compounds, hypochlorite, and aldehydes are effective, but most iodine-based disinfectants are not.[63]

E. Replication

The molecular aspects of calcivirus replication have only partially been determined.[42,43,48,51] Positive-strand 36S RNA serves as a template for transcription into progeny 36S RNA, subgenomic 22S RNA, and possibly 18S RNA. The 22S subgenomic RNA probably codes for the major capsid polypeptide. The 18S RNA may have a messenger function, or may be a degradation product.

Viral proteins accumulate in the cytoplasm and new virions are found as early as 4 hr after infection. Virions occur as single particles, in irregular accumulations, in linear arrays associated with microfibrils or with membrane bound cytoplasmic cisternae, and in paracrystalline arrays.[36,46,64-65]

F. Antigens

Virus neutralization, complement fixation, immunodiffusion, immunofluorescence, and immunoprecipitation have all been used in the serological investigation and classification of caliciviruses. No hemagglutinating activity for caliciviruses has been found. The number and location of calicivirus antigens has not been determined. The major polypeptide from San Miguel Sea Lion Virus (SMSV) obtained by heating in sodium dodecylsulfate and mercaptoethanol was immunogenic. It elicited antibodies reacting with homologous, and to a lesser extent, with heterologous intact virion antigens in St-RIP tests (radioimmune precipitation with protein — A — bearing staphylococci as immunoadsorbent).[66] A virion subunit from VESV consisting of a polypeptide trimer reacts with homologous antiserum in the complement fixation test.[48]

By virus neutralization, at least 11 and possibly 13 serotypes of VESV are recognized,[5] although between serotypes there are low levels of cross-reactivity, which is emphasized by complement fixation.[67] Eleven serotypes of SMSV have been determined.[68-69] These are distinct from VESV serotypes by neutralization tests, but antigenic relatedness between the two groups is indicated by immunodiffusion,[70] immunoelectron microscopy (IEM),[4] and St-RIP test.[66] This relationship is confirmed by RNA homology studies showing 70 to 90% homology among VESV and SMSV serotypes.[70]

Feline caliciviruses have given extensive cross-neutralization reactions making designation of serotypes difficult. The conclusion of recent studies[71-74] has been that FCV isolates are sufficiently closely related to be regarded as members of a single serotype. Some strains of FCV are more antigenically distinct and may be designated as serologic variants.

With regard to relationships between FCV and VESV/SMSV, cross-reactions have been found by IEM between two serotypes of VESV and FCV strain F-9.[4] Schaffer[52] reported St-RIP tests showing reaction between two serotypes of SMSV and three antiFCV cat sera, but hyperimmune goat sera to two other FCV strains showed no reaction. Burroughs et al.[70] found no cross-reaction by immunodiffusion between FCV strain K-1 and three serotypes of VESV and SMSV.

III. VIRUS-HOST INTERACTIONS OF FELINE CALICIVIRUS

A. In Vitro Cytopathogenicity

Cultivation of FCV is readily achieved in primary, secondary, and continuous cell cultures derived from domestic cats and lion kidney.[75] Feline tracheal and tongue organ cultures have also been used successfully.[76] Nonfeline cells are generally resistant, except that dolphin kidney cells[75] and green monkey (Vero) cells[77] will support FCV replication.

Feline caliciviruses are not particuarly fastidious as regards growth requirements. Stationary incubation at 37°C in a pH of 7.4 to 7.6 is as satisfactory as roller-drum incubation at 33°C at pH 6.8 to 7.0.[29] In growth cycle studies in kitten kidney monolayers, multiplication of KCD virus occurred within the first 4 to 6 hr with maximum titers of 10^7 to 10^8 median cell culture infective doses, ($CCID_{50}/m\ell$) at 24 to 30 hr depending on the size of the inoculum.[18] Strain FCV 10-66 inoculated into similar cells at a multiplicity of infection of 25 to 1 showed multiplication at 3 hr postinoculation and reached maximal titers of intra- and extracellular virus at 6 and 8 hr, respectively.[78]

The initial cytopathic effect (CPE) is increased refractility and swelling of individual cells. This is followed by contraction of the cells to round or oval shapes and detachment from the culture surface. The residual cytoplasm may maintain contact with neighboring uninfected cells by fibrillar processes. The rounding and detachment of cells progresses rapidly throughout the monolayer during a period of 24 hr.[18,24,79]

Giemsa stain shows infected cells to have pyknotic or karyorrhexic nuclei and basophilic cytoplasmic strands. No inclusion bodies or syncytia have been seen, although a paranuclear eosinophilic cytoplasmic mass similar to that seen with some picornaviruses has been described.[80-81] Electron microscopy of infected culture cells[36,46,65,82] shows rounding of the cell and nucleus, loss of pseudopodia, and the appearance of numerous smooth membrane-bound, cytoplasmic vesicles. Virus particles are observed only in the cytoplasm. Changes occurring within the nucleus, although thought to be secondary to the viral infection and nonspecific, include the formation of dense nuclear masses of chromatin.

Crandell[60] first described plaque formation by FCV. Additional information was provided using overlays containing methyl cellulose,[61] agar,[78] and agarose.[83] Generally plaques are visible 36 to 72 hr after inoculation when they range from 1 to 5 mm in diameter. Ormerod and Jarrett[84] showed that the sulfated polysaccharide (polyanion) fraction of agar was inhibitory for plaque size with some FCV isolates. They identified four distinct populations of plaque sizes: minute, small, large, and an uncommon extra-large group. There is evidence that the extra-large plaque formation is correlated with increased pathogenicity.[85]

Attempts to infect embryonated eggs with FCV by various routes have been unsuccessful.[18,60,86] The inoculation of sucking or weaned mice intracerebrally and intranasally, of day-old guinea pigs intracerebrally, or of half-grown rabbits intravenously, has not resulted in pathogenic infection.[18]

B. Association with Disease

Until very recently, FCV had been associated with disease only in domestic cats and cheetahs.[28] Now, Everman et al.[2] have isolated a calicivirus, neutralized by antiFCV serum, from a case of glossitis in a dog. In the cat, FCV infection may often be subclinical, but the typical disease syndrome is initial lethargy, roughened hair coat, anorexia and pyrexia, followed by respiratory tract signs and oral ulcerations.[87] The incubation period varies from 1 to 10 days, with 2 to 4 days as the most usual.

The respiratory signs are ocular and nasal discharges, sneezing, and dyspnea. The upper respiratory signs, even with virulent strains of FCV, are usually mild unless the infection is complicated with other agents such as rhinotracheitis virus, or secondary bacteria. Lower respiratory tract involvement often is undetected clinically, but extensive pneumonia, particularly if complicated by other organisms, causes exaggerated respiratory movements, mouth breathing, and possibly cyanosis.

The oral lesions begin as vesicles that rapidly ulcerate. Individual ulcers may coalesce to give extensive areas of epithelial loss. The anterior dorsum of the tongue is the main site, but the hard palate may also be involved. This latter site is more notably affected in cats receiving the abrasive dry diets.[88] The erosions may also be seen on the lips and nose, particularly the nasal philtrum and occasionally on the paws around the base of the footpads and between the toes.[89] Particles consistent in morphology with FCV were observed in electron microscopy samples obtained from multiple eosinophilic skin ulcers and granulomas.[90] The affected cat also had a single large ulcer of the hard palate.

Diarrhea has been noted in a number of experimental infections and in a few field cases of FCV infection in which vomition also occurred.[91]

Stiffness of gait, hyperesthesia, mild joint pain, and muscle soreness have also been observed during the acute phase of some FCV infections.[78,92-93] Some, at least, of these cats experience a transient partial or total loss of righting reflexes. This coincides with the height of fever and full recovery occurs within a few days. Virus has been isolated with this syndrome from the brain tissue of a kitten which died (with severe interstitial pneumonia), but no pathology was detected.[94] Love and Baker[91] mention isolation of FCV from the brain of a cat exhibiting severe convulsions followed by death in 48 hr. This cat also had tongue ulcers. Love and Baker[91] also describe isolation of FCV from the cerebrum as well as tongue, pharynx, lung, and liver of a 13-week-old kitten that died with acute interstitial pneumonia. A moderate degree of focal gliosis and perivascular cuffing was found in sections of the cerebrum and cerebellum.

The possible association between a strain of FCV designated Manx strain and the urethral obstruction syndrome in cats[95-96] has been further investigated, and its role is at most a secondary one.[97-98] Other investigators have failed to show any role for FCV in that syndrome.[99]

The mortality of FCV infections is generally low, although rates of 30% have been recorded.[82] As would be anticipated, the highest fatality occurs in the younger kittens, where death may be within 24 to 48 hr of the onset of illness.[91] A strain of FCV, A-3, relatively nonvirulent for mature cats,[100] produced fatal infections in 3-week-old antibody-free kittens[101] and weanling kittens.[102] Mortality is increased in cases complicated by concurrent viral or bacterial infection.

Recovered cats frequently remain persistently infected carriers of virus.[103] Most of these carriers are asymptomatic, but some have chronic oral lesions. The most frequent of these lesions is a line of gingival reddening at the alveolar margin of all the teeth, in the absence of tartar accumulation or obvious periodontal disease.[104] In other cases there is lymphoepithelial proliferation of a more focal nature. This is often located on the gums near the angle of the jaws. No direct association between these lesions and FCV has been demonstrated.

The diagnosis of FCV disease on clinical grounds alone is uncertain. The main differential diagnosis is FVR[22] in which the upper respiratory signs are usually more severe.[87] Laboratory confirmation of FCV is made by virus isolation in feline cell culture, the sample of choice being an oropharyngeal swab submitted in a viral transport medium containing antibiotics. Immunofluorescence may be used to confirm the identity of an isolate as FCV.[105] Dual infection with FCV and FVR virus can occur.[106,107]

C. Experimental Infection and Pathogenesis

Experimental FCV infections have been investigated using a variety of routes of inoculation. Intranasal and aerosol routes have been the most widely used, but intraocular, i.m., i.p., and i.v. inoculations have also been made. The route of infection has an influence on the pathogenesis of the disease. Direct intranasal instillation of virus usually results in lesions that are confined mostly to the upper respiratory tract, although pneumonia may occur.[85,100,102] Aerosol inoculation with a virulent strain typically leads to pneumonic disease with or without upper tract signs.[85,108-111] However, as infected cats rarely generate aerosols and aerogenous transmission of FCV does not occur over distances of greater than 1.24 m,[112] the direct intranasal route is probably a more natural model.

The spectrum of clinical signs following FCV infection varies from asymptomatic,[18,113] through a mild nonfatal respiratory disease[24] to more severe and often fatal infections.[108] This spectrum is largely dependent on the strain of virus as shown by Povey and Hale[102] working with 15 strains and by Hoover and Kahn[109] working with 10 strains. Burki[114] had previously drawn attention to a possible variable organ affinity between strains of FCV such that some strains show a preference for nose, others for oropharynx, yet others for lower intestinal tract and abdominal organs such as spleen. The hypothesis was based on the different concentrations of virus in different tissues and sites with different strains. However, intercurrent infections with other feline viruses such as rhinotracheitis, leukemia, panleukopenia,[115] and infectious peritonitis or the presence of potentially pathogenic bacterial flora may also influence apparent virulence. Thus, signs may be more severe and mortality higher in conventional than in specific pathogen-free cats.[82,108] Environmental climate may be important, particularly in initiating epidemics. It was observed that a series of 27 outbreaks of respiratory disease attributed to feline calicivirus and viral rhinotracheitis were regularly preceded by two rapid drops of ambient temperature in the cats' environment. It was considered important that the interval between the two stresses was approximately equal to the incubation period of the virus infections.[116] This hypothesis has not been developed further.

Following aerosol exposure to a virulent FCV, virus is detected on day 1 in upper respiratory tract epithelia including the tonsil and in the lung.[108] Ultrastructural studies[117,118] have revealed lung changes as early as 12 hr postinfection. There is degeneration and desquamation of Type I pneumocytes and crystalline arrays of FCV virions can be found in these cells. Degeneration of Type II pneumocytes is also present but to a much lesser extent. The lysis and necrosis of the pneumocytes are observed as acute exudative alveolitis between 12 and 96 hr. Although feline alveolar macrophages do not support complete viral replication,[119] infected cells produce a chemotactin that causes large numbers of neutrophils to accumulate in alveolar spaces. There is focal endothelial swelling and occasional perivascular and intraendothelial fibrin accumulation.

Viremia may occur during many FCV infections, but has only been reported on a few occasions.[24,106] Kahn and Gillespie[108] recovered virus from the plasma of one of six infected cats at day 3 postinfection and in the blood cells of two of three cats on day 4. At these times FCV may often be cultured from a variety of organs and tissues, including spleen, kidney, liver, small intestine, and cerebellum.

After day 4 postinfection virus is rarely detected outside the respiratory tract or its adnexa. This coincides with the appearance of virus-neutralizing antibody. Viral antigen is readily detected in epithelial cells of the tonsil and at the margins of oral ulcerations, and in the lower respiratory tract.[108,120] From day 7 onward, lung changes are dominated by proliferative responses of the Type II cells such that in some cases there is widespread adenomatoid epithelialisation of alveoli. Monocytes and alveolar macrophages dominate the free cell population, and there is peribronchiolar and interalveolar accumulation of lymphoplasmocytic cells.[117]

Virus is recovered regularly from day 1 onwards from the upper respiratory tract, in particular the oropharynx, and is persistently shed from this site, or can be cultured from tonsils, for extended periods of time.[100,102,106,108,121]

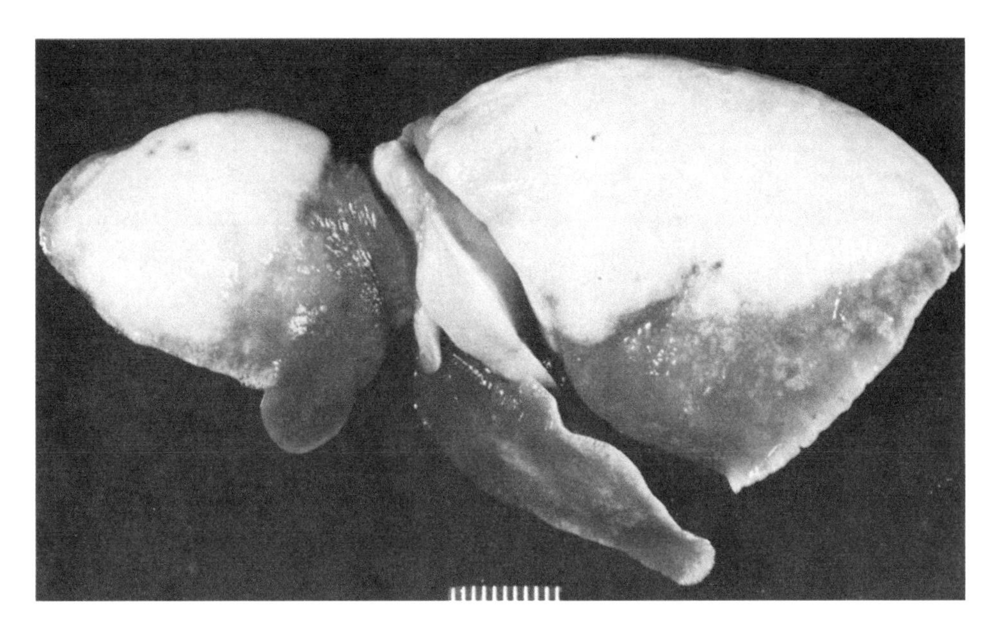

FIGURE 2. Pneumonia affecting the ventral portions of the left lobes of the lung of a cat infected with a virulent pneumotropic strain of feline calicivirus. (From Hoover, E. A. and Kahn, D. E., *J. Am. Vet. Med. Assoc.*, 166, 463, 1975. With permission.)

In the study reported by Wardley and Povey,[100] one cat was still excreting virus 176 days after infection and the halflife for viral elimination was 75 days. Rectal shedding of virus has been well-documented for the acutely ill cat.[61,100,106] This has been explained as swallowed virus, but following i.m. inoculation, FCV was detected in mesenteric lymph nodes and intestinal contents before it was found in the upper respiratory tract.[106] This finding indicated that local replication could be responsible for at least some fecal shedding of virus.

D. Gross Pathology

Holzinger and Kahn[122] described the pathological lesions in cats following aerosol exposure to the virulent FCV strain 255. Between postexposure days (ped) 1 and 5 there was diffuse red mottling of the lungs, which were heavy, wet, and slightly swollen. A foamy serosanguinous fluid could be expressed from the cut suface. The margins of the lungs were pale and slightly distended with gas. Between ped 4 and 10 lungs had firm, gray, slightly elevated subpleural foci of parenchymal consolidation 1 to 2 cm in diameter and extending 1 to 5 mm into the parenchyma. Initially most foci were close to the hilus of the lungs but then spread peripherally. In the aerosol infection study of Ormerod et al.,[85] the gross lesions were regularly found in the dependent portions of all lobes and in the dorsal area of the caudal lobes. Typical lung lesions are as in Figure 2.

The oral ulcers begin as vesicles that rapidly rupture and leave irregular ulcers with discrete margins. The ulcers are generally 2 to 5 mm in diameter. Several may coalesce to form much larger areas, particularly on the anterior dorsum of the tongue (Figure 3).

Apart from some cattarhal exudate in the rhinarium[123] there are no other gross lesions, although Holzinger and Kahn[122] described a transverse dark banding of the spleens in 20% of experimentally infected cats. These bands were neither elevated nor depressed. Their significance, if any, has not been determined.

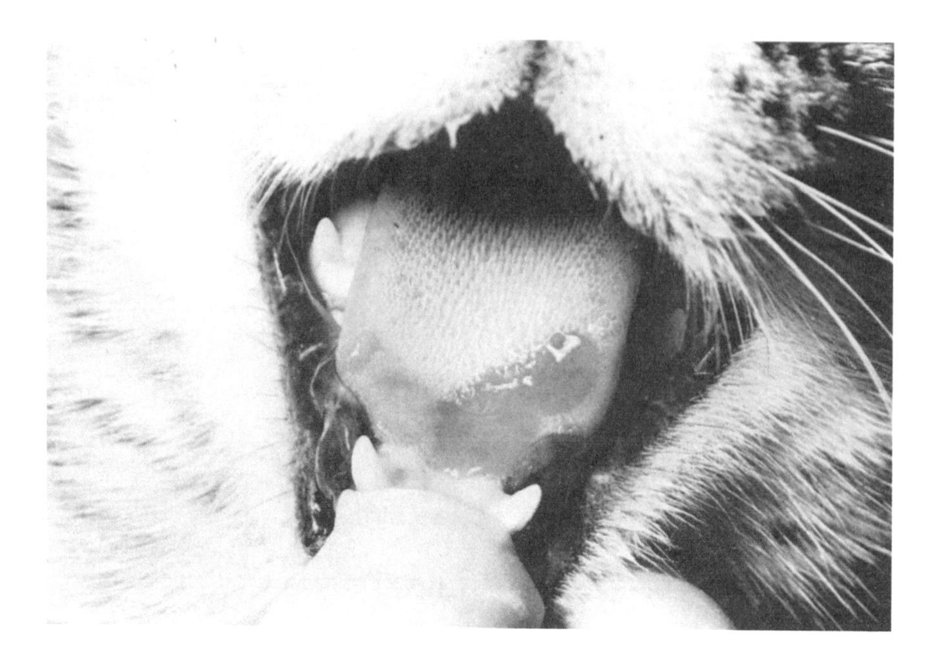

FIGURE 3. Typical ulcerative lesion of feline calicivirus involving the anterior dorsal margin of the tongue.

E. Microscopic Pathology

The microscopic pathology in the lungs resulting from aerosol infection with a virulent strain can be described in three phases.[85,109,117,119,122,124] Phase I occurs in the first 5 days after exposure and is an exudative phase. Following virus-induced degeneration and desquamation of Type I pneumocytes, and to a lesser extent degeneration of Type II pneumocytes, there is multifocal infiltration of leukocytes between and within alveoli. Migration of large numbers of neutrophils into alveolar sacs contributes to the formation of an exudate that extends into the alveolar ducts and terminal bronchioles. However, there is little necrosis of the epithelium of these ducts. Alveolar capillaries are congested, and there is some hemorrhage and leakage of proteinaceous fluid into alveolar septa and into alveoli with some fibrin deposition. Towards the end of this first phase, there is exudation into the alveoli of large macrophages in addition to neutrophils.

Phase II is a proliferative and interstitial pneumonia occurring between days 6 and 14 after exposure. As a probable result of the earlier viral damage to Type I pneumocytes, Type II alveolar pneumocytes undergo a hypertrophic and hyperplastic reaction (Figure 4). This results in alveolar spaces being lined by two or more layers of large polyhedral cells, many in mitosis or binucleate. There is extensive desquamation of these cells into the alveoli contributing to the pool of alveolar macrophages. The epithelial cells of the alveolar ducts and terminal bronchioles also proliferate to two or three cells in thickness and form exaggerated folds. The interstitial reaction consists of increasing numbers of lymphocytes and plasma cells and a decrease of neutrophils.

Phase III, from ped 14 onward, is a period of resolving lesions. There is still the adenomatoid appearance from the hyperplasia of pneumocytes and terminal duct epithelium, and this may take several weeks to resolve. Exudation is much reduced. The interstitial accumulation of large mononuclear cells with indistinct cytoplasm together with the lymphocytes and plasma cells result in alveolar septal thickening. Lymphocytes and plasma cells also are seen as focal aggregations, particularly adjacent to

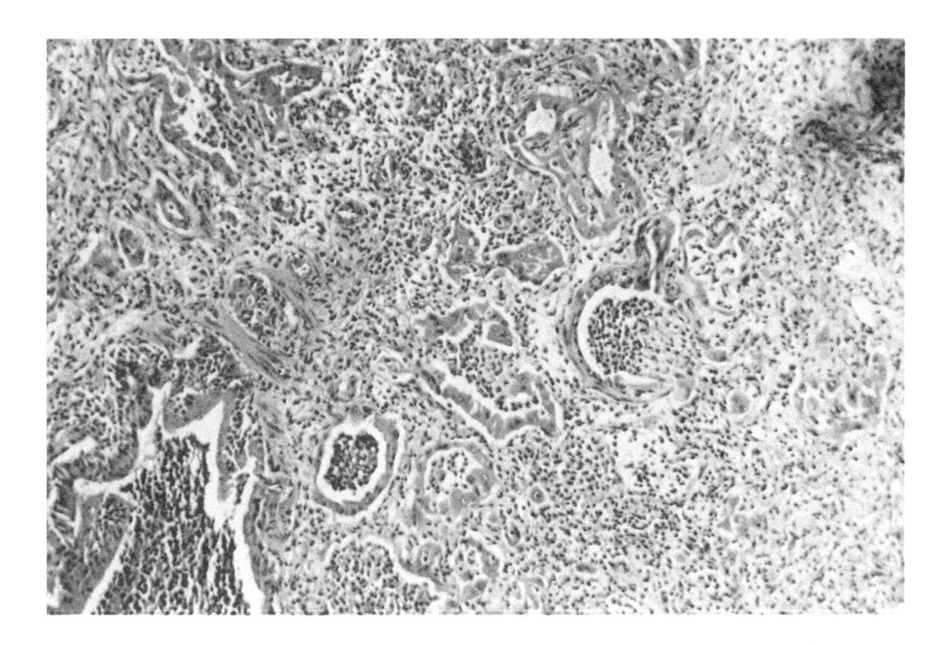

FIGURE 4. Light micrograph of lung 15 days post-exposure to FCV Exaggerated hypertrophy and hyperplasia of alveolar pneumocytes alveolar ducts and terminal bronchioles. Extensive mononuclear interstitial reaction.

bronchioles. There is some fibroblast proliferation during this phase, but intraalveolar fibrosis is sparse and nonprogressive.

Throughout the above phases, adjacent parts of the lungs not directly involved with the inflammatory process can show alternate areas of collapse and emphysema.

Lung histopathology after exposure to less virulent strains of FCV[109] showed mild focal lesions with minimal exudation. There was hyperplasia and exfoliation of pneumocytes and interstitial infiltration of alveolar septa by mononuclear and occasionally polymorphonuclear cells.

It is of interest that the description and illustrations of a suspected viral pneumonia of cats published by Blake et al. in 1942[125] have close similarities to FCV pneumonia and may well represent an early description of the disease.

Extra-pulmonary histopathology associated with experimental or natural FCV infections has been less studied. The oral lesions begin with nuclear pyknosis and increased cytoplasmic acidophilia in groups of cells in either the nonkeratinizing stratum corneum or the superficial stratum spinosum.[124] Necrosis of these cells leads to vesicle development and then loss of overlying epithelium leaving crateriform defects with neutrophilic inflammatory response at the periphery and base. There are no inclusion bodies. With antiFCV fluorescent reagents, strong fluorescence is detected in the epithelial cells at the periphery of these ulcers.[122]

Even with virulent FCV strains, there is little or no epithelial damage in the conjunctival, nasal, or pharyngeal mucosa. Mild subepithelial infiltrates of neutrophils and focal accumulations of lymphocytes and plasma cells are frequently found and may be quite persistent.[123,124,126]

F. Epizootiology

1. Prevalence

Following the initial isolations of the late 1950s, viruses that we now recognize as FCV were isolated with ease and high frequency in many parts of the world from the throat and less regularly the nose and conjunctiva of cats with and without signs of respiratory disease.[25,27,29,80,81,86,126-139]

In cats with respiratory disease, the prevalence rates for FVR and FCV have been very similar in most surveys. However, FVR was isolated from 28% and FCV from only 5.3% of 75 fatal cases of respiratory infection in kittens. In a case-matched control study of 65 cats with respiratory disease, FCV was isolated from 20%, FVR from 33.8%, and both viruses were present in three cats (4.6%).[138] In the age, sex, and residence-matched controls, FCV was isolated from 7.7% and FVR from 7.7%. Dual infections with FCV and FVR have now been reported by several authors.[106,137,138] Folkers and Hoogenboom[136] reported that in some catteries with endemic respiratory disease a high proportion were mixed infections.

In healthy cats, FCV are generally more frequently isolated than FVR. Thus, in an extensive survey[134] of the prevalence of FCV (and FVR) in clinically normal cats in Britain, 100 household pets, 974 cats attending a total of five cat shows, and 426 cats in three laboratory colonies were sampled by means of oropharyngeal swabs. FCV was isolated from 8% of the household pets, 24.02% of the cats attending shows, and from 41.5% of the cats from two of the laboratory colonies. The figures for FVR isolations were 1%, 1.75%, and 0.4%, respectively. When the isolations of FCV in the show-cat group were analyzed statistically, the infection rate was significantly higher ($p \leq 0.01$) in animals less than 1 year of age. This age relationship with FCV was also found, in both diseased and healthy matched control cats, by Bech-Nielson et al.[138] In the study of Wardley et al.,[134] neutered cats were less likely ($p \leq 0.05$) to be infected with FCV than entire animals. It was suggested this may reflect hormonal influence, or could result from reduced social exposure of the neutered group.

Serological surveys based on virus neutralization have been restricted by the antigenic variation encountered with FCV. However, neutralizing antibody to one or several strains has been found in up to 80% of adults cats.[119,139-141] The breadth of strains neutralized by a given sera and the titers of antibody tend to be greater in colony than individual household pets, and both tend to increase with age.

Although antigenic variation is regularly noted with FCV isolates, they are very closely related. Thus, an antiserum prepared by the sequential inoculation of cats with three FCV strains was capable of neutralizing all of 93 isolates tested. The majority of isolates were North American, but isolates from Britain, Australia, and New Zealand were included.

2. Sources of Infection

The sources of infectious FCV are the clinically diseased animal, the subclinically infected cat, or the persistently infected carrier. Experimental infections have shown a period of virus shedding, mainly oropharyngeal, not exceeding 21 days and more usually 7 to 14 days.[106,108] However, Wardley and Povey,[100] using a vigorous tonsillar swabbing technique, were able to detect virus, although intermittently, beyond 30 days in all of 15 cats experimentally infected with one of three FCV strains.

Subclinical infections may occur because of the relative avirulence of the infecting strain or because of partial immunity. The ability of homologous and heterologous strains of FCV to reinfect partially immune cats has been demonstrated on many occasions. For instance, Kahn and Walton[142] described the reinfection of pregnant queen cats with homologous virus in the face of virus neutralizing antibody titers as high as 1:180. Such queens shed virus for up to 3 weeks and were capable of infecting kittens born to them.

In the epidemiology of viral disease, the existence of "carrier" animals that remain infected with the particular virus over a much longer period than is normal for that disease, is a phenomenon of importance. The frequent isolations of FCV from healthy cats[126,131] had suggested carriers may occur. This was confirmed by Povey and Johnson,[143] who reported continuous shedding of FCV by cats, one of which had yielded virus from 35 of 37 oropharyngeal swabs taken over an 11-month period during which the cat was isolated from other cats. This cat was followed subsequently for a total of 124 weeks until it died from interstitial nephritis. Samples taken the day before death had FCV and the isolate was very close antigenically to that recovered more than 2 years previously.[97] Kahn and Walton[142] and Walton[144] referred to similar persistence of FCV infection, in one cat for 150 days. Both groups of workers demonstrated that the carriers were capable of producing disease in contact kittens.

In the experimental infections of Wardley and Povey,[100] persistence of virus for periods of between 34 and 186 days postinfection occurred in all 15 of the kittens maintained in isolation. Gradually the kittens eliminated virus, and overall virus shedding had a halflife of 75 days. Wardley[145] reported a detailed monitoring of FCV carrier cats over periods of 3 months to 2 years. During the 2-year observation of two carriers, significant ($p \leqslant 0.01$) variations in the mean monthly viral titers in the oropharynxes were found, but no definite pattern to these fluctuations was apparent. In six carrier cats, mean titers of virus in oropharyngeal fluids ranged from 1.9 to 4.3 $\log_{10}CCID_{50}$ FCV/mℓ of saliva. During the observations, oropharyngeal swabs were almost consistently positive (1459 positives out of 1507 swabbings), whereas nasal, conjunctival, urine, and blood samples were all negative. Of 73 fecal samples, 5 from the carriers were positive for FCV.

Wardley and Povey[120] examined a variety of tissues between 2 and 31 days postinfection with FCV and found that irrespective of infecting strain the tonsil yielded the highest concentrations of virus on almost all occasions. Titers varied from 2.0 to 5.5 $\log_{10}CCID_{50}$/g of tonsil tissue with a mean of 4.3. Similarly, in all long-term carrier animals autopsied after as long as 2 years after infection, the tonsil was the only one of 26 tissues consistently positive for virus. Persistence of FCV in the tonsil had previously been recorded.[102,108] However, tonsillectomy has not resulted in the subsequent elimination of virus from FCV carriers.[101]

3. Transmission

The major modes of transmission of FCV are by direct cat to cat contact or by indirect transfer on the hands of attendants. Infected cats do not generate infectious aerosols, and air-borne infection has not been achieved over distances greater than 1.24 m.[112] The virus in experimental aerosols is labile at relative humidities between 20 and 70%, but resists dryer conditions.[146] However, in surface survival tests, FCV was recoverable for up to 8 days in a dry environment, or 10 days from a moist surface at temperatures of 7 to 20°C.[143] Disinfectants are variable in their efficacy against FCV. Sodium hypochlorite at concentrations as low as 0.01% is very effective at inactivating FCV within 10 min.[63,147]

G. Immunity

Virus neutralizing antibody in serum has been the most frequently studied parameter of the immune response to FCV infection. The interaction of cell-mediated responses has not been elucidated, and there is only limited information on secretory responses.

Virus neutralizing antibody is detectable (titer 1:4) by 5 days after infection and by 7 days titers range from 1:4 to 1:64 (against 30 to 300 $CCID_{50}$ of homologous strain). Titers rise steadily over 5 weeks or so to attain levels of 1:512 or greater.[18,106,108,148,149] The class-specific immunoglobulin responses have not been determined, but Olsen et al.[148] found that early antibody, which was neutralizing and precipitating, did not fix complement, whereas antibody present from day 21 postinfection onwards was complement fixing as well as capable of neutralizing virus and precipitating soluble antigen.

The antigenic relatedness and yet diversity in cross-neutralization reactions of various FCV strains was discussed in an earlier section. Initial in vivo studies using a challenge strain of FCV different from the original infecting strain demonstrated lack of cross-protection.[61] However, Kahn et al.[150] showed that aerosol exposure of kittens to a strain of low virulence (F9) provided protection against subsequent challenge with a virulent strain, 255. This demonstration of cross-protection was extended to a total of eight strains of FCV by Povey and Ingersoll.[149] In these studies, as well as that of Olsen et al.,[148] it was shown that antibody capable of neutralizing some heterologous strains of FCV could be detected in sera of cats following a primary exposure. The appearance and rise of this heterotypic antibody was delayed (14 to 35 days) in comparison to homotypic antibody detection, and titers were low. Based on their results, Povey and Ingersoll[149] concluded that in their assay system a titer of 1:16 or greater was indicative of clinical protection against challenge ($10^{5.3}CCID_{50}$) with homologous and at least some heterologous FCV. A serum titer of 1:7 or less correlated with susceptibility to challenge with heterologous virus. The protection obtained in the above studies was incomplete in that it did not usually prevent viral replication and reisolation after challenge, although the duration and quantity of virus shed was reduced. This has been a consistent experience in various experiments[151-153] and even after homologous challenge in cats recovered from natural infection.[114]

The mucosal (local) immune responses to FCV infection have been paid scant attention until recently. Wardley[101] and Povey and Ingersoll[149] reported low levels (mean 1:5) of homotypic neutralizing antibody activity in nasal secretions of some cats 2 to 4 weeks after intranasal inoculation with FCV. Johnson[154] in a study that concentrated on immunity to FCV in the kitten following oculonasal vaccination found homotypic, but not heterotypic neutralizing activity in nasal secretions that was predominantly IgG mediated. Although IgA is the predominant immunoglobulin in normal feline secretion (tears and saliva) by a factor of 10:1 over IgG,[155] and although it was detected in equal frequency with IgG (80% of samples) in tears and nasal secretions of cats with FCV infection,[156] it would appear to be nonneutralizing for FCV.[154]

Maternally derived immunity to FCV has also been little studied considering its importance in epizootiology and vaccination program design. Virus neutralizing antibody levels in the sera of kittens at birth are only 3% of those of their mothers, but the kittens' titers rise rapidly after colostral intake to a mean of 65% of the mother's titer.[157] Thereafter the antibody is steadily catabolized and titers fall with a halflife of 15 days. In the 12 kittens from four litters followed by Johnson and Povey,[157] 66% of kittens had no detectable antibody by 12 weeks of age and the remaining four kittens lacked antibody by 14 weeks of age. This compared with the persistence of maternal antibody of up to 13 weeks noted as previously.[106]

The capability of maternal antibody to protect kittens against homotypic FCV has been in some doubt. Kahn and Gillespie[108] found that aerosol challenge of kittens 6 to 10 weeks of age with passive antibody titers of 1:4 to 1:16 produced moderated disease. Kahn and Walton[142] noted FCV disease as early as 6 days of age in kittens suckling seropositive virus shedding dams. Johnson[154] has recently demonstrated that kittens born to carrier mothers initially resisted infection, but became infected between 3 and 6 weeks of age. Signs were very mild unless maternal antibody titers had fallen to 1:4 or below before infection, in which case more typical disease occurred.

The success of the cross-protection studies encouraged the development of monotypic FCV vaccines that have always been marketed as a combination with other antigens, such as FVR and panleukopenia. Two routes of administration were used — either intramuscular/subcutaneous[151-153,158-160] or intranasal/intraocular.[161,162] Apart from the vaccines described by Povey and Wilson,[152] Povey,[163] Chappuis et al.[164] and Povey et al.,[153] in which the calicivirus fraction was inactivated, the vaccines have contained live attenuated FCV.

Live virus vaccines given by the intranasal/intraocular route have been demonstrated to provide good protection[152] with onset by 48 hr after vaccination.[165] Disadvantages of this method of immunization include the transient occurrence of sneezing and ocular[152] and nasal discharges in vaccinated cats,[162,166,167] and persistent shedding of virus that may cause clinical illness in contact cats.[166] Protection conferred by the parenteral routes has been adequate to good.[151-153,158-160,163]

IV. CONCLUSIONS

The Caliciviridae are a virus family of expanding importance as they are recognized in an increasing number of species. The feline caliciviruses have been the most extensively studied, and they serve as a model for the pathobiology of the family. They also have good comparative potential in the study of pneumonia and in particular diffuse alveolar damage. The lesions of FCV pneumonia have been compared and contrasted with those of nitrogen dioxide toxicity and of human influenza.[117] Although most attention has been given to the association of FCV with respiratory disease, the relatively frequent reports of diarrhea, particularly in experimental infections, requires clarification as to pathogenesis. Similarly, the incoordination, spasticity, and loss of righting reflexes noted recently with some strains of FCV is intriguing.

The frequency of persistent infections with FCV provides good opportunities for comparative studies on virus-host interaction. This is a carrier situation where there is continuous production of virus in the presence of an apparently normal immune system. The events that take place immediately preceding the sudden and permanent clearance of virus from such carriers are worthy of study.

Progress in the control of FCV disease has been made with commercial vaccines, but the problem of strain heterogeneity has not been fully overcome. The identification of protective antigens within the FCV particle and of cross-protective antigens would open the way towards genetically engineered or synthetic approaches to vaccine development.

REFERENCES

1. **Matthews, R. E. F.**, Classification and nomenclature of viruses: third report of the International Committee on Taxonomy of Viruses, *Intervirology*, 12, 129, 1979.
2. **Evermann, J. F., Bryan, G. M., and McKeirnan, A. J.**, Isolation of a calicivirus from a case of canine glossitis, *Canine Pract.*, 8(2), 36, 1981.
3. **Smith, A. W., Skilling, D. E., Dardiri, A. H., and Latham, A. B.**, Calicivirus pathogenic for swine: a new serotype isolated from opaleye Girella nigricans, an ocean fish, *Science*, 209, 940, 1980.
4. **Smith, A. W., Skilling, D. E., and Ritchie, A. E.**, Immunoelectron microscopic comparisons of caliciviruses, *Am. J. Vet. Res.*, 39, 1531, 1978.
5. **Bankowski, R. A.**, Vesicular exanthema, *Adv. Vet. Sci. Comp. Med.*, 10, 23, 1965.
6. **Madeley, C. R. and Cosgrove, B. P.**, Caliciviruses in man, *Lancet*, 1, 199, 1976.
7. **Woode, G. N. and Bridger, J. C.**, Isolation of small viruses resembling astroviruses and caliciviruses from acute enteritis of calves, *J. Med. Microbiol.*, 11, 441, 1978.
8. **Bridger, J. C.**, Detection by electron microscopy of caliciviruses, astroviruses and rotavirus-like particles in the faeces of piglets with diarrhoea, *Vet. Rec.*, 107, 532, 1980.
9. **Wyeth, P. J., Chettle, N. J., and Labram, J.**, Avian calicivirus (correspondence), *Vet. Rec.*, 109(21), 477, 1981.
10. **Gough, R. E. and Spackman, D.**, Virus-like particles associated with disease in guinea fowl (correspondence), *Vet. Rec.*, 109(22), 497, 1981.
11. **Flewett, T. H. and Davies, H.**, Caliciviruses in man, *Lancet*, 1, 311, 1976.
12. **Spratt, H. C., Marks, M. I., Gomersall, M., Gill, P., and Pia, C. H.**, Nosocomial infantile gastroenteritis associated with minirotavirus and calicivirus, *J. Pediatr.*, 93, 922, 1978.
13. **Cubitt, W. D., McSwiggan, D. A., and Moore, W.**, Winter vomiting disease caused by calicivirus, *J. Clin. Pathol.*, 32, 786, 1979.

14. Chiba, S., Sakuma, Y., Kogasaka, R., Akihara, M., Horino, K., Nakao, T., and Fukui, S., An outbreak of gastroenteritis associated with calicivirus in an infant home, *J. Med. Virol.*, 4(4), 249, 1979.
15. Cubitt, W. D., McSwiggan, D. A., and Arstall, S., An outbreak of calicivirus infection in a mother and baby unit, *J. Clin. Pathol.*, 33, 1095, 1980.
16. Cubitt, W. D. and McSwiggan, D. A., Calicivirus gastroenteritis in North West London, *Lancet*, 2(8253), 975, 1981.
17. Long, G. G., Evermann, J. F., and Gorham, J. R., Naturally occurring picornavirus infection of domestic mink, *Can. J. Comp. Med.*, 44, 412, 1980.
18. Fastier, L. B., A new feline virus isolated in tissue cultures, *Am. J. Vet. Res.*, 18, 382, 1957.
19. Bolin, V. S., The cultivation of panleukopenia virus in tissue culture, *Virology*, 4, 389, 1957.
20. Cohen, P., Yohn, D. S., Pavia, R. A., and Hammon, W. M., The relationship of a feline virus isolated by Bolin to feline panleukopenia kidney cell degenerating virus and 2 feline respiratory viruses, *Am. J. Vet. Res.*, 22, 637, 1961.
21. Sinha, S. K. and Burger, D., Some findings concerning a report on the cultivation of feline panleukopenia virus in cell culture, *Can. Vet. J.*, 2, 62, 1961.
22. Crandell, R. A. and Maurer, F. D., Isolation of a feline virus associated with intranuclear inclusion bodies, *Proc. Soc. Exp. Biol. Med.*, 97, 487, 1958.
23. Sinha, S. K., Feline viruses, *Calif Vet.*, 6, 18, 1958.
24. Crandell, R. A. and Madin, S. H., Experimental studies on a new feline virus, *Am. J. Vet. Res.*, 21, 551, 1960.
25. Crandell, R. A., Nieman, W. H., Ganaway, J. R., and Maurer, F. C., Isolation of cytopathic agents from the nasopharyngeal region of the domestic cat, *Virology*, 10, 283, 1960.
26. Hersey, D. F. and Maurer, F. D., Immunological relationship of selected feline viruses by complement fixation, *Proc. Soc. Exp. Biol. Med.*, 107, 645, 1961.
27. Piercy, S. E. and Prydie, J., Feline influenza, *Vet. Rec.*, 75, 86, 1963.
28. Sabine, M. and Hyne, R. H. J., Isolation of feline picornavirus from cheetahs with conjunctivitis and glossitis, *Vet. Rec.*, 87, 794, 1970.
29. Burki, F., Picornaviruses of cats, *Arch. Gesamte Virusforsch.*, 15, 690, 1965.
30. Zwillenberg, L. O. and Burki, F., On the capsid structure of some small feline and bovine viruses, *Arch. Gesamte Virusforsch.*, 19, 373, 1966.
31. Almeida, J. D., Waterson, A. P., Prydie, J., and Fletcher, E. W. L., The structure of feline picornavirus and its relevance to cubic viruses in general, *Arch. Gesamte Virusforsch.*, 25, 105, 1968.
32. Wawrzkiewicz J., Smale, C. J., and Brown, F., Biochemical and biophysical characteristics of vesicular exanthema virus and the viral ribonucleic acid, *Arch. Gesamte Virusforsch.*, 25, 337, 1968.
33. Oglesby, A. S., Schaffer, F. L., and Madin, S. H., Biochemical and biophysical properties of vesicular exanthema of swine virus, *Virology*, 44, 329, 1971.
34. Wildy, P., Classification and nomenclature of viruses, in *Virology*, 5th ed., Melnick, J. L., Ed., S. Karger, Basel, 1971, 55.
35. Burroughs, J. N. and Brown, F., Physico-chemical evidence for the reclassification of the caliciviruses, *J. Gen. Virol.*, 22, 281, 1974.
36. Studdert, M. J. and O'Shea, J. D., Ultrastructural studies of the development of feline calicivirus in a feline embryo cell line, *Arch. Virol.*, 48, 317, 1975.
37. Ehresmann, D. W., Studies on Calicivirus Genomic Replication and Transcription: Characterization of RNA from Virions and Infected Cells, Ph.D. thesis, University of California, Berkeley, 1978.
38. Ehresmann, D. W. and Schaffer, F. L., Calicivirus intracellular RNA: fractionation of 18 to 22S RNA, and lack of typical 5′-methylated caps on 36S and 22S San Miguel sea lion virus RNAs, *Virology*, 95, 251, 1979.
39. Black, D. N., Burroughs, J. N., Harris, T. J. R., and Brown, F., The structure and replication of calicivirus RNA, *Nature (London)*, 274, 614, 1978.
40. Burroughs, J. N. and Brown, F., Presence of a covalently linked protein on calicivirus RNA, *J. Gen. Virol.*, 41, 443, 1978.
41. Black, D. and Brown, F., A major difference in the strategy of the calici- and picornaviruses and its significance in classification, *Intervirology*, 6, 57, 1975/6.
42. Ehresmann, D. W. and Schaffer, F. L., RNA synthesized in calicivirus-infected cells is atypical of picornaviruses, *J. Virol.*, 22, 572, 1977.
43. Black, D. N. and Brown, F., Proteins induced by infection with caliciviruses, *J. Gen. Virol.*, 38, 75, 1978.
44. Fretz, M. and Schaffer, F. L., Calicivirus proteins in infected cells: evidence for a capsid polypeptide precursor, *Virology*, 89, 318, 1978.

45. Schaffer, F. L., Bachrach, H. L., Brown, F., Gillespie, J. H., Burroughs, J. N., Madin, S. H., Madeley, C. R., Povey, R. C., Scott, F., Smith, A. W., and Studdert, M. J., Caliciviridae. Report of the Caliviridae Study Group, Vertebrate Virus Subcommittee, International Committee on Taxonomy of Viruses, *Intervirology,* 14, 1, 1980.
46. Peterson, J. E. and Studdert, M. J., Feline picornavirus: structure of the virus and electron microscopic observations on infected cell cultures, *Arch. Gesamte Virusforsch.,* 32, 249, 1970.
47. Schaffer, F. L. and Soergel, M. E., Single major polypeptide of a calicivirus: characterization of calicivirus isolates from pinnipeds, *Intervirology,* 1, 210, 1976.
48. Burroughs, J. N., Doel, T. R., Smale, C. J., and Brown, F., A model for vesicular exanthema virus, the prototype of the calicivirus group, *J. Gen. Virol.,* 40, 161, 1978.
49. Adldinger, H. K., Lee, K. M., and Gillespie, J. H., Extraction of infectious ribonucleic acid from a feline picornavirus, *Arch. Gesamte Virusforsch.,* 28, 245, 1969.
50. Bachrach, H. L. and Hess, W. R., Animal picornaviruses with a single major species of capsid protein, *Biochem. Biophys. Res. Commun.,* 55, 141, 1973.
51. Love, D. N., Feline calicivirus: purification of virus and extraction and characterization of its ribonucleic acid, *Cornell Vet.,* 66, 498, 1976.
52. Schaffer, F. L., Ehresmann, D. W., Fretz, M. K., and Soergel, M. E., A protein VPg, covalently linked to 36S calicivirus RNA, *J. Gen. Virol.,* 47, 215, 1980.
53. Schaffer, F. L., Caliciviruses, in *Comprehensive Virology,* Vol. 14, Fraenkel-Conrat, H. and Wagner, R. R., Eds., Plenum Press, New York, 1979, 249.
54. Rowlands, D. J., Sangar, D. V., and Brown, F., Buoyant density of picornaviruses in caesium salts, *J. Gen. Virol.,* 13, 141, 1971.
55. Pichler, L., Buoyant density in CaCl of infectious feline picornavirus virions, *Zentralbl. Bakteriol. Hyg. Orig. A,* 222, 162, 1972.
56. Soergel, M. E., Smith, A. W., and Schaffer, F. L., Biophysical comparisons of calicivirus serotypes isolated from pinnipeds, *Intervirology,* 5, 239, 1975.
57. Love, D. N., and Jones, R. F., Studies on the buoyant density of a feline calcivirus, *Arch. Gesamte Virusforsch.,* 44, 142, 1974.
58. Schaffer, F. L. and Soergel, M., Biochemical and biophysical characterization of calicivirus isolates from pinnipeds, *Intervirology,* 1, 210, 1973.
59. Lee, K. M. and Gillespie, J. R., Thermal and pH stability of feline calicivirus, *Infect. Immun.,* 7, 678, 1973.
60. Crandell, R. A., A description of eight feline picornaviruses and an attempt to classify them, *Proc. Soc. Exp. Biol. Med.,* 126, 240, 1967.
61. Bartholomew, P. T. and Gillespie, J. H., Feline Viruses. I. Characterization of four isolates and their effect on young kittens, *Cornell. Vet.,* 58, 248, 1968.
62. Burki, F. and Pichler, L., Further biochemical testing of feline picornaviruses, *Arch. Gesamte Virusforsch.,* 33, 126, 1971.
63. Scott, F. W., Virucidal disinfectants and feline viruses, *Am. J. Vet. Res.,* 41, 410, 1980.
64. Zee, Y. C., Hackett, A. J., and Madin, S. H., Electron microscopic studies on vesicular exanthema of swine virus: intracytoplasmic viral crystal formation in cultured pig kidney cells, *Am. J. Vet. Res.,* 29, 1025, 1968.
65. Love, D. N. and Sabine, M., Electron microscopic observation of feline kidney cells infected with a feline calicivirus, *Arch. Virol.,* 48, 213, 1975.
66. Soergel, M. B., Schaffer, F. L., Sawyer, J. C., and Prato, C. M., Assay of antibodies to caliciviruses by radioimmune precipitation using staphylococcal protein A as IgG adsorbent, *Arch. Virol.,* 57, 271, 1978.
67. Bankowski, R. A., Wichmann, R., and Kummer, R. M., A complement-fixation test for identification and differentiation of immunological types of the virus of vesicular exanthema of swine, *Am. J. Vet. Res.,* 14, 145, 1953.
68. Smith, A. W., Prato, C. M., and Skilling, D. E., Characterization of two new serotypes of San Miguel sea lion virus (SMSV), *Intervirology,* 8, 30, 1977.
69. Smith, A. W., Skilling, D. E., and Latham, A. B., Isolation and identification of five new serotypes of calicivirus from marine mammals, *Am. J. Vet. Res.,* 42(4), 693, 1981.
70. Burroughs, N., Doel, T., and Brown, F., Relationship of San Miguel sea lion virus to other members of the calicivirus group, *Intervirology,* 10, 51, 1978.
71. Povey, R. C., Serological relationships among feline caliciviruses, *Infect. Immun.,* 10, 1307, 1974.
72. Chappuis, G. and Stellman, C., Biomathematical system of relationship and dominance for the classification of feline picornavirus, *J. Biol. Stand.,* 2, 319, 1974.
73. Kalunda, M., Lee, K. M., Holmes, D. F., and Gillespie, J. H., Serologic classification of feline caliciviruses by plaque-reduction neutralization and immunodiffusion, *Am. J. Vet. Res.,* 36, 353, 1975.

74. Burki, F., Starustka, B., and Ruttner, O., Attempts to serologically classify feline caliciviruses on a national and an international basis, *Infect. Immun.*, 14, 876, 1976.
75. Lee, K. M., Kniazeff, A. J., Fabricant, C. G., and Gillespie, J. H., Utilization of various cell culture systems for propagation of certain feline viruses and canine herpes virus, *Cornell Vet.*, 59, 534, 1969.
76. Love, D. N. and Donaldson-Wood, C., Replication of a strain of feline calicivirus in organ culture, *Arch. Virol.*, 47, 167, 1975.
77. Studdert, M. J., Caliciviruses. Brief review, *Arch. Virol.*, 58, 157, 1978.
78. Studdert, M. J., Martin, M. C., and Peterson, J. E., Viral diseases of the respiratory tract of cats: isolation and properties of viruses tentatively classified as picornaviruses, *Am. J. Vet. Res.*, 31, 1723, 1970.
79. Scott, F. W., Csiza, C. K., and Gillespie, J. H., Feline viruses. IV. Isolation and characterization of feline panleukopenia virus in tissue culture and comparison of cytopathogenicity with feline picornavirus, herpes virus, and reovirus, *Cornell Vet.*, 60, 165, 1970.
80. Torlone, V., Agents citopathogeno isolato da una forma rinocongiunfivale del gatto, *Vet. Ital.*, 11, 915, 1960.
81. Spradbrow, P. B., Bagust, T. J., Burgess, G., and Portas, B., The isolation of picornaviruses from cats with respiratory disease, *Aust. Vet. J.*, 46, 105, 1970.
82. Kahn, D. E. and Gillespie, J. H., Feline viruses. X. Characterization of a newly-isolated picornavirus causing interstitial pneumonia and ulcerative stomatitis in the domestic cat, *Cornell Vet.*, 60, 669, 1970.
83. Love, D. N., The utilization of a zwitterionic buffer system in the plaque assay of a feline calicivirus, *Aust. J. Exp. Biol. Med. Sci.*, 51, 263, 1973.
84. Ormerod, E. and Jarrett, O., A classification of feline calicivirus isolates based on plaque morphology, *J. Gen. Virol.*, 39, 537, 1978.
85. Ormerod, E., McCandlish, I. A. P., and Jarrett, O., Diseases produced by feline caliciviruses when administered to cats by aerosol or intranasal instillation, *Vet. Rec.*, 104, 65, 1979.
86. Bittle, J. L., York, C. J., Newberne, J. W., and Martin, M., Serologic relationship of new feline cytopathogenic viruses, *Am. J. Vet. Res.*, 21, 547, 1960.
87. Povey, R. C., Feline respiratory infections — a clinical review, *Can. Vet. J.*, 17, 93, 1976.
88. Johnson, R. P. and Povey, R. C., Effect of diet on oral lesions of feline calicivirus infection, *Vet. Rec.*, 110, 106, 1982.
89. Cooper, L. M. and Sabine, M., Paw and mouth disease in a cat, *Aust. Vet. J.*, 48, 644, 1972.
90. Neufeld, J. L., Burton, L., and Jeffery, K. R., Eosinophilic granuloma in a cat. Recovery of virus particles, *Vet. Pathol.*, 17, 97, 1980.
91. Love, D. N. and Baker, K. D., Sudden death in kittens associated with feline picornavirus, *Aust. Vet. J.*, 48, 643, 1972.
92. Pedersen, N., personal communication, 1981.
93. Laliberte, L., Johnson, R., and Povey, C., unpublished data, 1982.
94. Povey, C., unpublished data, 1982.
95. Rich, L. J. and Fabricant, C. G., Urethral obstruction in male cats: transmission studies, *Can. J. Comp. Med.*, 33, 164, 1969.
96. Fabricant, C. G. and Rich, L. T., Microbial studies of feline urolithiasis, *J. Am. Vet. Med. Assoc.*, 158, 976, 1971.
97. Fabricant, C. G., Herpesvirus-induced urolithiasis in specific-pathogen-free male cats, *Am. J. Vet. Res.*, 38, 1837, 1977.
98. Fabricant, C. G., Serological responses to the cell associated herpesvirus and the Manx calicivirus of SPF male cats with herpesvirus-induced urolithiasis, *Cornell Vet.*, 71, 59, 1981.
99. Gaskell, R. M., Gaskell, C. J., Page, W., Dennis, P., and Voyle, C. A., Studies on a possible viral aetiology for the feline urological syndrome, *Vet. Rec.*, 105, 243, 1979.
100. Wardley, R. C. and Povey, R. C., The clinical disease and patterns of excretion associated with three different strains of feline caliciviruses, *Res. Vet. Sci.*, 23, 7, 1977.
101. Wardley, R. C., Studies on Feline Caliciviruses with Particular Reference to Persistent Infections, Ph.D. thesis, University of Bristol, England, 1974.
102. Povey, R. C. and Hale, C. J., Experimental infections with feline caliciviruses (picornaviruses) in specific-pathogen-free kittens, *J. Comp. Pathol.*, 84, 245, 1974.
103. Povey, R. C., Wardley, R. C., and Jessen, H., Feline picornavirus infection. The *in vivo* carrier state, *Vet. Rec.*, 92, 224, 1973.
104. Povey, R. C., Differential diagnosis and control of infectious respiratory diseases in cats, *Small Anim. Vet. Med. Update*, 1(1), 1, 1977.
105. Gillespie, J. H., Judkins, B., and Kahn, D. E., Feline viruses. XIII. The use of immunofluorescent test for the detection of feline picornavirus, *Cornell Vet.*, 61, 172, 1971.
106. Povey, R. C., Studies on Viral Induced Respiratory Diseases of Cats, Ph.D. thesis, University of Bristol, England, 1970.

107. Stevenson, B. J. and Burgess, G. W., Feline respiratory disease of multiple aetiology, *N.Z. Vet. J.*, 26, 257, 1978.
108. Kahn, D. E. and Gillespie, J. H., Feline viruses: pathogenesis of picornavirus infection in the cat, *Am. J. Vet. Res.*, 32, 521, 1971.
109. Hoover, E. A. and Kahn, D. E., Experimentally induced feline calicivirus infection: clinical signs and lesions, *J. Am. Vet. Med. Assoc.*, 166, 463, 1975.
110. Love, D. N., Pathogenicity of a strain of feline calicivirus for domestic kittens, *Aust. Vet. J.*, 51, 541, 1975.
111. Povey, R. C., Effect of orally administered ribavirin on experimental feline calicivirus infection in cats, *Am. J. Vet. Res.*, 39, 1337, 1978.
112. Wardley, R. C. and Povey, R. C., Aerosol transmission of feline caliciviruses. An assessment of its epidemiological importance, *Br. Vet. J.*, 133, 504, 1977.
113. Burki, F., Viren des Respirationsapparates bei Katzen, 17th World Vet. Congr., Hannover, W. Germany, 1963, 1.
114. Burki, F., Zur organaffinitat feliner Picornaviren, *Zentralbl. Bakteriol. I. Orig.*, 200, 281, 1966.
115. Bittle, J. L., Emergy, J. B., York, C. J., and McMillen, J. K., Comparative study of feline cytopathogenic viruses and feline panleukopenia virus, *Am. J. Vet. Res.*, 22, 374, 1961.
116. Smith, L. P., Hugh-Jones, M. E., and Jackson, O. F., Weather conditions and disease (correspondence), *Vet. Rec.*, 91, 642, 1972.
117. Langloss, J. M., Hoover, E. A., and Kahn, D. E., Diffuse alveolar damage in cats induced by nitrogen dioxide or feline calicivirus, *Am. J. Pathol.*, 89, 637, 1977.
118. Langloss, J. M., Hoover, E. A., and Kahn, D. E., Ultrastructural morphogenesis of acute viral pneumonia produced by feline calicivirus, *Am. J. Vet. Res.*, 39, 1577, 1978.
119. Langloss, J. M., Hoover, E. A., Kahn, D. E., and Kniazeff, A. J., Elaboration of chemoactive substances by alveolar cells: possible mechanisms for the initial neutrophilic response in feline caliciviral pneumonia, *Am. J. Vet. Res.*, 40, 186, 1979.
120. Wardley, R. C. and Povey, R. C., The pathology and sites of persistence associated with three different strains of feline calicivirus, *Res. Vet. Sci.*, 23, 15, 1977.
121. Pichler, L. and Burki, F., Zur Pathogenese der feliner Picornavirus infektion, *Wein. Tierarztl. Mschr.*, 57, 246, 1970.
122. Holzinger, E. A. and Kahn, D. E., Pathologic features of picornavirus infections in cats, *Am. J. Vet. Res.*, 31, 1632, 1970.
123. Lindt, S., Zur Morphologie und Atiologie der Erkrankungen des oberen Respirationstraktes bei katzen, *Schweiz. Arch. Tierheilk.*, 107, 196, 1965.
124. Hoover, E. A. C. and Kahn, D. E., Lesions produced by feline picornavirus of different virulence in pathogen-free cats, *Vet. Pathol.*, 10, 307, 1973.
125. Blake, F. G., Howard, M. E., and Tatlock, H., Feline virus pneumonia and its possible relation to some cases of primary atypical pneumonia in man, *Yale J. Biol. Med.*, 15, 139, 1942.
126. Povey, R. C. and Johnson, R. H., A survey of feline viral rhinotracheitis and feline picornavirus infection in Britain, *J. Small Anim. Pract.*, 12, 233, 1971.
127. Burki, F., Lindt, S., and Freudiger, U., Katzenshnupfen in einem Tierheim. II. Mitt. Virologischer und experimenteller, *Teil Zentralbl. Vet. Med.*, 11, 110, 1964.
128. McEwan, P. J. and Miles, J. A. R., An electron microscope study of viruses associated with upper respiratory tract infections in cats, *Proc. Univ. Otagao Med. Sch.*, 45, 21, 1967.
129. Brehaut, L., Jones, R. H., McEwan, P. J., and Miles, J. A. R., Viruses associated with feline respiratory disease in Dunedin, *N.Z. Vet. J.*, 17, 82, 1969.
130. Kamizono, M., Konishi, S., Ogata, M., and Kobori, S., Studies on cytopathogenic viruses isolated from cats with respiratory infection, I., *Jpn. J. Vet. Sci.*, 30, 197, 1968.
131. Walton, T. E. and Gillespie, J. H., Feline Viruses. VI. Survey of the incidence of feline pathogenic agents in normal and clinically-ill cats, *Cornell Vet.*, 60, 215, 1970.
132. Takahashi, E., Konishi, S., and Ogata, M., Studies on cytopathogenic viruses from cats with respiratory infections. II. Characterization of feline picorna viruses, *Jpn. J. Vet. Sci.*, 33, 81, 1971.
133. Flagstaad, A., Isolation and classification of feline picornavirus and herpesvirus in Denmark, *Acta Vet. Scand.*, 13, 462, 1972.
134. Wardley, R. C., Gaskell, R. M., and Povey, R. C., Feline respiratory viruses — their prevalence in clinically healthy cats, *J. Small Anim. Pract.*, 15, 579, 1974.
135. Jensen, M. M., Buell, D. J., and McKim, R. M., Isolation rates of feline respiratory viruses in local cat populations, *J. Small Anim. Pract.*, 18, 659, 1977.
136. Folkers, C. and Hoogenboom, A. M. M., Intranasal vaccination against upper respiratory tract disease (URD) in the cat. I. Virological and serological observations in cats suffering from URD, *Comp. Immunol. Microbiol. Infect. Dis.*, 1, 37, 1978.
137. MacLachlan, N. J. and Burgess, G. W., A survey of feline viral upper respiratory tract infections, *N.Z. Vet. J.*, 26, 260, 1978.

138. Bech-Nielsen, S., Fulton, R. W., Cox, H. U., Hoskins, J. D., Malone, J. B., and McGrath, R. K., Feline respiratory tract disease in Louisiana, *Am. J. Vet. Res.*, 41, 1293, 1980.
139. Goto, H., Horimoto, M., Shimizu, K., Hiraga, T., Matsuoka, T., Nakano, T., Morohoshi, Y., Maejima, K., and Urano, T., Prevalence of feline viral antibodies in random-source laboratory cats, *Exp. Anim.*, 30, 283, 1981.
140. Studdert, M. J. and Martin, M. C., Virus diseases of the respiratory tract of cats. I. Isolation of feline rhinotracheitis virus, *Aust. Vet. J.*, 46, 99, 1970.
141. Laurent, J. C., Oudar, J., and Chappuis, G., Affections respiratoires felines d'origine virale. Sondage serologique, *Rev. Med. Vet.*, 128, 1099, 1977.
142. Kahn, D. E. and Walton, T. E., Epizootiology of feline respiratory infections, *J. Am. Vet. Med. Assoc.*, 158, 955, 1971.
143. Povey, R. C. and Johnson, R. H., Observations on the epidemiology and control of feline viral respiratory disease, *J. Small Anim. Pract.*, 11, 485, 1970.
144. Walton, T. E., Comments on epizootiology of feline respiratory infections, *J. Am. Vet. Med. Assoc.*, 158, 960, 1971.
145. Wardley, R. C., Feline calicivirus carrier state. A study of the host/virus relationship, *Arch. Virol.*, 52, 243, 1976.
146. Donaldson, A. I. and Ferris, N. P., The survival of some air-borne animal viruses in relation to relative humidity, *Vet. Microbiol.*, 1, 413, 1976.
147. Yagami, K., Ando, S., Omata, Y., Furukawa, T., and Fukui, M., Studies on viral respiratory disease in laboratory cats. I. Isolation of feline herpesvirus and choice of proper disinfectant, *Exp. Anim.*, 31, 27, 1982.
148. Olsen, R. G., Kahn, D. E., Hoover, E. A., Saxe, N. J., and Yohn, D. S., Differences in acute and convalescent-phase antibodies of cats infected with feline picornaviruses, *Infect. Immun.*, 10, 375, 1974.
149. Povey, R. C. and Ingersoll, J., Cross-protection among feline caliciviruses, *Infect. Immun.*, 11, 877, 1975.
150. Kahn, D. E., Hoover, E. A., and Bittle, J. L., Induction of immunity to feline caliciviral disease, *Infect. Immun.*, 11, 1003, 1975.
151. Kahn, D. E. and Hoover, E. A., Feline caliciviral disease: experimental immunoprophylaxis, *Am. J. Vet. Res.*, 37, 279, 1976.
152. Povey, R. C. and Wilson, M. R., A comparison of inactivated feline viral rhinotracheitis and feline caliciviral disease vaccines with live-modified viral vaccines, *Fel. Pract.*, 8, 35, 1978.
153. Povey, R. C., Koonse, H., and Hayes, M. B., Immunogenicity and safety of an inactivated vaccine for the prevention of rhinotracheitis, caliciviral disease, and panleukopenia in cats, *J. Am. Vet. Med. Assoc.*, 177, 347, 1980.
154. Johnson, R. P., Immunity to Feline Calicivirus in Kittens, Ph.D. thesis, University of Guelph, Guelph, Ontario, Canada, 1980.
155. Vaerman, J. P., Studies on IgA Immunoglobulins in Man and Animals, Ph.D. thesis, Catholic University of Louvain, Belgium, 1970.
156. Schultz, R. D., Scott, F. W., Duncan, J. R., and Gillespie, J. H., Feline immunoglobulins, *Infect. Immun.*, 9, 391, 1974.
157. Johnson, R. P. and Povey, R. C., Transfer and decline of maternal antibody to feline calicivirus, *Can. Vet. J.*, 24, 6, 1983.
158. Bittle, J. L. and Rubic, W. J., A feline calicivirus vaccine combined with feline viral rhinotracheitis and feline panleukopenia vaccine, *Fel. Pract.* 5(6), 13, 1975.
159. Bittle, J. L. and Rubic, W. J., Immunization against feline calicivirus infection, *Am. J. Vet. Res.*, 37, 275, 1976.
160. Scott, F. W., Evaluation of a feline viral rhinotracheitis-feline calicivirus disease vaccine, *Am. J. Vet. Res.*, 38, 229, 1977.
161. Davis, E. V. and Beckenhauer, W. H., Studies on the safety and efficacy of an intranasal feline rhinotracheitis-calici vaccine, *Vet. Med. Small Anim. Clin.*, 71, 1405, 1976.
162. Wilson, J. H. G., Intranasal vaccination against upper respiratory tract disease (U.R.D.) in the cat. II. Results of field studies under enzootic conditions in the Netherlands with a combined vaccine containing live attenuated calici- and herpesvirus, *Comp. Immunol. Microbiol. Infect. Dis.*, 1, 43, 1978.
163. Povey, R. C., The efficacy of two commercial feline rhinotracheitis-calicivirus-panleukopenia vaccines, *Can. Vet. J.*, 20, 253, 1979.
164. Chappuis, G., Brun, A., Precausta, F., and Terre, J., Immunization against respiratory disease in cats, *Comp. Immunol. Microbiol. Infect. Dis.*, 1, 221, 1979.
165. Davis, E. V. and Beckenhauer, W. H., Time required for "Felomune CVR" to stimulate protection, *Norden News*, p.30, 1977.
166. Kahn, D. E., Letter: report on intranasal feline rhinotracheitis-calici virus vaccine criticized, *Vet. Med. Small Animal Clin.*, 72, 8, 1977.
167. Povey, R. C., Feline respiratory disease — which vaccine?, *Fel. Pract.*, 7(5), 12, 1977.

Chapter 6

BOVINE RESPIRATORY SYNCYTIAL VIRUS

Lawrence E. Mathes and Michael K. Axthelm

TABLE OF CONTENTS

I. INTRODUCTION AND HISTORY

Bovine respiratory syncytial virus (BRSV) is believed to be a major cause of viral respiratory disease in cattle.

The first isolation of a respiratory syncytial virus was from an outbreak of upper respiratory illness in a chimpanzee colony.[1] A human isolate was later recovered from infants and children suffering from bronchiolitis and pneumonia.[2] The human respiratory syncytial virus (HRSV) has subsequently been shown to be a major cause of respiratory disease in infants. The first indication that a bovine respiratory syncytial virus (BRSV) may exist came from research studies where HRSV was being propagated in tissue culture. In these studies, investigators noticed that pooled calf serum used in growth medium inhibited HRSV replication.[3] The inhibitory factor was later shown to be antibody, suggesting that the bovine used as the source of serum had been exposed to HRSV or an antigenically related bovine virus.[4] The first BRSV isolation was made by Paccaud and Jacquier from an outbreak of bovine respiratory disease in Switzerland.[5] Shortly thereafter, outbreaks of BRSV disease were reported in Japan[6] and Belgium.[7] These early reports based their identification of BRSV on cross-neutralization with HRSV. Today, bovine respiratory syncytial virus is present in cattle populations world-wide. Antibodies to BRSV have been identified in cattle in Africa,[8] the British Isles,[4,9] Canada,[10-12] and the U.S.[13-16]

II. CLASSIFICATION

Bovine and human respiratory syncytial virus are tentatively grouped with pneumonia virus of the mouse (PVM) to make up the genus *Pneumovirus* of the family Paramyxoviridae.[17] Within this genus all species are morphologically similar. HRSV and BRSV are closely related biophysically and antigenically to each other, and cattle are susceptible to HRSV infection. PVM is only distantly related to HRSV and BRSV.[17]

Another syncytial inducing virus of cattle, bovine syncytial virus, belongs to the family Retroviradae and is unrelated to BRSV.[18,19]

III. GEOGRAPHICAL AND SPECIES DISTRIBUTION

Based on serologic surveys and isolations, bovine respiratory syncytial virus is prevalent in cattle populations in several localities around the world and must be considered endemic in these areas. In England, 83% of cattle tested had neutralizing antibody.[4] In Switzerland, one third of the serum samples tested were antibody positive.[5] In Quebec, 35% of cattle,[11] 67% in Alabama,[13] and 76% in Saskatchewan[10] were BRSV antibody positive. BRSV has been isolated from cattle with upper respiratory distress in Switzerland,[5] Britain,[20] Czechoslovakia,[21] Japan,[22] and, in the U.S. in Missouri,[16] Iowa,[13,15] and Nebraska.[23] Berthiaume and colleagues found that 25 of 31 sheep surveyed in Canada had complement-fixing BRSV, suggesting that sheep may be infected with BRSV or an antigenically related virus under natural conditions.[24] Sheep are susceptible to BRSV experimentally, and clinical signs and pulmonary lesions are similar to those in cattle.[25,26] Additionally, BRSV (or HRSV) antibody has been detected in horses,[24] cats,[27] dogs,[28] swine,[27] and goats,[27] but not in rabbits, chickens, or ferrets.[4]

Recently, a goat respiratory syncytial virus (GRSV) closely related but not identical to BRSV was isolated from pigmy goats with respiratory disease.[29] Serologic surveys of healthy sheep and goats indicated that about 50% of the animals tested are seropositive for GRSV.[30,31]

European and American BRSV outbreaks are believed to be the result of an endemic condition where small numbers of animals occasionally displayed symptoms. The original Japanese outbreak was believed to be a new infection among nonimmune animals.[19] The outbreak took on epidemic proportions with greater than 40,000 animals becoming infected.[32]

IV. PROPERTIES

Pneumoviruses are medium size (80 to 130 μm)[33] RNA viruses with helical segments that resemble paramyxovirus in many respects.[34] The RNA is single-stranded with a molecualar weight of 5×10^6 daltons[35,36] and is nonmessenger or negative-stranded.[35,37] The envelope is amorphous with spike-like projections.[19] Complete virions are assembled and bud from the plasma membrane of infected cells. Pneumoviruses are morphologically different from other paramyxoviruses having nucleocapsids of 12 to 14 nm rather than 18 nm.[33,38,39] Bovine respiratory syncytial virus and HRSV have no demonstrable hemagglutinin or neuraminidase activity.[6,16,32,40] By contrast, PVM has both neuraminidase and hemagglutination activities.[38,41,42]

The HRSV genome codes for ten polyadenylated mRNAs.[43] The corresponding virus-specific proteins have been characterized by in vitro translation of individual mRNAs.[44,45] Of these, seven are known to be virion structural proteins and the remaining three are candidates for nonstructural proteins.[44] Letter designations and molecular weights in daltons assigned to the respective proteins are as follows: L, the large nucleocapsid protein (200 K); G, the large envelope glycoprotein (84 K); F, the small envelope glycoprotein (68 K); N, the major nucleocapsid protein (42 K); P, the phosphoprotein (34 K); M, the matrix protein (26 K); VP25, a (25 K) protein; and the (9.5 K), (11 K), and (14 K) proposed nonstructural proteins.[45] The F protein is cleaved to yield the disulfide-linked subunits F_1 (22 K) and F_2 (50 K) similar to the fusion protein of other paramyxoviruses.[46] Bovine respiratory syncytial virus has virion proteins that migrate similar to F_1, N, M, and VP25 proteins of HRSV.[47] A 10,000 dalton component has also been reported.[47] Because of its close similarity to HRSV, it is probable that further analysis of BRSV will discover additional proteins analogous to those of HRSV.

Pneumoviruses similar to other paramyxoviruses are sensitive to low pH, ether, chloroform, and to temperatures of 56°C for 30 min.[32,48]

The host range of BRSV has been investigated in several studies. BRSV induces symptoms of respiratory disease in cattle,[22,49-53] sheep,[25] and subclinical infections in guinea pigs and mice.[54] Tissue culture cells from a wide range of hosts are suitable for BRSV. The tissue culture host range, however, seems to vary with isolate. While the Swiss isolate replicates in bovine origin cell (primary fetal kidney and lung), other cell types, including KB, HeLA, BHK 21, secondary rhesus kidney cells, and primary fetal sheep kidney cells, are insensitive to the agent.[5,49] The Japanese isolate, by contrast, propagates well in a variety of bovine origin cells (kidney, testicle, thyroid, thymus, duodenum, rectum) and in cells from other species including swine (fetal kidney), hamster (lung, kidney), monkey kidney (Vero), and human (fetal lung and kidney, HeLa HEp-2).[54] The Belgian strain reportedly grows in bovine origin cells and human (HeLa) cells.[49] In other reports, BRSV was propagated in primary fetal bovine lung,[5] kidney,[5,22] trachea,[55] testicle,[22] bovine turbinate cells,[56] and goat turbinate cells.[57] Infection of tracheal explants with BRSV did not affect ciliary movement,[55,58] and infected cells appeared to be the peri-tracheal connective tissue rather than ciliated epithelium.[55]

BRSV grows to 2×10^5 PFU/mℓ over a period of between 11 and 21 days at 37°C in fetal tracheal cells.[55] Titers were increased three times by incubation at 33°C.

The BRSV virion is extremely sensitive to mechanical disruption such as freeze-thawing, and even under ideal conditions loses infectivity rapidly.[56] This instability may account partly for the early difficulties encountered in trying to isolate infectious

BRSV from biological samples. $MgSO_4$, 1*M*, is reported to increase the stability of RSV substantially.[59]

BRSV cytopathic effect appears early as syncytial cell formation, often containing eosinophilic inclusions.[5] Treatment of infected cultures with tunicamycin, a drug that blocks glycosylate, inhibits syncytial formation and release of extracellular infected virus. Fusion is probably associated with the F protein as with other paramyxovirus. Late stage infection cells undergo complete degeneration. Fluorescent antibody testing of BRSV-infected cells shows maximum fluorescence at 24 to 48 hr after which FA is not consistent.[57] Fluorescence appears as a delicate granular network[57] and is limited to the cytoplasm.

V. CLINICAL SIGNS

The clinical signs and age range for BRSV infections vary in severity between outbreaks. The Swiss outbreak[5] started suddenly with fever, nasal discharge, and coughing. Bronchopneumonia developed in some cows and older animals appeared to take a more severe course. The disease subsided after 3 to 5 days in calves and 8 to 10 days in cows. Animals over 7 years of age were not affected. No deaths were reported.

The Japanese outbreak[6] appeared as an epizootic throughout the country. Symptoms included anorexia, depression, fever, respiratory distress, cough, nasal discharge, lacrimation, and salivation. The disease fatality rate was 0.26%.

In Nebraska,[23] severe pulmonary edema and emphysema of weaned calves was attributed to BRSV infection. This disease was characterized by early symptoms that included anorexia, increased frequency and severity of coughing, nasal and ocular discharge, accelerated respiration rates, and temperatures of 105 to 108°F. In the acute phase, animals suffered from extreme respiratory distress, rapid shallow breathing and a dry, hacking cough. Subcutaneous edema was reported especially around throat and neck.

Other outbreaks reported coughing, fever, anorexia, nasal discharge, and rapid respiration rates as the major symptoms.[12,20,21,60] Animals experimentally infected by transnasal inoculation generally display mild symptoms of pyrexia, leukopenia, and nasal discharge.[10,15,49,50,51,56,58,61] Symptoms more closely resembling the natural outbreaks have been induced only by using repeated inoculation via intranasal/intratracheal route.[52,53] A pneumonia-like disease was induced, characterized by coughing, tachypnea, and hyperpnea.[52] This wide variation in severity of clinical signs may be partly accounted for by partially immune animals and concurrent bacterial, mycoplasma, and infections with other viruses in addition to virus strain.

VI. PATHOGENESIS

The natural route of transmission of BRSV has not been conclusively determined but is thought to be through the respiratory tract. Experimental inoculation by intranasal/intratracheal routes have been successful in inducing symptoms similar to the natural disease,[52,53] suggesting inhalation as the route of infection. Infectious virus is apparently difficult to isolate after the initial stage of the disease, although viral antigen can be detected by FA in infected tissues in later stages.[58] The disease is primarily localized in the lower respiratory system.

Gross lesions from naturally occurring cases are principally found in the respiratory tract and consist of pneumonia characterized by lobular consolidation of primarily anterior lung lobes and interstitial emphysema.[12,60,62] Small areas of congestion and hemorrhage are evident in pneumonic lobules.[60] Subcutaneous edema of the cervical tissue may be seen in animals with severe respiratory distress.[23]

Histological examination reveals bronchitis, bronchiolitis, and alveolitis[12,60,62] characterized by epithelial cell necrosis, and bronchiolar and alveolar epithelial hyperplasia. Giant cells (syncytia) are seen in bronchiolar and alveolar epithelial cells in some, but not all cases.[58,60] Macrophage, lymphocyte, plasma cell, and neutrophil accumulation may be present in alveolar septae and peribronchiolar tissues.[12,58,60] Bronchioles are often occluded with exudate composed of neutrophils, macrophage, and necrotic cell debris (bronchiolitis oblitrans).[12] In calves with both parainfluenza type 3 and BRSV infection, severe intraalveolar edema with hyaline membrane formation and widespread epithelial hyperplasia have been noted.[63] Virus antigen can be seen by IF staining in bronchiolar and alveolar epithelium.[53]

Experimental reproduction of BRSV in cattle and sheep has shown the virus to be primarily a lower respiratory tract pathogen. This is further substantiated by the fact that while BRSV will infect tracheal epithelium in vitro, impairment of ciliary function and cytopathic change in tracheal epithelial cell does not occur. The experimentally induced disease generally has yielded only mild clinical symptoms, often showing no gross lesions.[14,49,64] Microscopically, however, bronchiolitis, alveolitis, and giant cell formation are seen in the lung.[99]

Recently, moderate to severe pulmonary disease was obtained experimentally by intranasal and intratracheal aerosolization and the pulmonary changes studied sequentially. In this study, calves given multiple doses of BRSV by intranasal/intratracheal routes showed clinical signs similar to the naturally occurring disease.[52,53] Gross changes in the lung consisted of large areas of consolidation in the cranial, middle, accessory, and cranial lung lobes in absence of coagulative necrosis. These changes were prominent at 5 to 8 days after infection, and thereafter reduce in severity.

Histologically, early lesions consisted of principally bronchiolitis with moderate cellular infiltration of alveolar spaces and septae, and lymphocytic cuffing of small septal blood vessels. Between 4 to 6 days after infection, bronchial and bronchial-epithelial changes were severe with necrosis and loss of epithelial cells. Epithelial syncytial giant cells were widespread in bronchioles and alveolar wells and contained eosinophilic intracytoplasmic viral inclusions. Alveolar and septal cellular infiltrates increased and were characterized by a dimorphic population of neutrophils and mononuclear cells. Exudate in small bronchi and bronchioles was extensive and occasionally completely occluded the lumina (bronchiolitis oblitrans). Severely damaged epithelial surfaces had evidence of early reepithelization. Marked squamous metaplasia and hyperplasia of the bronchiolar epithelium contributed to the bronchiolar occlusion. Viral antigen could be demonstrated by IF staining in nasopharyngeal cell and epithelial cells of bronchi, bronchioles, and alveoli. After 10 days, the major changes were those of repair. Reepithelization was prominent, cellular exudate in alveoli, bronchi, and bronchioles became organized, and there was an increase in lymphocytes and macrophages. Bronchial and peribronchiolar fibrosis and lymphoid hyperplasia accompanied these changes.

The clinical and pathologic findings in experimentally induced BRSV pneumonia are similar in many ways to those found in pneumonia experimentally induced by PI-3.[52] Differential diagnosis between these two agents based on epithelial syncytia formation as a prominent feature of BRSV is tentative. Syncytia appear as a prominent feature only at certain stages of BRSV,[52] and PI-3 also can induce occasional syncytia.[65]

BRSV outbreaks are commonly associated with other pathogens and/or opportunistic agents such as Mycoplasma spp.,[60,66] Pasturella spp.,[66] PI-3,[66] BVD,[67] IBR,[67] and BAV.[67] However, when antibodies to various respiratory viruses are measured in samples from cattle suffering from respiratory disease, BRSV was the most prevalent agent,[11,66-68] showing between 23 and 55% occurrence in animals. This suggests that BRSV is a major contributor to the bovine respiratory disease complex.

VII. DIAGNOSIS

Because the clinical signs and lesions are similar to other respiratory diseases, diagnosis of BRSV infection must depend on: (1) isolation and identification of virus, (2) detection of virus antigen in biologic samples from suspect animals, or (3) serologic conversion of paired serum samples from acute and convalescent stages of disease. Due to the sensitivity of virus to freeze/thawing and to prolonged storage, it is important that specimens, usually from conjunctiva or nasal material, be cultured for infectivity as rapidly as possible. Best results have been reported when samples are processed within 1 hr.[56] Even under ideal conditions, recoverable infectious virus appears to be present only during the early stages of disease,[58,69] although histology and IF indicate the presence of virus in tissue for prolonged periods.[58] Serologic testing of paired serum samples may be performed by SN,[5,6,22,32] CF,[5,7,24,70] FA,[8,11,21,50,57,71] and ELISA[27,72] methods.

VIII. IMMUNITY

In BRSV-challenged gnotobiotic or colostrum-deprived calves, serum neutralizing antibody is detectable by day 6 to 13.[50,53] CF and SN antibody titers are reported to correlate.[5,70] Antibodies detected by FA appeared sooner and remained longer than SN antibodies.[71] In the same study, high interferon titers were detected by 3 days after infection, but fell to undetectable levels by day 6. Interferon levels remained low for about a week, then reappeared, but only to moderate levels.[71]

Although there is some suggestion that preexisting antibodies to BRSV predispose the animal to disease as is apparently the case with HRSV in children with maternal antibody,[15] experimental studies have not been confirmatory.[49,56,71] The presence of maternally derived circulating antibodies does not appear to protect calves from disease.[53] Nasal secretory antibody, however, is protective suggesting a role of local immunity.[56] Established immunity due to previous exposure protects animals from subsequent challenge.[56]

Attempts to produce protective immunity by vaccination have been only partially successful.[23,73,74] A vaccine regimen using formalin-killed BRSV emulsified in incomplete Freund's adjuvant in the primary inoculation, and in aluminum hydroxide adjuvant in the booster produced moderate levels of SN antibody.[73] After challenge, four of five animals showed no signs of infection.[73]

Animals given two doses of an experimentally attenuated BRSV vaccine showed high SN antibody titers.[23] Field trials reduced the incidence of undifferentiated respiratory disease from 34% in unvaccinated animals to 15% in animals given one or two doses of the experimental vaccine. In both vaccine experiments, antibodies induced by immunization did not appear to cause exacerbation of experimentally induced disease[73] or predispose animals to natural infection.[23]

REFERENCES

1. Morris, J. A., Blount, R. E., and Savage, R. E., Recovery of cytopathogenic agent from chimpanzees with coryza, *Proc. Soc. Exp. Biol. Med.*, 92, 544, 1956.
2. Chanock, R., Roizman, B., and Myers, R., Recovery from infants with respiratory illness of a virus related to chimpanzee coryza agent (CCA). I. Isolation, properties and characterization, *Am. J. Hyg.*, 66, 281, 1957.
3. Taylor-Robinson, D. and Doggett, J. E., An assay method for respiratory syncytial virus, *Br. J. Exp. Pathol.*, 64, 473, 1963.

4. Doggett, J. E., Taylor-Robinson, D., and Gallop, R. G. C., A study of an inhibitor in bovine serum active against respiratory syncytial virus, *Arch. Gesamte Virusforsch.*, 23, 126, 1968.
5. Paccaud, M. F. and Jacquier, C., A respiratory syncytial virus of bovine origin, *Arch. Gesamte Virusforsch.*, 30, 327, 1970.
6. Inaba, Y., Tanaka, Y., Sato, K., Ito, H., and Omori, T. M. M., Nomi virus, a virus isolated from an apparently new epizootic respiratory disease of cattle, *Jpn. J. Microbiol.*, 14, 246, 1970.
7. Wellemans, G., Leunen, J., and Luchsinger, E., Isolement d'un virus (220/69) serologiquement semblable au virus respiratoire syncytial (RS) humain, *Ann. Med. Vet.*, 114, 89, 1970.
8. Mahin, L. and Welleman, G., Serological evidence for the intervention of bovine respiratory syncytial virus in a respiratory disease outbreak in Moroccan cattle, *Zentralbl. Vet. Med.*, 29, 76, 1982.
9. Pirie, H. M., Acute fatal pneumonia in calves due to respiratory syncytial virus, *Vet. Rec.*, 108, 411, 1981.
10. Moteane, M., Babiuk, L. A., and Schiefer, B., Studies on the occurrence and significance of bovine respiratory syncytial virus in Saskatchewan, *Can. J. Comp. Med.*, 42, 246, 1977.
11. Elazhary, M. A. S. Y., Roy, R. S., Champlin, R., Higgins, R., and Marsolais, G., Bovine respiratory syncytial virus in Quebec: antibody prevalence and disease outbreak, *Can. J. Comp. Med.*, 44, 299, 1980.
12. Elazhary, M. A. S. Y., Silim, A., and Morin, M., A natural outbreak of bovine respiratory disease caused by bovine respiratory syncytial virus, *Cornell Vet.*, 72, 325, 1982.
13. Rossi, C. R. and Diessel, G. K., Serological evidence for the association of bovine respiratory syncytial virus with respiratory tract disease in Alabama cattle, *Infect. Immunol.*, 10, 293, 1974.
14. Smith, M. H., Frey, J. L., and Dierks, R. E., Isolation and characterization of a bovine respiratory syncytial virus, *Vet. Rec.*, 94, 599, 1974.
15. Smith, M. H., Frey, M. L., and Dierks, R. E., Isolation, characterization, and pathogenicity studies of a bovine respiratory syncytial virus, *Arch. Virol.*, 47, 237, 1975.
16. Rosenquist, B. D., Isolation of respiratory syncytial virus from calves with acute respiratory disease, *J. Infect. Dis.*, 130, 177, 1974.
17. Kingsbury, E., Paramyxoviridae, *Intervirol.*, 10, 137, 1978.
18. Woods, G. T., Isolation of bovine syncytial virus in Britain, *Vet. Rec.*, 91, 363, 1972.
19. Woods, G. T., Bovine parvovirus. I. Bovine syncytial virus, and bovine respiratory syncytial virus and their infections, *Adv. Vet. Sci. Comp. Med.*, 18, 273, 1974.
20. Jacobs, J. W. and Edington, N., Isolation of respiratory syncytial virus from cattle in Britain, *Vet. Rec.*, 88, 694, 1971.
21. Pospisil, Z., Mensik, J., and Valicek, L., Isolation and identification of a bovine respiratory syncytial virus in Czechoslovakia, *Acta Vet. Brno.*, 47, 79, 1978.
22. Inaba, Y., Tanaka, Y., Sato, K., Omori, T., and Matumoto, M., Bovine respiratory syncytial virus; studies on an outbreak in Japan, 1968—1969, *Jpn. J. Microbiol.*, 16, 373, 1972.
23. Bohlender, R. E., McCune, M. W., and Frey, M. L., Bovine respiratory syncytial virus infection, *Mod. Vet. Pract.*, 63, 613, 1982.
24. Berthiaume, L., Joncas, J., Boulay, G., and Pavilanis, V., Serological evidence of respiratory syncytial virus infection in sheep, *Vet. Rec.*, 93, 337, 1973.
25. Lehmkuhl, H. D. and Cutlip, R. C., Experimentally-induced respiratory syncytial viral infection in lambs, *Am. J. Vet. Res.*, 40, 512, 1979.
26. Cutlip, R. C. and Lehmkuhl, H. D., Lesions in lambs experimentally infected with bovine respiratory syncytial virus, *Am. J. Vet. Res.*, 40, 1479, 1979.
27. Richardson-Wyatt, L. S., Belshe, R. B., London, W. T., Sly, D. L., Camargo, E., and Chanock, R. M., Respiratory syncytial virus antibodies in nonhuman primates and domestic animals, *Lab. Anim. Sci.*, 31, 413, 1981.
28. Lundgren, D. L., Magnuson, M. G., and Clapper, W. E., A serological survey in dogs for antibody to human respiratory virus, *Lab. Anim. Care,* 19, 352, 1969.
29. Smith, M. H., Lehmkuhl, H. D., and Phillips, S. M., Isolation and characterization of a respiratory syncytial virus from goats, *Am. Assoc. Vet. Lab. Diag.*, 22, 259, 1979.
30. Brako, E. E., Fulton, R. W., Nicholson, S. S., and Amborski, G. F., Prevalence of bovine herpesvirus-1, bovine viral diarrhea, parainfluenza-3, goat respiratory syncytial, bovine leukemia and bluetongue viral antibodies in sheep, *Am. J. Vet. Res.*, 45, 813, 1984.
31. Fulton, R. W., Downing, M. M., and Hagstad, H. V., Prevalence of bovine herpesvirus-1, bovine viral diarrhea, parainfluenza-3, bovine adenoviruses-3 and -7, and goat respiratory syncytial viral antibodies in goats, *Am. J. Vet. Res.*, 43, 1454, 1982.
32. Inaba, Y., Tanaka, Y., Omori, T., and Matumoto, M., Isolation of bovine respiratory syncytial virus, *Jpn. J. Exp. Med.*, 40, 473, 1970.
33. Ito, Y., Tanaka, Y., Inaba, Y., and Omori, T., Structure of bovine respiratory syncytial virus, *Arch. Gesamte Virusforsch.*, 40, 198, 1973.

34. McIntosh, K. and Fishaut, J. M., Immunopathologic mechanisms in lower respiratory tract disease of infants due to respiratory syncytial virus, *Prog. Med. Virol.*, 26, 94, 1980.
35. Lambert, D. M., Pons, M. W., Mbuy, G. N., and Dorsch-Hasler, K., Nucleic acids of respiratory syncytial virus, *J. Virol.*, 36, 837 1980.
36. Zhdanov, V. M., Dreizin, R. S., Yankevici, O. D., et al., Biophysical characteristics of respiratory syncytial virus, *Rev. Roum. Virol.*, 25, 277, 1974.
37. Huang, Y. T. and Wertz, G. W., The genome of respiratory syncytial virus is a negative-stranded RNA that codes for at least seven mRNA species, *J. Virol.*, 43, 150, 1982.
38. Berthiaume, L., Joncas, J., and Pavilanis, V., Comparative structure, morphogenesis and biological characteristics of the respiratory syncytial (RS) virus and the pneumonia virus of mice (PVM), *Arch. Gesamte Virusforsch.*, 45, 39, 1974.
39. Bussell, R. H., Waters, D. J., and Seals, M. K., Measles, canine distemper and respiratory syncytial virians and nucleocapsids. A comparative study of their structure, polypeptide and nucleic acid composition, *Med. Microbiol. Immunol.*, 160, 105, 1974.
40. Richman, A. V., Pedreira, F. A., and Tauraso, N. M., Attempts to demonstrate hemagglutination and hemadsorption by respiratory syncytial virus, *Appl. Microbiol.*, 21, 1099, 1971.
41. Mills, K. C. and Dochez, A. R., Specific agglutination of murine erythrocytes by a pneumonitis virus in mice, *Proc. Soc. Exp. Biol. Med.*, 57, 140, 1944.
42. Harter, D. H. and Chopping, P. W., Studies on pneumonia virus on mice (PVM) in cell culture. I. Replication in baby hamster kidney cells and properties of the virus, *J. Exp. Med.*, 126, 251, 1967.
43. Collins, P. L., Huang, Y. T., and Wertz, G. W., Identification of a tenth mRNA of respiratory syncytial virus and assignment of polypeptides to the 10 viral genes, *J. Virol.*, 49, 572, 1984.
44. Huang, Y. T. and Wertz, G. W., Respiratory syncytial virus mRNA coding assignments, *J. Virol.*, 46, 667, 1983.
45. Dickens, L. E., Collins, P. L., and Wertz, G. W., Transcriptional mapping of human respiratory syncytial virus, *J. Virol.*, 52, 364, 1984.
46. Lambert, D. M. and Pons, M. W., Respiratory syncytial virus glycoproteins, *Virology*, 130, 204, 1983.
47. Cash, P., Wunner, W. H., and Pringle, C. R., A comparison of the polypeptides of human and murine pneumonia virus, *Virology*, 82, 369, 1977.
48. Inaba, Y., Tanaka, Y., Omori, T., and Matumoto, M., Physicochemical properties of bovine respiratory syncytial virus, *Jpn. J. Microbiol.*, 17, 211, 1973.
49. Jacobs, J. W. and Edington, N., Experimental infection of calves with respiratory syncytial virus, *Res. Vet. Sci.*, 18, 299, 1975.
50. Elazhary, M. A. S. Y., Galina, M., Roy, R. S., Fontaine, M., and Lamothe, P., Experimental infection of calves with bovine respiratory syncytial virus (Quebec strain), *Can. J. Comp. Med.*, 44, 390, 1980.
51. Mohanty, S. B., Ingling, A. L., and Lillie, M. G., Experimentally-induced respiratory syncytial viral infection in calves, *Am. J. Vet. Res.*, 36, 417, 1975.
52. Bryson, D. G., McNulty, M. S., Logan, E. F., and Cush, P. F., Respiratory syncytial virus pneumonia in young calves: clinical and pathologic findings, *Am. J. Vet. Res.*, 44, 1648, 1983.
53. McNulty, M. S., Bryson, D. G., and Allan, G. M., Experimental respiratory syncytial virus pneumonia in young calves: microbiologic and immunofluorescent findings, *Am. J. Vet. Res.*, 44, 1656, 1983.
54. Matumoto, M., Inaba, Y., Kurogi, H., Sato, K., Omori, T., Goto, Y., and Hirose, O., Bovine respiratory syncytial virus: host range in laboratory animals and cell cultures, *Arch. Gesamte Virusforsch.*, 44, 280, 1974.
55. Thomas, L. H., Stott, E. J., Jebbett, J., and Hamilton, S., The growth of respiratory syncytial virus in organ cultures of bovine foetal trachea, *Arch. Virol.*, 52, 251, 1976.
56. Mohanty, S. B., Lillie, M. G., and Ingling, A. L., Effect of serum and nasal neutralizing antibodies on bovine respiratory syncytial virus infection in calves, *J. Infect. Dis.*, 134, 409, 1976.
57. Potgieter, N. D. and Aldridge, P. L., Use of the indirect fluorescent antibody test in the detection of bovine respiratory syncytial virus antibodies in bovine serum, *Am. J. Vet. Res.*, 38, 1341, 1977.
58. Thomas, L. H. and Stott, E. J., Diagnosis of respiratory syncytial virus infection in the bovine respiratory tract by immunofluorescence, *Vet. Rec.*, 108, 432, 1981.
59. Fernie, B. F. and Gerin, J. L., The stabilization and purification of respiratory syncytial virus using MgSO4, *Virology*, 106, 141, 1980.
60. Pirie, H. M., Petrie, L., Pringle, C. R., Allan, E. M., and Kennedy, G. J., Acute fatal pneumonia in calves due to respiratory syncytial virus, *Vet. Rec.*, 108, 411, 1981.
61. Thomas, L. H., Stott, E. J., Collins, A. P., Jebbett, N. J., and Stark, A. J., Evaluation of respiratory disease in calves: comparison of disease response to different viruses, *Res. Vet. Sci.*, 23, 157, 1977.
62. Bryson, D. G., McFerran, J. B., and Ball, H. J. N. S. D., Observations on outbreaks of respiratory disease in housed calves. II. Pathological and microbiological findings, *Vet. Rec.*, 103, 503, 1978.

63. Bryson, D. G., McFerran, J. B., Ball, H. J., and Neill, S. D., Observations on outbreaks of respiratory disease in calves with parainfluenza type 3 virus and respiratory syncytial virus infection, *Vet. Rec.*, 104, 45, 1979.
64. Edington, N. and Jacobs, J. W., Respiratory syncytial virus in cattle, *Vet. Rec.*, 87, 762, 1970.
65. Bryson, D. G., McNulty, M. S., Ball, H. J., and Neill, S. D., The experimental production of pneumonia in calves by intranasal inoculation of parainfluenza Type III virus, *Vet. Rec.*, 105, 566, 1979.
66. Bryson, D. G., McFerran, J. B., Ball, H. J., and Neill, S. D., Observations on outbreaks of respiratory disease in housed calves. I. Epidemiological, clinical and microbiological findings, *Vet. Rec.*, 103, 485, 1978.
67. Lehmkuhl, M. S. and Gough, P. M., Investigations of causative agents of bovine respiratory tract disease in beef cow-calf herd with an early weaning program, *Am. J. Vet. Res.*, 38, 1717, 1977.
68. Wellemans, G. and Leunen, J., L'importance du virus V220/69 dans les troubles respiratoires des veaux en belgique, *Ann. Med. Vet.*, 115, 35, 1971.
69. Omar, A. R., Jennings, A. R., and Betts, A. O., The experimental disease produced in calves by the J121 strain of parainfluenza virus type 3, *Res. Vet. Sci.*, 7, 379, 1966.
70. Taskahasi, E., Inaba, Y., and Kurogi, H., Diagnosis of bovine respiratory syncytial virus infection by complement fixation test, *Natl. Inst. Anim. Health*, 15, 179, 1975.
71. Elazhary, M. A. S. Y., Silim, A., and Roy, R. S., Interferon, fluorescent antibody, and neutralizing antibody responses in sera of calves inoculated with bovine respiratory syncytial virus, *Am. J. Vet. Res.*, 42, 1378, 1981.
72. Gillette, K. G., Enzyme-linked immunosorbent assay for serum antibody to bovine respiratory syncytial virus: comparison with complement-fixation and neutralization tests, *Am. J. Vet. Res.*, 44, 2251, 1983.
73. Mohanty, S. B., Rockemann, D. D., and Davidson, J. P., Effect of vaccinal serum antibodies on bovine respiratory syncytial viral infection in calves, *Am. J. Vet. Res.*, 42, 881, 1981.
74. Zygraich, N. and Welleman, G., Immunological markers of an attenuated bovine respiratory syncytial (BRS) virus vaccine, *Zentralbl. Vet. Med.*, 28, 355, 1981.

Chapter 7*

FELINE INFECTIOUS PERITONITIS

Niels C. Pedersen

TABLE OF CONTENTS

* Color Figures 1 to 9 appear following p. 120.

I. INTRODUCTION

Feline infectious peritonitis (FIP) is a disease of domestic and wild felidae.[45] It was first described as a specific disease entity by Wolfe and Griesemer.[70] Earlier clinical reports date back only to the early 1960s, and it is doubtful that the disease existed before the 1950s.

The name "feline infectious peritonitis" refers to the principal form of the disease, an inflammatory condition of the visceral serosa and omentum.[70] Subsequently, Montali and Strandberg[38] described a second form of the disease. This form of FIP was characterized by granulomatous involvement of parenchymatous organs such as the kidneys, mesenteric lymph nodes, liver, pancreas, central nervous system (CNS) and spine, and the uveal tract of the eyes.[41] The granulomatous form of FIP was called "dry" or "noneffusive" because there was not an inflammatory exudation into the body cavities. Classical FIP, which comprised about three fourths of the cases, was termed "wet" or "effusive."

II. ETIOLOGIC AGENT

Wolfe and Griesemer,[70] on the basis of filtration experiments, were the first to suggest that FIP was caused by a virus. Zook and co-workers[74] observed viral particles in the tissues of experimentally infected cats, but were unable to characterize the agent. Ward[65] recognized the close similarity of the FIP virus (FIPV) in tissues with members of the newly classified family of Coronaviridae. The FIPV was first cultivated in vitro in peritoneal macrophages,[42] and later in suckling mouse, rat, and hamster brains.[30,40] Recently, several different isolates have been propagated in vitro in conventional tissue cultures.[2,12,39,50] The FIPV is closely related to transmissible gastroenteritis virus (TGEV) of swine and canine coronavirus (CCV).[26] In fact, the relationship between these viruses is so close that it has been proposed that FIPV, TGEV, and CCV are really strains of a single species.[26] FIPV, TGEV, and CCV have recently been placed together in group one of the mammalian coronaviruses.[60]

In 1976 a serological test was developed for the detection of serum antibodies to FIPV.[43] When a large number of healthy and ill cats were tested, it was noted that both groups of animals had antibodies to FIPV, although antibody titers were more consistently present and were usually higher in cats with FIP. It was postulated, therefore, that FIP was an uncommon manifestation of a common infection. The common infection appeared to be largely asymptomatic or so mild as to be undefinable. Later on it was discovered that many seropositive reactions in healthy cats were due to a second group of coronavirus that caused mild or inapparent enteritis and not FIP.[50,51] These feline enteric coronaviruses (FECVs) were antigenically and morphologically indistinguishable from FIPV.[3,50,51]

FIPV and FECV isolates of cats have been recently classified according to tissue culture growth characteristics and serological cross-reactions with CCV.[48] Type I feline coronaviruses are fastidious growers in tissue culture and are less closely related to CCV by virus neutralization. This group includes FIPV-UCD1, FIPV-UCD3, FIPV-UCD4, FIPV-TN406,[2] and FECV-UCD. Type II feline coronaviruses include FIPV-79-1146, FIPV-UCD2, and FECV-79-1683. Viruses in this group grow very readily and to high titer in cell culture, and resemble CCV more closely than viruses of Type I.[48]

A. Host Range

FIP is mainly a disease of domestic cats. It has been recognized, however, in a number of wild felidae. Among larger cats, FIP has been diagnosed in lions,[6,8] mountain lions,[14] leopards,[8,64] cheetah,[56] and the jaguar.[8] Smaller wild cats that have been infected with FIPV include the lynx and caracal,[57] sand cat,[62] and pallas cat.[45]

III. EPIZOOTIOLOGY

FIP has been diagnosed in cats of all ages, but the peak incidence is between 6 months and 5 years of age.[41] There is no significant sex predisposition, but the disease appears to occur more frequently in purebred than in domestic cats. This may not be due to genetic predisposition, but rather to enhancing factors associated with the cattery environment. The disease is world-wide in distribution.[27]

The incidence of FIP is highly variable. In most cases the disease is sporadic and attacks isolated animals. A cat population may not suffer any deaths from FIP for years and then several cases might be seen in rapid succession. The disease can then seemingly disappear only to reappear months or years later. Much higher infection rates have been seen, however, in isolated catteries. A 3 to 49% yearly infection rate was observed in kittens raised in one cattery over a 4-year period.[59] Similar explosive outbreaks have been seen in several other catteries.[45] In catteries, FIP will often occur in several kittens from the same litter, with deaths occurring over a period of weeks to many months.

A. Clinical Signs of Effusive FIP

The clinical course of effusive FIP is from 1 to 6 weeks or longer. The onset of disease is heralded by the appearance of a chronic, fluctuating fever. Associated with the fever there is usually a progressive decline in weight, activity, and appetite. Terminally, the cats go into shock and die. Peritonitis is seen in over 90% of the cats with effusive FIP (Figure 1) and pleuritis in around 40%.[45] Peritonitis is associated with distension of the abdomen from ascites, while pleuritis usually causes pleural effusion and dyspnea. The peritoneal and pleural fluids are characteristic for the disease. Pleural and abdominal fluid have protein levels near blood levels and contain numerous macrophages, neutrophils, and lymphocytes.[41] The fluid ranges from almost colorless to deep straw yellow and is frequently viscous in nature. Involvement of other organ systems such as the eyes and CNS is clinically apparent in only 10% of the cats with effusive disease,[45] although a somewhat higher proportion may have clinically silent lesions in these and other nonserosal sites.

B. Clinical Signs of Noneffusive FIP

Cats with noneffusive FIP are ill for 1 to 12 weeks or more. Similar to the effusive form, a chronic fluctuating fever accompanies the disease. There is also a progressive decline in general body condition and appetite. Added upon these features, however, are signs referable to specific organ systems. Peritoneal cavity lesions are found in 50% of the cats with noneffusive FIP and pleural cavity lesions in 10%.[45] Unlike the effusive form, however, there is a high incidence of ocular or CNS signs in cats with noneffusive FIP. One third of cats with noneffusive FIP demonstrate signs referable to the CNS and a similar number have clinically apparent ocular disease.

Peritoneal cavity signs in noneffusive FIP usually consist of irregular solitary or multiple masses within the kidneys, or hepatic or mesenteric lymph nodes (Figures 2 and 3). Granulomatous lesions in the liver, spleen, pancreas, omentum, and serosal surfaces are less frequent (Figure 3). Thoracic cavity lesions of noneffusive FIP are usually clinically silent, and when present, they are usually found on the pleural surfaces or on the heart (Figure 4). CNS involvement is varied in its clinical expression. Spinal signs such as posterior paresis, incoordination, hyperesthesia, brachial, trigeminal, facial, and sciatic nerve palsies have all been described.[24,33,35,41,61] Hydrocephalus, secondary to disease of the chroid and ependyma, has also been reported.[13,22,34] Cranial involvement can lead to dementia, personality changes (rage, withdrawal), or convulsive disorders. Cerebeller-vestibular signs such as nystagmus, head tilt or circling, have also been associated with FIP. Ocular lesions can be associated with lesions in the CNS or peritoneal cavity, or can occur by themselves.[47] Uveitis and chrorioretinitis are the predominant clinical manifestations of ocular disease (Figure 5).[4,5,10,15,61]

C. Clinicopathologic Findings

Complete blood counts show similar changes regardless of the form of the disease. Leukocytosis with neutrophilia and lymphopenia is a common abnormality. With chronicity, a low grade to moderate anemia of the depression type is also seen. Inclusion bodies within circulating neutrophils have been described in some cats with FIP.[66] It has been suggested that these inclusions represent immune complexes.[27] Plasma proteins are elevated in one half of the cats with effusive FIP and three fourths of the cats with noneffusive FIP. Plasma protein elevations are caused by variable increases in the alpha-2, beta, and gamma globulins.[41] Serum haptoglobin elevations contribute to part of the increase in alpha-2 globulins.[21]

A disseminated intravascular coagulopathy occurs in cats with effusive FIP. It is usually clinically inapparent, however, but may contribute to the characteristic features of the pleural and abdominal effusions. The coagulopathy is associated with increases in the prothrombin and partial thromboplastin times, increased levels of fibrin degradation products, prolonged activated clotting times, and decreased levels of intrinsic clotting factors.

Ascitic and pleural fluid from cats with effusive FIP is light to dark yellow and has a sticky viscous consistency, somewhat like synovial fluid or egg white. The exudate is very high in protein and contains from 1,600 to 25,000 or more leukocytes per microliter. Neutrophils, lymphocytes, macrophages, and mesothelial cells are the predominant cell type found in the fluid. Cerebral spinal fluid and aqueous humor in cats with CNS or ocular disease also show similar increases in proteins and leukocytes.

Following the introduction of tests for the detection of feline leukemia virus infection, it was noted that from 40 to 50% of cats with FIP were concomitantly infected with FeLV.[9] With the elimination of FeLV from many catteries, there has been a decrease both in the number of cases of FIP and in the proportion of cats with FIP that also have concurrent FeLV infection.

IV. PATHOLOGY

The pyogranuloma is the typical lesion of effusive FIP.[25,69,71] A pyogranuloma consists of necrotic debris and neutrophils, surrounded by a dense accumulation of phagocytic cells interspaced with a few lymphocytes and plasma cells. In addition, considerable amounts of fibrin and protein-rich fluid are deposited within and around the lesions.[69] Pyogranulomas appear as distinct or coalescing serosal plaques 0.5 to 2 mm or more in diameter (Figure 1). The visceral serosa of the chest and abdomen is more likely to be involved; the omentum is often thickened, edematous, and retracted into a compact mass (Figure 1). Although the pyogranulomatous process is usually surface oriented, a similar inflammatory reaction may extend into underlying tissues along penetrating veins (Figure 6). Focal lesions, often associated with phlebitis, may be seen deep in underlying muscle or organ parenchyma. These are also associated with a mixed inflammatory cell infiltrate.

The lesions of noneffusive FIP are more typically granulomatous in nature, but nevertheless bear a basic resemblance to the pyogranulomatous lesions of effusive disease. Granulomatous lesions are variable in size depending on the organ involved.[25,38,61] Lesions in the eyes and central nervous system more closely resemble the microscopic or small pyogranulomatous reactions seen in effusive FIP. Serosal, mesenteric, and omental lesions also consist mainly of small whitish plaques or nodules. Kidney, liver, and mesenteric lymph node lesions, however, are often very large, sometimes exceeding 5 cm in diameter. Similar to the pyogranulomas of effusive FIP, lesions of noneffusive FIP are usually surface oriented and extend into the underlying parenchyma along blood vessels (Figure 7). Histologically, the lesions of noneffusive FIP are more typically granulomatous, the outer zone is characteristically more fibrous, and the number of plasma cells and lymphocytes are much greater; edema, hyperemia, and fibrin and protein exudation are also not as pronounced as in the pyogranulomatous lesions of effusive FIP.

Lymphoid lesions are common in both effusive and noneffusive forms of FIP. Splenic enlargement may be due to histiocytic and plasmacytic infiltration of the red pulp, hyperplasia of the lymphoid elements in the white pulp, necrotizing splenitis with fibrin deposition and polymorphonuclear cell infiltrates, or by more organized pyogranulomatous reactions. Gross lymph node involvement is usually limited to thoracic and abdominal nodes. Lymph node enlargement is due to a range of lesions resembling those described for the spleen.

Fluorescent antibody staining of tissue sections from cats with both forms of the disease demonstrate the presence of FIPV in the lesions. In effusive FIP, a large amount of viral antigen is contained in phagocytic cells that make up the periphery of the pyogranulomas[49,67-69] (Figure 8), in Kupfer cells within and adjacent to hepatic sinusoids, and in macrophages within parenchymal lesions in the liver.[69] There is less virus antigen in the lesions of noneffusive FIP, and it is usually found within a few macrophages adjacent to veins in the center of the lesions (Figure 9).

A. Pathogenesis of FIP

The precise routes by which FIPV enters the body are not known. Direct, as well as circumstantial, evidence suggests that at least two routes are important, in utero and oral ingestion. The evidence for in utero infection is largely circumstantial. The 79-1146 feline coronavirus isolate, which induces FIPV when inoculated into cats[44,46] was isolated from a 4-day-old fading kitten.[37] The pleural and lung lesions in this kitten were of sufficient chronicity to suggest that the infection began prior to birth. The queen had kindled two normal litters followed by two abortions prior to this litter. FIP-like lesions were also observed in a litter of newborn kittens born to a queen that developed effusive FIP during pregnancy.[46] It is understandable in this latter situation how virus entered the fetuses, because cats with active effusive FIP have virus in their circulating leukocytes.[67] FIPV transmission from healthy queens to their fetuses is not as clearly understood. Transmission of FIPV from healthy queens might mimic the situation that occurs in latent FeLV[52] and feline syncytium forming virus (FeSFV)[16] infections. The FeLV and FeSFV genomes are present in an unexpressed form in leukocytes, and infection of the fetuses probably occurs when maternal leukocytes enter the fetus through placental breaks. The expression of virus is suppressed within the millieu of the queen's body, but not in the fetus. Only a proportion of the fetuses are infected with FeLV and FeSFV, and normal and affected kittens can exist side by side with noninfected ones.

In utero transmission of FIPV may also explain two additional phenomena. FIP frequently affects several kittens in the same litter, and certain queens will give birth to multiple litters that develop FIP.

The major route of infection is probably by ingestion of virus after birth. The obvious source of virus is from carrier cats. To explain the sporadic and random incidence of disease, carrier cats would have to shed virus either intermittently or at very low levels. Pedersen and co-workers[50] studied the incidence of disease that occurred following oral or oronasal administration of either high or low levels of virus. Following chronic ingestion of 20 to 100 $TCID_{100}$ of FIPV daily for many weeks, 50% of the kittens did not even become infected. This indicates that FIPV is of relatively low infectivity. Of the remaining kittens, 30% developed FIP and became seropositive,

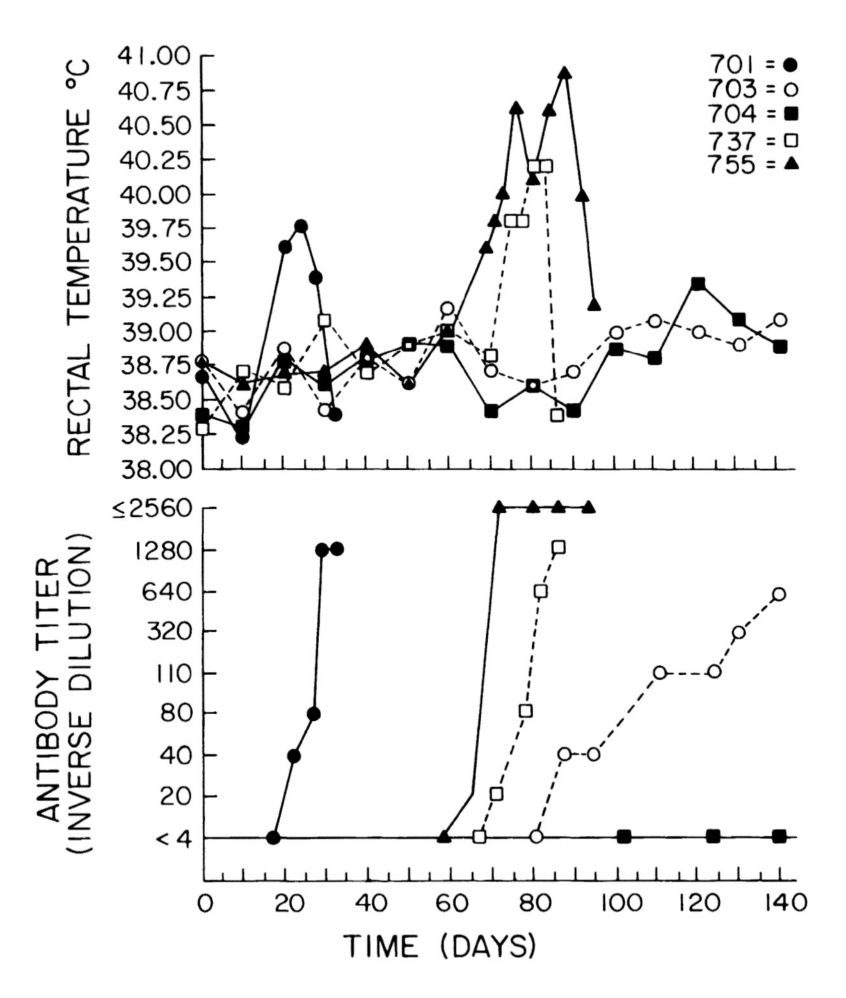

FIGURE 10. Febrile and humoral immune responses of kittens to continuous oral inoculation (daily feeding) with FIPV-UCD1. These 5 kittens, selected from a group of 15, were considered representative of the three different clinical outcomes, i.e., (1) no seroconversion and no illness (#704), (2) seroconversion without illness (#703), and (3) seroconversion with illness (#701, 737, 755). (From Pedersen, N. C., Boyle, J. F., and Floyd, K., et al., *Am. J. Vet. Res.*, 42, 363, 1981. With permission.)

while 20% developed antibodies without any other sign of illness (Figure 10). When the oronasally administered dose of virus was raised to 4000 $TCID_{100}$ of FIPV, still only about 30% of the kittens were infected after a single dose.[50] When the dose was increased tenfold, 50 to 75% became infected, and after a 100-fold increase virtually all of the kittens developed disease. With increasing doses of virus, the infection rate not only increased, but a greater proportion of cats developed fatal FIP and fewer seroconverted without illness. A similar observation was made by Pedersen and Black.[47] Cats inoculated with sublethal amounts of FIP either did not become infected, seroconverted without developing disease, or developed FIP. These experiments suggest several things. First, the sporadic and low incidence of disease in nature more closely resembles the experimental disease produced by low dose challenge, which suggests that cats were exposed to relatively small amounts of virus. Second, for every cat in that field that dies of FIP, one or less will seroconvert without developing any signs of illness. The chronic carriers are most likely to be seropositive healthy cats, because cats that become ill do not survive for long periods of time. In addition, naturally

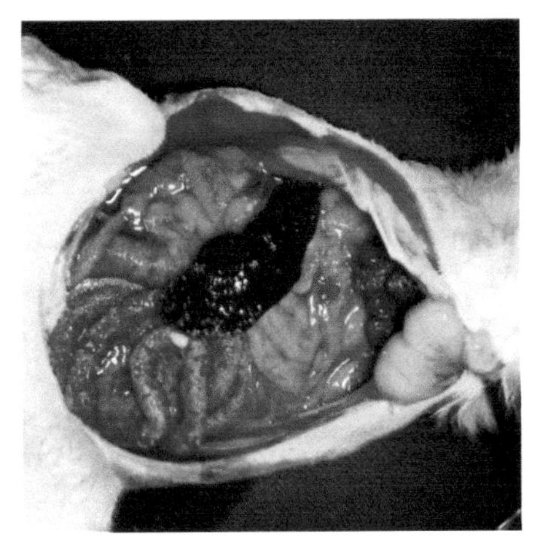

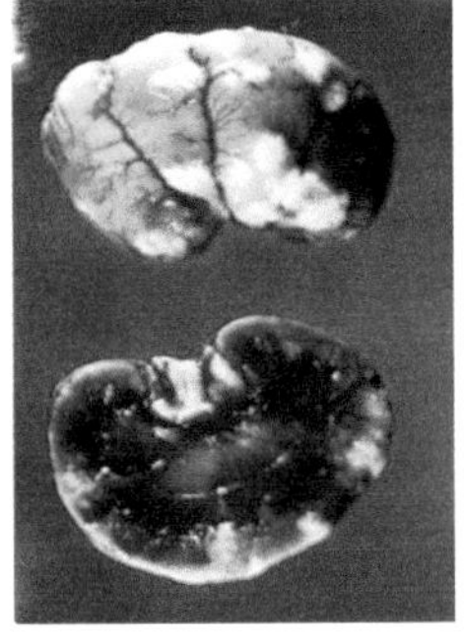

FIGURE 1 FIGURE 2 FIGURE 3

FIGURE 1. Gross appearance of the abdominal viscera of a cat with the classical (peritoneal) form of effusive FIP. The viscera are floating in yellowish gelatinous fluid, and numerous whitish fibrinous plaques are seen on the serosal surfaces of the intestine, liver, and spleen. The omentum is thickened, edematous, and retracted back towards the stomach.

FIGURE 2. Gross appearance of the kidney of a cat with noneffusive FIP. Numerous whitish granulomatous masses are seen extending from the capsular surface on cut section, these capsular granulomas extend downward into the renal parenchyma.

FIGURE 3. Gross appearance of the liver, hepatic lymph node (smaller mass), and mesenteric lymph node (larger mass) of a cat with noneffusive FIP. The liver contains numerous capsular oriented whitish-yellow masses. The hepatic and mesenteric lymph nodes are greatly enlarged, and are hard and irregular in shape and on palpation.

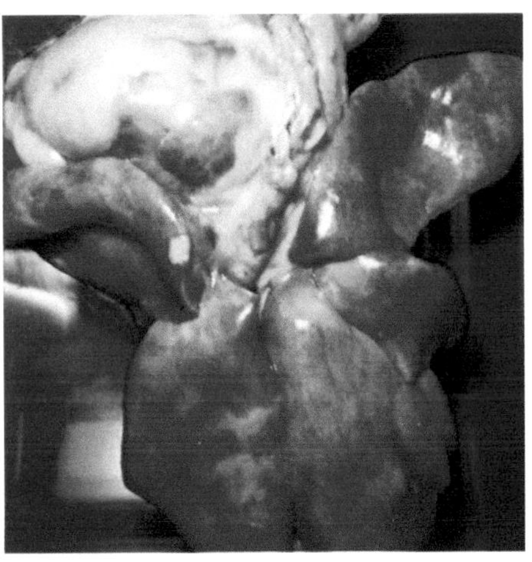

FIGURE 4 FIGURE 5

FIGURE 4. Gross appearance of the heart and lungs of a cat with noneffusive FIP. A small pleural granuloma is seen on the edge of the left cranial lung lobe. No other gross lesions were seen in the thoracic cavity.

FIGURE 5. Gross appearance of the eye of a cat with noneffusive FIP. There are keratic precipitates made up of filbrin and inflammatory cells on the inner surface of the cornea. The iris is swollen, indicating an associated anterior uveitis.

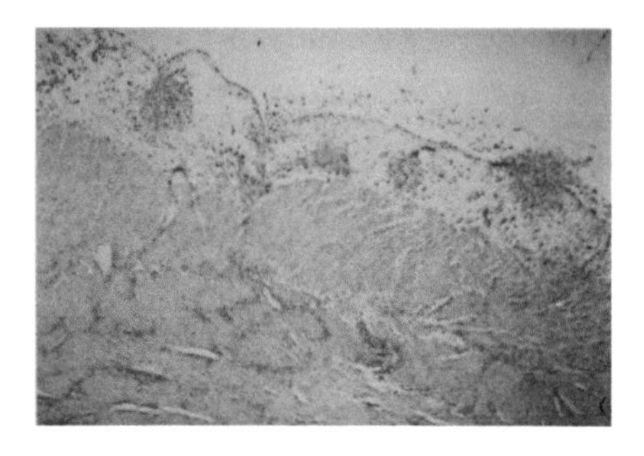

FIGURE 6

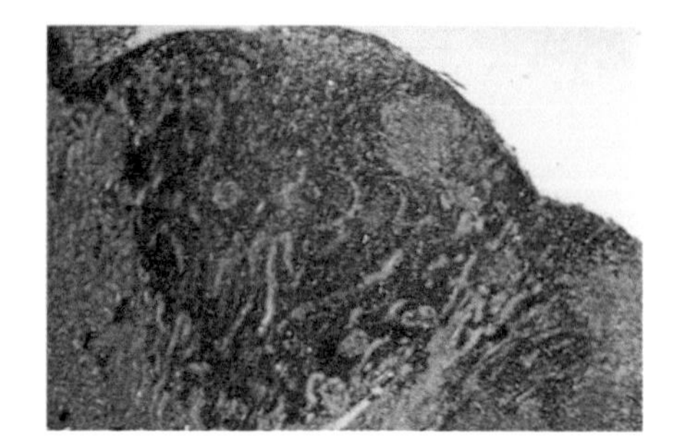

FIGURE 7

FIGURE 6. Photomicrograph of the serosal surfaces of the urinary bladder of a cat with effusive FIP. The serosa over the underlying muscularis is thickened and very edematous. There is a pronounced inflammatory response in the serosa, largely made up of focal aggregates of macrophages and polymorphonuclear neutrophils. The inflammatory response in the serosa extends into the underlying muscularis along penetrating veins. (Hematoxylin and eosin stain.)

FIGURE 7. Photomicrograph of the kidney of a cat with noneffusive FIP. Several paler focal pyogranulomatous reactions are seen in the cortex. These are surrounded by a dense accumulation of lymphocytic and plasmacytic cells that are infiltrating inward around the renal tubules. (Hematoxylin and eosin stain.)

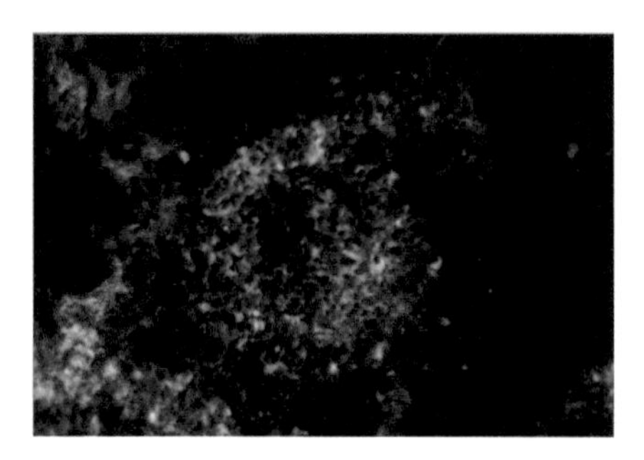

FIGURE 8

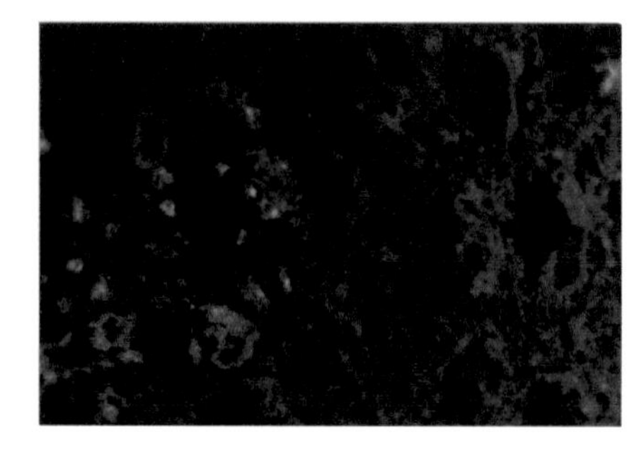

FIGURE 9

FIGURE 8. Photomicrograph of the omentum of a cat with effusive FIP. Specific immunofluorescent staining for FIPV antigen is seen in phagocytic cells forming the periphery of a pyogranuloma. Cryostat prepared, acetone fixed tissue section. (Cat anti-FIPV globulin-FITC stain, Evan's blue counterstain.)

FIGURE 9. Fluorescent photomicrograph of the kidney of a cat with noneffusive FIP. A large granulomatous mass makes up the left half of the field with normal kidney tubules on the right. FIP viral antigen is present only within a few phagocytic cells in the central portions of the granuloma. Cryostat prepared, acetone fixed tissue section. (Cat anti-FIPV globulin-FITC stain, Evan's blue counterstain.)

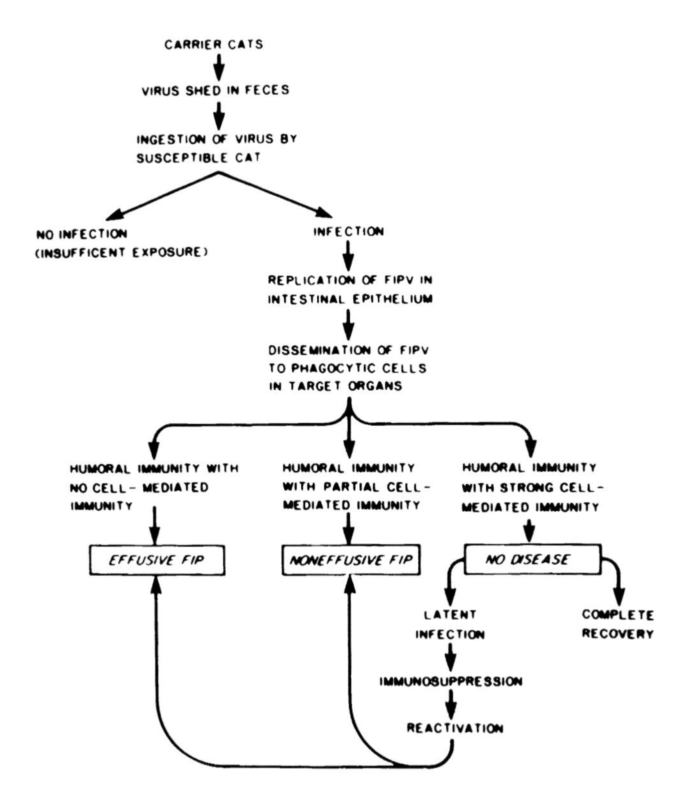

FIGURE 11. The proposed pathogenesis of FIP following the oral ingestion of virus.

occurring cases of FIP frequently occur in cats with no known exposure to cats with the disease.

In addition to FIPV carrier cats, it has been postulated, but not proven, that feline enteric coronavirus (FECV) carriers might be a source of FIPV.[50,51] In this situation, FIPV represents a common mutant form of the basic enteric coronavirus. Mutant viruses, when they are generated, could then infect either the host cat or other cats that come in contact with their secretions. The existence of closely related doublets of FIPVs and FECVs gives this postulate some credence. FIPV-UCD1 and FECV-UCD are antigenically indistinguishable, but cause very different types of illness; the FIPV-79-1146 and FECV-79-1683 isolates are another similar doublet.[53]

Regardless of the source of the virus, the initial event following oral or oronasal infection probably involves replication of virus in the intestinal epithelium (Figure 11). Hayashi and associates[23] have demonstrated the presence of FIPV in the intestinal epithelium of 2 of 4 cats with naturally occurring FIP and 12 of 14 cats with experimentally induced disease. Orally administered FIPV also replicates mainly in the mature apical epithelium of the intestinal villi of neonatal pigs and causes an enteritis virtually indistinguishable from TGE virus infection.[73] Clinical disease appears to be associated with dissemination of virus from the mucous membranes, probably by way of blood-borne phagocytic cells.[67] This is one essential difference between FIPV and closely related FECV strains. Feline enteric coronaviruses do not spread much further than the mesenteric lymph nodes.[53] Once in the bloodstream, FIPV is preferentially taken up by phagocytic cells in the liver, visceral peritoneum and pleura, uveal tract of the eyes, and the meninges and ependyma of the brain and spinal cord. This high tropism for phagocytic cells is a characteristic of FIPV and is another important differentiating feature from the FECVs. The presence of FIPV in macrophages is not due merely to phagocytosis; FIPV also replicates extremely well in such cells.[42] Following dissemination, the ultimate course of the disease appears to be dependent on the type and degree of immunity that develops (Figure 11). Virus containment is probably due to strong cell-mediated immunity. Humoral immunity by itself is not protective, and in some cases may even enhance the severity of the disease. Effusive FIP occurs in cats that mount only humoral immunity. Noneffusive FIP, on the other hand, is thought to occur when partial cell-mediated immunity is developed. In this sense, noneffusive FIP is an intermediate stage between nonprotective humoral immunity alone and protective cellular immunity. Evidence for immune phenomena in the various forms of FIPV infection will be discussed in the subsequent section.

Immunity to FIPV can be induced in a small proportion of cats. After being given very small doses of virus, about one third or less of the cats will seroconvert without ever developing signs of illness. These cats are immune to subsequent challenge with high levels of virus. Immunity in such cats does not, however, mean that the virus is eliminated from the body. On the contrary, there is fragmentary evidence that virus persists in either a latent or subclinical focus in the body, possibly for the remainder of the cat's life.

1. Immunologic Phenomena in the Effusive Form of FIP

Of cats with effusive FIP, 50% show characteristic elevations in all antibody containing globulin fractions.[41] Elevated globulin levels are usually, but not invariably, associated with high titers of indirect fluorescent reacting and virus neutralizing antibodies.[47] Serum electrophoresis demonstrates elevations of both beta and gamma globulin classes. Infrequently, the gamma fraction will be so elevated as to suggest a monoclonal gammopathy. Specific FIPV antibodies, however, are found almost exclusively in the IgG portion of the beta and gamma fractions. IgM and IgA fractions are also elevated, indicating that much of the antibody globulin produced by cats with FIP is not viral specific. It is possible that the virus infection has a nonspecific enhancing effect on the total level of antibody synthesis.

Although both IFA and virus neutralizing antibodies appear following infection with FIPV,[30,47] there is no evidence that humoral immunity is protective. In fact, earlier studies demonstrated an enhancing effect of preexisting humoral coronavirus immunity on FIPV-UCD1 infection.[49,67-69] At the time of the original studies, it was thought that the preexisting coronavirus antibody titers were due to previous FIPV infection. It is now known that these antibodies resulted from previous infections with antigenically similar nonFIP inducing enteric coronaviruses.[51]

The enhancing role of preexisting homologous coronavirus immunity was first demonstrated by Pedersen and Boyle.[49] Their findings were subsequently confirmed by Weiss and Scott.[67-69] In the experiments of Pedersen and Boyle,[49] specific pathogen free (spf) kittens that were negative for coronavirus antibodies, and conventionally reared kittens that had preexisting enteric coronavirus immunity, were inoculated i.p. with a cell-free tissue extract of FIPV-UCD1. Rectal temperatures were recorded daily and the kittens were observed for clinical signs of FIP. Some of the spf kittens inoculated i.p. with tissue-derived FIPV-UCD1 did not develop fever or any signs of illness (Figure 12). The remainder of the spf kittens developed fever around the 10th day postinoculation (Figure 13), and died 7 to 10 days later from effusive FIP. In contrast, kittens with preexisting homologous coronavirus immunity developed a high fever within 24 to 48 hr and became depressed and icteric by the 7th to 9th day (Figure 14).

When the humoral immune responses of these kittens were measured, it was noted that spf cats that did not become ill after FIPV challenge also did not develop detectable levels of FIPV antibodies. In contrast, the spf kittens that died of effusive FIP after primary inoculation developed measurable concentrations of serum antibodies

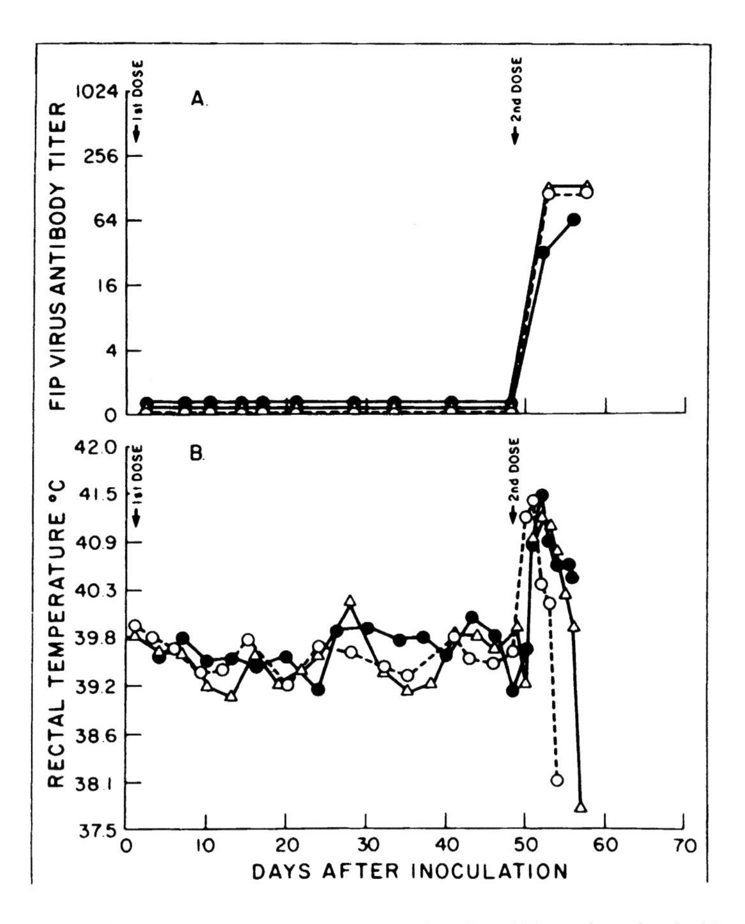

FIGURE 12. (A) Serum FIPV antibody titers in spf kittens inoculated with a cell-free FIPV inoculation. These kittens had no measurable antibody in their sera 7 weeks after the first inoculation with FIPV. After they were reinoculated, FIP virus antibody rapidly appeared in the serum in detectable titers. (B) Rectal temperatures of the same kittens. No febrile response was observed after the initial inoculation. After reinoculation, however, the kittens became febrile within 24 to 48 hr, and died 7 to 8 days later of effusive FIP. (From Pedersen, N. C. and Boyle, J. F., *Am. J. Vet. Res.*, 41, 808, 1980.

around the 10th day postinoculation, which was also about the time fever and illness first was detected (Figure 13). Specific pathogen free kittens that did not seroconvert or become ill after initial infection with FIPV were rechallenged with FIPV 6 weeks later (Figure 12). At the time of secondary challenge these kittens had no detectable levels of IFA reacting antibody to FIPV. Following secondary challenge with the same amount of tissue-derived FIPV, all of the kittens developed a fever within 48 hr and were moribund 6 to 10 days postinoculation. The subsequent serum antibody response in these kittens appeared to be anamnestic in character (Figure 12), which indicated that they had been immunologically primed after the inoculation even though they did not develop disease. This phenomenon was also described by Jacobse-Geels and coworkers.[32] They found that only one of six seronegative cats became infected after the first challenge with tissue derived FIPV, and that multiple injections with virus were required to infect the others.

Preexisting homologous or acquired autologous coronavirus immunity in this group of cats affected the course of FIPV challenge in several ways. The most obvious effect

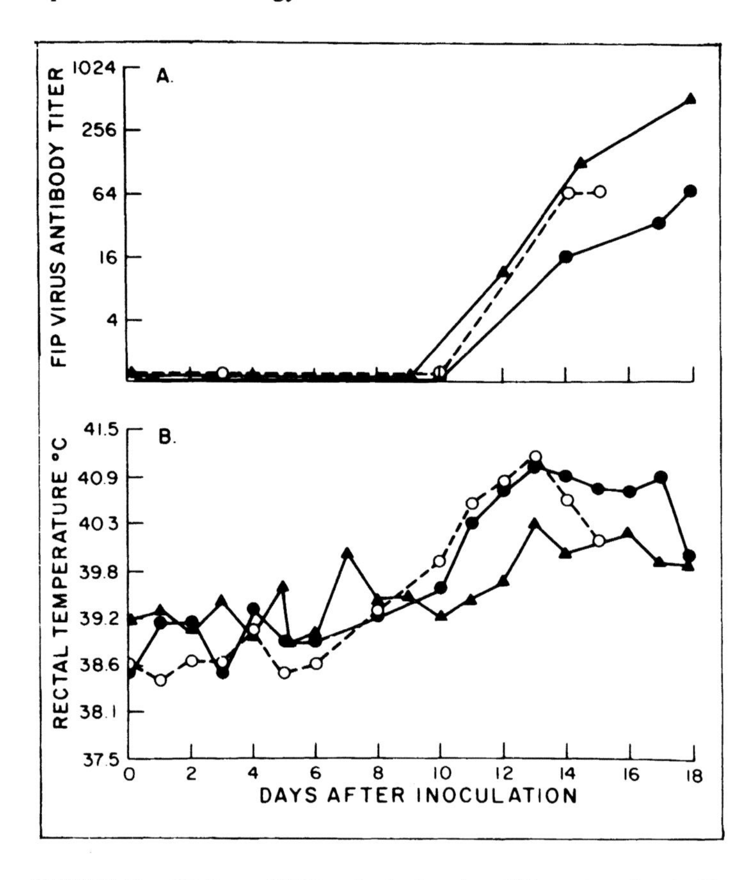

FIGURE 13. (A) Serum FIPV antibody titers in spf kittens inoculated with a cell free FIPV inoculum. These kittens developed measurable concentrations of FIPV antibody 9 to 10 days after primary inoculations. (B) Rectal temperatures of the same kittens. A fever appeared at approximately the same time that FIPV antibody was detected in the serum. The kittens died 5 to 8 days after effusive FIP. (From Pedersen, N. C. and Boyle, J. F., *Am. J. Vet. Res.*, 41, 868, 1980. With permission.)

was to greatly enhance the infectivity of the tissue-derived FIPV inoculum. Tissue-derived FIPV used in these early studies was probably complexed with IgG and complement because it was obtained from cats that were seropositive at the time their tissues were harvested. Viral antigen in such tissues is known to coexist with IgG and complement, and the IgG and complement can be eluted from the virus with low pH buffers.[49] Possibly for this reason, only about one half of the specific pathogen-free kittens inoculated with this material became ill or developed measurable levels of FIPV antibodies. In contrast, this same inoculum was virtually 100% infectious when administered to kittens that had preexisting homologous coronavirus immunity. Although about one half of the seronegative spf cats did not show any signs of infection after the first inoculation, they were nevertheless immunologically primed. Following a second identical inoculation, there was an extremely rapid rise in antibody titer suggesting an anamnestic type immune response. Just like cats with preexisting homologous coronavirus immunity, these cats also developed fever and other signs of FIP within 24 hr of the time of challenge. It was concluded, therefore, that infection with tissue-derived FIPV was enhanced when virus remained viable in the body long enough for the host to mount an immune response. If immunity was not induced soon enough, the tissue-derived virus would not remain viable and the infection would be aborted.

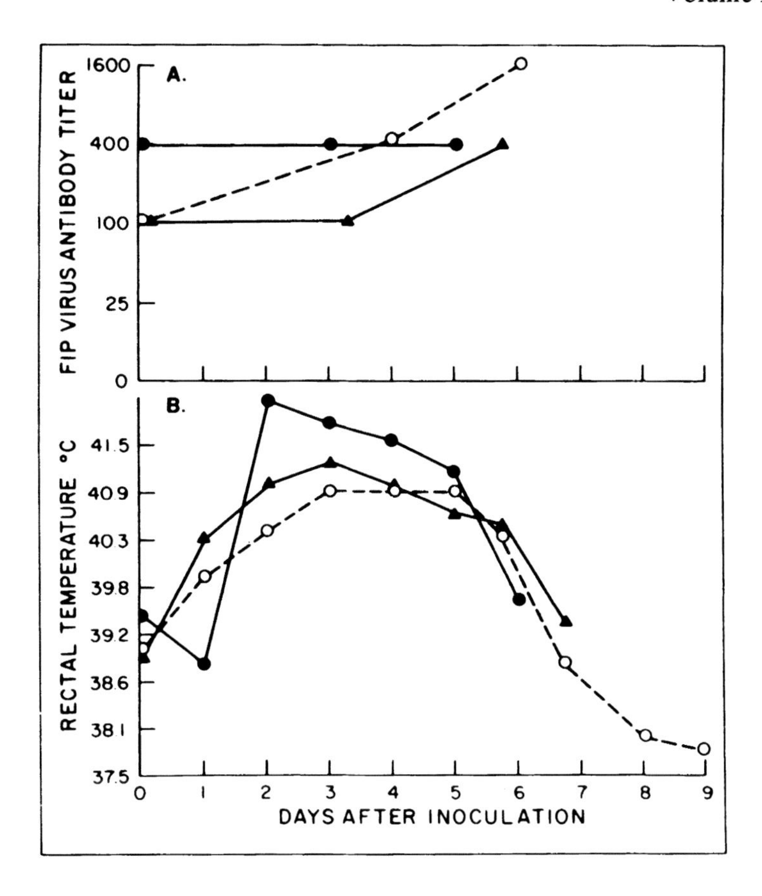

FIGURE 14. (A) Serum FIPV antibody titers in seropositive conventionally reared kittens inoculated with a cell-free FIPV inoculum. The preexisting coronavirus antibodies were elicited by an earlier FECV infection. There was an anamnestic antibody response following challenge. (B) Rectal temperatures of the same kittens. Fever appeared 24 to 48 hr after inoculation and the cats died 6 to 9 days later of effusive FIP. (From Pedersen, N. C. and Boyle, J. F., *Am. J. Vet. Res.*, 41, 868, 1980. With permission.)

Such an aborted infection would result in some immunologic priming, and would sensitize the cat to a second identical challenge inoculation. Low levels of preexisting immunity, or the rapid appearance of a secondary immune response, would then assure the infection would take hold the second time.

Pedersen and Boyle[49] postulated that homologous coronavirus antibodies were responsible for enhancement of FIPV infectivity. To test this premise, they inoculated a group of seronegative spf kittens i.p. with 40 m*l* of high titered serum collected from normal cats with high coronavirus antibody titers. Antibody that cross-reacted with FIPV-UCD1 on a titer of 1:25 to 1:100 was detected in the serum 2 days later. At this time, each kitten was inoculated i.p. with 20, 2, 0.2, 0.02, or 0.002 gEq of FIPV inoculum. A group of nontreated kittens was given the same amount of FIPV inoculum. Kittens that had been passively immunized with serum and inoculated with 0.02 to 20.0 gEq of material developed fever within 24 to 48 hr and were moribund by the 8th to 14th day postinoculation (Figure 5). Seronegative kittens that were not treated with serum did not develop disease after FIPV inoculation regardless of the dose of virus given (Figure 15).

Although it was apparent that passive sensitization to FIPV-UCD1 was related to something in the homologous coronavirus immune serum, it remained to be deter-

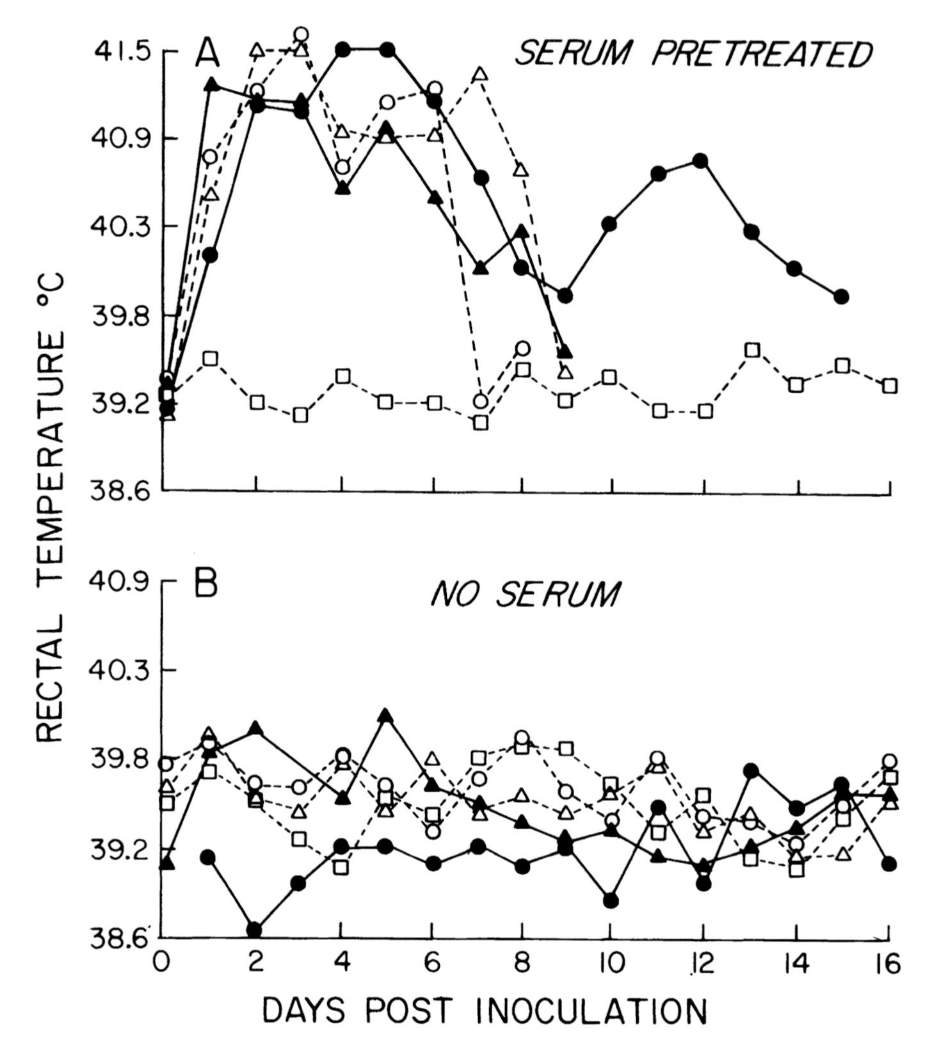

FIGURE 15. (A) Rectal temperature of spf kittens who were given FIPV antibody positive serum before being inoculated with a cell-free inoculum of FIPV. (B) Rectal temperatures of spf kittens who were not pretreated with FIPV antibody positive serum before they were inoculated with FIPV. (From Pedersen, N. C. and Boyle, J. F., *Am. J. Vet. Res.*, 41, 868, 1980. With permission.)

mined if the serum factor was antibody. To test this, Pedersen and Boyle[49] pretreated a spf kitten with 40 m*l* of highly purified IgG obtained from the serum of normal cats with cross-reacting coronavirus antibodies. A second spf kitten was left untreated. The IgG treated kitten developed fever within 48 hr, and was near death from effusive FIP by the 8th day after inoculation (Figure 16). The nontreated kitten developed fever on the 8th day after inoculation and died 5 days later of effusive FIP.

The mechanism by which humoral antibody enhances infectivity is not known. It can be argued that homologous immunity to enteric coronavirus infection generates nonneutralizing antibodies to FIPV and that these antibodies are responsible for disease enhancement. This is not the case, however, because cats infected with nonFIP inducing enteric coronaviruses produce antibodies that neutralize FIP-inducing coronaviruses as well.[3] Cats passively immunized with virus neutralizing serum from FIPV immune cats are also sensitized to FIPV challenge.[47] Furthermore, cats that are in-

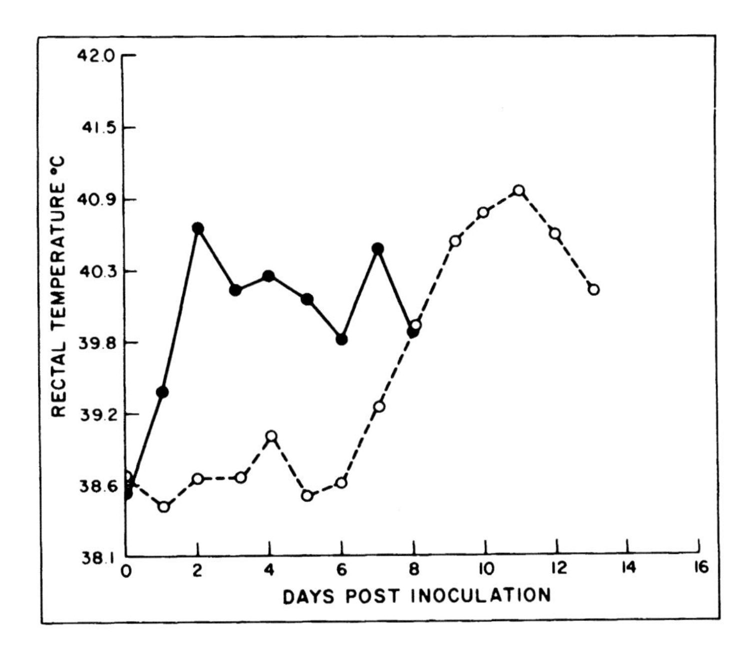

FIGURE 16. Rectal temperatures of spf kittens inoculated with a cell-free inoculum of FIPV. •—• Treated with purified IgG containing FIPV antibody. O—O Not pretreated with purified IgG. (From Pedersen, N. C. and Boyle, J. F., *Am. J. Vet. Res.*, 41, 868, 1980. With permission.)

fected with FIPV, and die as a result, generate large amounts of virus neutralizing antibody during the course of the disease.

The mechanism of antibody mediated enhancement of FIPV infectivity is not known. A similar phenomena is described, however, for dengue virus infection.[58] People with preinfection homotypic immunity to one serotype of dengue fever virus can develop a more severe disease (dengue hemorrhagic shock syndrome) if they are infected later with another serotype.[18] It has been shown both in vitro[19] and in vivo[58] that homotypic antibody may complex with virus and potentiate the infection of macrophages. This in turn increases yields of virus, increases the number of infected cells, and increases disease. Heterotypic flavivirus immunity also enhances dengue virus replication both in vitro[54] and in vivo.[58] Enhancement of dengue virus replication can be blocked by pretreatment of macrophage cultures with monoclonal antibodies to fc receptors,[55] suggesting that immune complexed virus is selectively bound to macrophages and that this binding facilitates macrophage infection.

Similar to the situation in dengue virus infection,[18] homotypic enhancement varies according to the strains of coronaviruses involved. Preimmunization with FECV-UCD consistently enhances infection with FIPV-UCD1 and FIPV-Black.[50,51,53] It does not, however, enhance infection with FIPV-79-1146.[53] Preimmunization with FECV-79-1683 induces a more variable enhancing immunity to FIPV-UCD1, and does not enhance infection with FIPV-79-1146.[53] Autotypic enhancement to virulent FIPV-Black occurs in cats immunized with attenuated FIPV-Black.[47] Immunization with inactivated FIPV will usually sensitize cats to subsequent challenge with its living counterpart.[46] Similarly, preimmunization of cats with immune complexed FIPV-UCD1 will enhance a subsequent infection with immune complexed FIPV-UCD1. The most interesting aspect of such comparative studies, however, is not the variability of enhancing immunity, but the lack of cross-protection. We have not identified one strain of feline coronavirus that would give solid cross-protection against another.

In addition to enhancing the infectivity of antibody complexed virus and facilitating

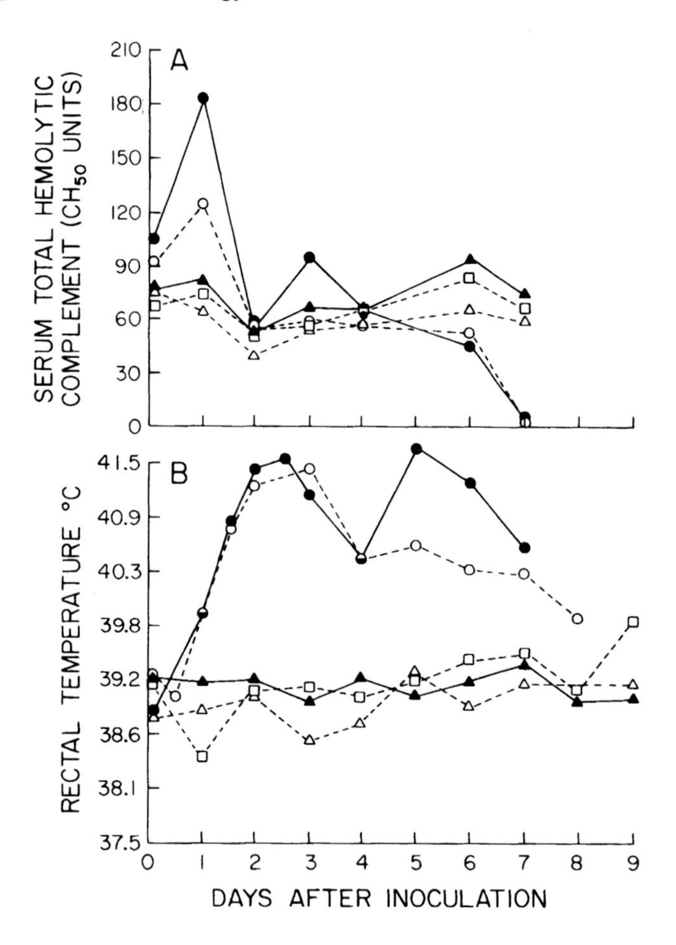

FIGURE 17. (A) Total hemolytic complement levels in the serum of cats with preexisting homologous coronavirus immunity. Two kittens were inoculated intraperitoneally with FIPV, (•—• and O—O). The remaining three kittens were not infected with FIPV. (B) Rectal temperatures of the same kittens. There was an abrupt rise in complement levels following infection with FIPV, followed by a terminal decrease. The initial increase in serum complement preceded the rise in rectal temperature. (From Pedersen, N. C. and Boyle, J. F., *Am. J. Vet. Res.*, 41, 868, 1980. With permission.)

the overall level of virus replication within the reticuloendothelial system, antibody may play a direct role in the pathogenesis of the lesions. Pedersen and Boyle[49] identified viral antigens, IgG, and complement within lesions of cats with experimentally induced FIP. They also demonstrated a dramatic increase in serum complement prior to the onset of clinical signs followed by a marked decrease in the terminal phases of the disease (Figure 17). Complement activation followed by complement depletion was also demonstrated by Jacobse-Geels and co-workers.[32] Jacobse-Geels and associates[32] presented compelling evidence that circulating immune complexes were also involved in the disease. Using a C1q binding assay, they demonstrated a marked increase in the levels of circulating immune complexes associated with the appearance of humoral antibodies and fever in cats experimentally infected with FIPV (Figure 18). Some of these circulating immune complexes also became deposited within the glomerular endothelium.[31]

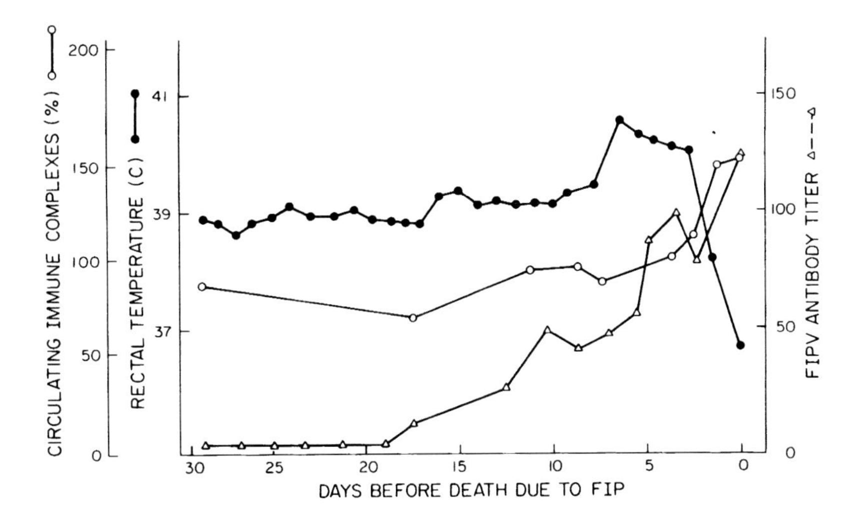

FIGURE 18. Mean values for rectal temperature, circulating immune complexes, and FIPV antibody titers in six kittens experimentally infected with FIPV.[32]

A predominant role for complement factors in FIP has been proposed by Jacobse-Geels and associates.[32] They viewed selective growth of FIPV in macrophages as a crucial factor in the pathogenesis of the disease. Because macrophages are also a site of complement synthesis, they postulated that the initial increase in complement levels reflected macrophage proliferation, infection, and virolysis. Once antibody appeared in the serum, immune complexes would be formed. These complexes would bind complement by more classical routes. At this point, they postulated that a feedback mechanism comes into play with virion bound activated C3 acting as an infection-promoting molecule, causing even more macrophages to be infected, increased synthesis or release of complement, and immune adherence of the complexes. Presumably, the magnitude of complement binding at this stage would lead to a pronounced drop in circulating complement factors.

The role of complement in the pathogenesis of the lesions of effusive FIP was studied in our laboratory. Kittens with homologous coronavirus immunity were decomplemented by parenteral infections of C3 cleaving enzyme purified from cobra venom. When the serum total hemolytic complement activity was decreased to unmeasurable levels, the kittens were infected with FIPV. Decomplementation had no appreciable effect on the resulting disease. The cats developed a fever within 1 to 2 days and died from fulminating effusive FIP in about the same time as nondecomplemented cats. If these experiments were valid, it can be concluded that complement probably does not play a central role in the overall pathogenesis of FIP. If complement is not an important factor, what other factors might be involved in creating the lesions of FIP? It is possible that tissue damage is precipitated by lymphokine-like or enzymatic factors released from infected and activated macrophages. Indeed, macrophage and T lymphocyte cytotoxic factors are released as a result of dengue-virus infection.[17] Eling and associates[11] have also postulated that activated macrophages induce endocardial lesions in murine malaria.

A common feature between effusive FIP of cats and dengue hemorrhagic shock syndrome of man is the occurrence of a systemic coagulopathy. The pathogenesis of the coagulopathy in these two diseases is manifested in different ways. Frank hemorrhage in many organs and tissues is seen in dengue hemorrhagic fever, but are not seen in cats with effusive FIP. The coagulopathy in FIP seems to contribute, however, to the amount and character of the fluid that exudes into the abdomen and/or chest cavities. The coagulopathy of effusive FIP has been extensively studied by Weiss and associates. They found that cats with experimentally induced effusive FIP had greatly increased prothrombin and partial thromboplastin times, increased levels of fibrin degradation products, increased activated clotting times, and decreased levels of intrinsic clotting factors. The coagulopathy seen in dengue hemorrhagic fever and effusive FIP may be caused by multiple factors. Activation of terminal complement factors may also cause activation of terminal clotting factors. This in turn would lead to a depletion of clotting factors and coagulopathy.

2. Immunologic Phenomena in the Noneffusive Form of FIP

Much less is known about immune mechanisms of noneffusive FIP. What is known is based mainly on circumstantial evidence and similarities between this and other granulomatous diseases of man and animals.

Total plasma proteins are elevated in more than three fourths of the cats with noneffusive FIP.[43] Similar to the effusive form, protein elevations are due mainly to elevations in immune globulins. Plasma proteins as high as 11 g/ℓ have been observed in this form of FIP, and 60 to 80% of this protein is immunoglobulin.[41] Although serum coronavirus antibody titers in noneffusive FIP are frequently very high, much of this immunoglobulin is not directed against viral antigens. Antibody to viral proteins are almost exclusively in the IgG fraction, and can be detected by indirect fluorescent antibody, virus neutralization, ELISA, or immunoprecipitation.

Circumstantial evidence suggests that noneffusive FIP represents an intermediate state of host responsiveness lying between that of cats with effusive FIP and cats that are immune to reinfection. In three cases that have been followed from infection to death, noneffusive FIP began as a brief episode of effusive type disease. This indicates an evolution in the immune response from humoral to some other form, probably cell-mediated. The initial effusive nature of the disease may explain why fibrinous plaques and scars are often seen on the pleural and serosal surfaces of cats with noneffusive FIP. The lesions of noneffusive FIP, like the lesions of effusive FIP, usually began in the serosal, meningeal, or ependymal surfaces and extend downward into the parenchyma. The essential difference in the two forms of FIP is that parynchymal involvement in noneffusive FIP often dominates over more superficial lesions.

There is strong circumstantial evidence that noneffusive FIP results from cell-mediated immune phenomena. The granulomatous nature of the lesions and the presence of small amounts of coronaviral antigen deep in centrally located phagocytic cells (Figure 9) are highly reminiscent of tuberculosis and the deep mycotic infections. The most telling evidence, however, is the failure to show any protective effects for humoral immunity. Much like humoral immunity in tuberculosis and deep mycotic infections, antibody seems to be merely a consequence of continuous antigenic stimulation. This almost pathologic attempt to make antibodies is reflected by the overall increase in antibody synthesis.

If noneffusive FIP is caused by cellular immune phenomena, then there must be some qualitative or quantitative deficiency in the immunity of the cats with noneffusive FIP compared to cats that have been rendered immune to FIPV infection. Some cats will become solidly immune to FIPV following infection, and this immunity has no relationship to humoral immunity.[47] Protective immunity must, therefore, represent a specific state where cellular immune phenomena work in concert to bring about desruction or sequestration of cell-bound virus.

The predominance of lesions in the CNS and eyes in noneffusive FIP is in itself evidence that partial immunity is in force in these animals. The CNS and eyes are protected from the rest of the body by a blood-brain barrier. This barrier limits the access of immune cells as well as humoral antibody. It is reasonable to assume that partial immunity will be most effective against virus or virus-infected cells that have not crossed the blood-brain barrier. It is also logical to assume that lesions in animals with partial immunity will persist in areas where systemic immunity is least likely to penetrate. Canine distemper, another viral disease where cell-mediated and humoral immunity are both important, demonstrates the same phenomena. Puppies that develop rapid systemic immunity will abort the infection before it has time to spread from the reticuloendothelial system to epithelial structures and the CNS.[1] Dogs that fail to mount immunity will develop a widespread epithelial disease and will often die of enteritis and pneumonia before CNS involvement becomes pronounced. Dogs that mount intermediate immunity, however, develop very little epithelial disease, but ultimately die from CNS involvement. This presumably occurs when partial immunity clears the virus from epithelial tissues, but fails to gain access to the CNS.

The incidence of noneffusive vs. effusive FIP can be increased in several ways. First, cats preimmunized with cross-reacting enteric coronaviruses may show a higher relative incidence of noneffusive FIP than cats that have no previous coronavirus exposure.[51] Second, the smaller the dose of virus that is given the more likely it becomes for cats to develop protective immunity and the higher the relative incidence of noneffusive FIP. Conversely, cats given high doses of virus are much less likely to develop protective immunity or noneffusive FIP. Most of such cats will succumb to noneffusive disease.

B. FIP and Feline Leukemia Virus Infection

Cotter and associates[9] noticed that from 40 to 50% of the cats with naturally occurring FIP were concurrently infected with feline leukemia virus (FeLV). This figure was subsequently confirmed by other researchers.[41] With the elimination of FeLV infection from many catteries, there has been a dramatic decrease in both the incidence of FIP and in the percentage of naturally occurring cases of FIP that are FeLV infected.

The relationship of FeLV infection and FIP can best be evaluated in controlled studies. During the last 15 years we have raised over 2600 cats for FeLV research. These kittens came from a colony where FIPV was apparently enzootic. During this period of time, 28 cats from this colony developed FIP. Of these 28 cats, 25 developed FIP within 2 to 12 weeks of the time that they became infected with FeLV and only 3 cats developed FIP without any exposure to FeLV. In most cases, the FeLV infection was active at the time the cats developed FIP. The few cats that were FeLV negative at the time they developed FIP were usually transiently FeLV viremic a few weeks before. In a related study, three cats that had been "immunized" against FIPV using sublethal amounts of virulent virus were infected with FeLV. One of these cats developed FIP 3 weeks later in spite of having resisted a number of challenges with FIPV prior to this time.[46]

The mechanism by which FeLV infection potentiates the incidence of FIP is largely unknown, but from controlled observations such as these, several assumptions can be made. Since only 10% of the cases of FIP occurred in siblings that were not exposed to FeLV, it can be assumed that many more normal cats harbor FIPV as an asymptomatic or latent infection. It can also be assumed that FeLV infection in some way interferes with established FIPV immunity. The mechanism of this interference is not known, but could involve any of a number of immunosuppressive effects that have been described for FeLV infection.

V. SEROLOGIC STUDIES

Specific pathogen-free kittens that are infected by oral or intratracheal instillation of FIPV serological react in several ways.[50] Some cats do not develop any signs of infection after prolonged exposure, and they remain antibody negative (Figure 10). Infected cats that do not develop signs of illness demonstrate a plateau-shaped antibody response, while cats that develop FIP demonstrate a progressive antibody titer rise (Figure 10). Virus neutralizing antibodies tend to correlate with the IFA titers in both groups of cats.[47] Sometimes, however, an infected cat will only develop virus-neutralizing antibodies, and the IFA titers will be negligible.

Serologic responses are much more difficult to interpret in the field. That is because of the great amount of antigenic similarity between FIPV and nonFIP inducing FECVs, and the ubiquitousness of FECV infection in nature. Boyle and co-workers,[3] using sensitive immunoblotting techniques, were unable to show differences in the antibody specificity of serum from cats infected with FECVs or with FIPV isolates. For this reason, serodiagnosis of FIP in the field is fraught with a great deal of inaccuracy. Currently used serologic procedures, however, still have some usefulness. Antibody titers by IFA greater than 1:3200 are usually associated with FIP, frequently of the noneffusive type. Titers this high are uncommon in cats infected with FECVs. Titers of 1:100 to 1:3200 are commonly seen in cats with effusive FIP and in a portion of cats with noneffusive disease. Unfortunately IFA titers of 1:25 to 1:3200 are also seen in cats that have undergone the enteric coronavirus infections and in cats that have previously been infected with FIPV but did not become ill. The diagnosis of FIP in cats with titers in this range depends, therefore, on a careful review of the entire clinical and clinicopathologic picture. Positive coronavirus titers should alert clinicians to the possibility of FIP, while negative titers are often helpful in ruling it out. It has been noted, however, that some cats with pathologically confirmed FIP have been seronegative by IFA so that a negative IFA titer is not always helpful.

A. Immunization

Initial attempts to immunize cats with TGE virus of swine have been unsuccessful.[63,72] Although FIPV would immunize baby pigs against TGE, TGE immunization provided no protection against FIP. Immunization with killed FIPV has also proven uniformly unsuccessful.[31] The immunity derived from autologous killed vaccines almost always renders the cat more susceptible to challenge with the virulent living virus, and the disease that results is usually more severe and fulminating.

Attenuated live FIPV has also failed to immunize cats.[47] Kittens immunized oronasally with attenuated FIPV-TN406 developed both IFA and virus-neutralizing antibodies. Following challenge with virulent FIPV-TN406, however, the infection rate was enhanced, incubation period reduced, and disease severity increased in vaccinated kittens compared to nonvaccinates. Apparently, avirulent virus does not confer a protective type of immunity, but may have elicited humoral immunity that was actually deliterious. Pedersen and Black[47] postulated that avirulent virus failed to provide protective immunity because it did not persist in the body long enough. This suggests that immunity to FIPV may involve the establishment of a latent or asymptomatic carrier state.

Cats can be immunized against FIP using very small amounts of virulent virus.[47] This was not of clinical relevance, however, because the dose required to immunize some cats caused fatal FIP in others. Furthermore, when three immunized cats were infected later with FeLV, one of them developed fatal effusive FIP.[47] This also suggested that immunity in some of these cats was a result of a persistent latent or asymptomatic FIPV infection.

B. Treatment

There is no treatment that has proven uniformly and consistently effective. Cats that develop FIP usually die in 1 to 16 weeks. Nevertheless, there have been several reports of cats that have gone into remission after being treated with various drugs. Colgrove and Parker[7] described cats that went into remission after being treated with the antibiotic Tylocine and prednisolone. They credited the antibiotic for the apparent cures, which sparked a decade of use of Tylocine for the treatment of this disease. It is now conceded, however, that the antibiotic is of no value whatsoever in the treatment of FIP, and the fortuitous response in the original cats was probably due to self-cures or to the prednisolone. Pedersen[41,45] also described some cats that went into remission after the use of prednisolone and phenylalanine mustard or cyclophosphamide. Madewell and co-workers[36] also successfully treated a cat with effusive FIP with prednisolone and phenylalanine mustard. This too has proven to be of limited effectiveness.

In the author's experience, less than 10% of cats with FIP will go into a brief or sustained remission after treatment with immunosuppressive drugs. Successfully treated cats had several things in common. They usually had milder signs of illness, and they were still eating and not overly debilitated at the time treatment was undertaken; debilitated animals inevitably die and drug therapy actually hastens their demise. In addition, the owners were more apt to administer continuous supportive care in the form of fluid therapy, force feeding, and other such attentions.

A complicating factor in evaluating treatment success is the occurrence of spontaneous remission. Not every cat with FIP will die. Post-mortem examination on older cats without signs of overt FIP have occasionally demonstrated fibrous lesions on the spleen and liver that indicate a past FIP infection. Cats with ocular FIP and no other systemic manifestations have occasionally gone into remission just with topical treatment. We have also observed cats with chronic fever, enlarged mesenteric lymph nodes that were histologically compatible with FIP, and high coronavirus titers, that have spontaneously gone into disease remission with time and no treatment. Finally, cats with small quiescent lesions in their spleens and mesenteric lymph nodes have been discovered during the course of routine ovariohysterectomies. What this all means is that it is sometimes difficult to ascertain whether a treatment is successful or if remission was naturally induced.

REFERENCES

1. Appel, M. J. G., Pathogenesis of canine distemper, *Am. J. Vet. Res.*, 30, 1167, 1969.
2. Black, J. W., Recovery and *in vitro* cultivation of a coronavirus from laboratory-induced cases of feline infectious peritonitis (FIP), *Vet. Med. Small Anim. Clin.*, 75(5), 811, 1980.
3. Boyle, J. F., Pedersen, N. C., Evermann, J. F., McKeirnan, A. J., Ott, R. L., and Black, J. W., Plaque assay, polypeptide composition and immunochemistry of feline infectious peritonitis virus and feline enteric cornavirus isolates, *Adv. Exp. Med. Biol.*, 173, 133, 1984.
4. Campbell, L. H. and Reed, C., Ocular signs associated with feline infectious peritonitis in two cats, *Feline Pract.*, 5(3), 32, 1975.
5. Campbell, L. H. and Schiessl, M. M., Ocular manifestations of toxoplasmosis, infectious peritonitis and lymphosarcoma in cats, *Mod. Vet. Pract.*, 59, 761, 1978.
6. Colby, E. D. and Low, R. J., Feline infectious peritonitis, *Vet. Med. Small Anim. Clin.*, 65, 783, 1970.
7. Colgrove, D. J. and Parker, A. J., Feline infectious peritonitis, *J. Small Anim. Pract.*, 12, 225, 1971.
8. Colly, L. P., Johannesburg, S. Africa, cited in, *J. Am. Vet. Med. Assoc.*, 158, 981, 1971.
9. Cotter, S. M., Gilmore, C. E., and Rollins, C., Multiple cases of feline leukemia and feline infectious peritonitis in a household, *J. Am. Vet. Med. Assoc.*, 162, 1054, 1973.
10. Doherty, M. J., Ocular manifestations of feline infectious peritonitis, *J. Am. Vet. Med. Assoc.*, 159, 317, 1971.
11. Eling, W. M. C., Heinen-Borries, U., VanRun-VanBreda, J. J., and Jerusalem, C. R., Activated macrophages: inductors of endocardial lesions in murine malaria, *Cell Biol. Int. Rep.*, 7(5), 321, 1983.

12. Evermann, J. F., Baumgartner, L., Ott, R. L., Davis, E. V., and McKeirnan, A. J., Characteristics of feline infectious peritonitis virus isolate, *Vet. Pathol.*, 18, 256, 1981.
13. Fankhauser, R. and Fatzer, R., Meningitis und Chorioependymitis Granulomatosa der Katze: Mogliche Beziehungen Zur Felinen Infectiosen Peritonitis (FIP), *Klientierpraxis*, 22, 19, 1977.
14. Fowler, M. E., *Zoo and Wild Animal Medicine*, W. B. Saunders, Philadelphia, 1978, 660.
15. Gelatt, K. M., Iridocyclitis-panophthalmitis associated with FIP, *Vet. Med. Small Anim. Clin.*, 68, 56, 1973.
16. Gillespie, J. H. and Scott, F. W., Feline viral infections, *Adv. Vet. Sci.*, 17, 163, 1973.
17. Gulati, L., Chaturvedi, U. C., and Mathur, A., Dengue virus-induced cytotoxic factors induce macrophages to produce a cytotoxin, *Immunology*, 49, 121, 1983.
18. Halstead, S. B., Rojanasuphot, S., and Sangkawibha, N., Original antigenic sin in dengue, *Am. J. Trop. Med. Hyg.*, 32, 154, 1983.
19. Halstead, S. B., Larsen, K., Kliks, S., Peiris, J. J. M., Cardosa, J., and Porterfield, J. S., Companion of P388D, mouse macrophage cell line and human monocytes for assay of dengue-2 infection-enhancing antibodies, *Am. J. Trop. Med. Hyg.*, 32, 151, 1983.
20. Halstead, S. B., *In vivo* enhancement of dengue infection with passively transferred antibody, *J. Infect. Dis.*, 140, 527, 1979.
21. Harvey, J. W. and Gaskin, J. M., Feline haptoglobin, *Am. J. Vet. Res.*, 39, 549, 1978.
22. Hayashi, T., Utsumi, F., Takahashi, R., and Fujiwara, K., Pathology of noneffusive type feline infectious peritonitis and experimental transmission, *Jpn. J. Vet. Sci.*, 42, 197, 1980.
23. Hayashi, T., Watabe, Y., Nakayama, H., and Fujiwara, K., Enteritis due to feline infectious peritonitis virus, *Jpn. J. Vet. Sci.*, 44, 97, 1982.
24. Holliday, T. A., Clinical aspects of some encephalopathies of domestic cats, *Vet. Clin. N. Am.*, 1, 367, 1971.
25. Holmberg, C. A. and Gribble, D. H., Feline infectious peritonitis: diagnostic gross and microscopic lesions, *Feline Pract.*, 3(4), 11, 1973.
26. Horzinek, M. C., Lutz, H., and Pedersen, N. C., Antigenic relationship among homologous structural polypeptides of porcine, feline and canine coronaviruses, *Infect. Immun.*, p.1148, 1982.
27. Horzinek, M. C. and Osterhaus, A. D. M. E., Feline infectious peritonitis: a world-wide serosurvey, *Am. J. Vet. Res.*, 40, 1487, 1979.
28. Horzinek, M. C. and Osterhaus, A. D. M. E., The virology and pathogenesis of feline infectious peritonitis. Brief review, *Arch. Virol.*, 59, 1, 1979.
29. Horzinek, M. C., Osterhause, A. D. M. E., and Ellens, D. J., Feline infectious peritonitis virus, *Zentralbl. Vet. Med. B*, 24, 398, 1977.
30. Horzinek, M. C. and Osterhaus, A. D. M. E., Wirahadiredja, R. M. S., and deKreeck, P., Feline infectious peritonitis virus. III. Studies on the multiplication of FIP virus in the suckling mouse, *Zentralbl. Vet. Med. B*, 25, 806, 1978.
31. Jacobse-Geels, H. E. L., Daha, M. R., and Horzinek, M. C., Isolation and characterization of feline C3 and evidence for the immune complex pathogenesis of feline infectious peritonitis, *J. Immunol.*, 125, 1606, 1980.
32. Jacobse-Geels, H. E. L., Daha, M. R., and Horzinek, M. C., Antibody, immune complexes and complement activity. Fluctuations in kittens with experimentally induced feline infectious peritonitis, *Am. J. Vet. Res.*, 43, 666, 1982.
33. Kornegay, J. N., Feline infectious peritonitis, *J. Am. Anim. Hosp. Assoc.*, 14, 580, 1978.
34. Krum, S., Johnson, K., and Wilson, J., Hydrocephalus associated with the noneffusive form of feline infectious peritonitis, *J. Am. Vet. Med. Assoc.*, 167, 746, 1975.
35. Legendre, A. M., and Whitenack, D. L., Feline infectious peritonitis with spinal cord involvement in two cats, *J. Am. Vet. Med. Assoc.*, 167, 931, 1975.
36. Madewell, B. R., Crow, S. E., and Nickerson, T. R., Infectious peritonitis in a cat that subsequently developed a myeloproliferative disorder, *J. Am. Vet. Med. Assoc.*, 172, 169, 1978.
37. McKeirnan, A. J., Evermann, J. F., Hargis, A., Miller, L. M., and Ott, R. L., Isolation of feline coronaviruses from two cats with diverse disease manifestations, *Feline Pract.*, 11(3), 16, 1981.
38. Montali, R. J. and Strandberg, J. D., Extraperitoneal lesions in feline infectious peritonitis, *Vet. Pathol.*, 9, 109, 1972.
39. O'Reilly, K. J., Fishman, L. M., and Hitchcock, L. M., Short communication: feline infectious peritonitis; isolation of a coronavirus, *Vet. Rec.*, 104, 348, 1979.
40. Osterhaus, A. D. M. E., Horzinek, M. C., Wirahadiredja, R. M. S., and Kroon, A., Feline infectious peritonitis (FIP) virus. IV. Propagation in suckling rat and hamster brain, *Zentralbl. Vet. Med.*, 5(25), 816, 1978.
41. Pedersen, N. C., Feline infectious peritonitis: something old, something new, *Feline Pract.*, 6(3), 42, 1976.
42. Pedersen, N. C., Morphologic and physical characteristics of feline infectious peritonitis virus and its growth in autochthonous peritoneal cell cultures, *Am. J. Vet. Res.*, 37, 1449, 1976.

43. Pedersen, N. C., Serologic studies of naturally occurring feline infectious peritonitis, *Am. J. Vet. Res.*, 37, 1449, 1976.
44. Pedersen, N. C., Feline infectious peritonitis and feline enteric coronavirus infections. I. Feline enteric coronavirus, *Feline Pract.*, 13(4), 13, 1983.
45. Pedersen, N. C., Feline infectious peritonitis and feline enteric coronavirus infections. II. Feline infectious peritonitis, *Feline Pract.*, 13(5), 5, 1983.
46. Pedersen, N. C., unpublished observations, 1983.
47. Pedersen, N. C. and Black, J. W., Attempted immunization of cats against feline infectious peritonitis using either avirulent live virus or sublethal amounts of virulent virus, *Am. J. Vet. Res.*, 44, 229, 1983.
48. Pedersen, N. C., Black, J. W., Boyle, J. F., Evermann, J. F., McKeirnan, A. J., and Ott, R. L., Pathogenic differences between various feline coronavirus isolates, *Adv. Exp. Med. Biol.*, 173, 365, 1984.
49. Pedersen, N. C. and Boyle, J. F., Immunologic phenomena in the effusive form of feline infectious peritonitis, *Am. J. Vet. Res.*, 41, 868, 1980.
50. Pedersen, N. C., Boyle, J. F., and Floyd, K., Infection studies in kittens utilizing feline infectious peritonitis virus propagated in cell culture, *Am. J. Vet. Res.*, 42, 363, 1981.
51. Pedersen, N. C., Boyle, J. F., Floyd, K., Fudge, A., and Barker, J., An enteric coronavirus infection of cats and its relationship to feline infectious peritonitis, *Am. J. Vet. Res.*, 42, 368, 1981.
52. Pedersen, N. C., Ho, E., Johnson, L., Plucker, S., and Theilen, G. H., The clinical significance of latent feline leukemia virus infection in cats, *Feline Pract.*, 14(2), 32, 1984.
53. Pedersen, N. C., Evermann, J. F., McKeirnan, A. J., and Ott, R. L., Pathogenicity studies of feline coronavirus isolates 79-1146 and 79-1683, *Am. J. Vet. Res.*, 45, 2580, 1984.
54. Peiris, J. S. M. and Porterfield, J. S., Antibody mediated enhancement of flavivirus replications in macrophage-like cell lines, *Nature (London)*, 28, 507, 1979.
55. Peiris, J. S. M., Gordon, S., Unkeless, J. C., and Porterfield, J. S., Monoclonal anti-Fc receptor IgG blocks antibody enhancement of viral replication in macrophages, *Nature (London)*, 289, 189, 1981.
56. Pfeifer, M. L., Evermann, J. F., Roelke, M. E., Gallina, A. M., Ott, R. L., and McKeirnan, A. J., Feline infectious peritonitis in a captive cheetah, *J. Am. Vet. Med. Assoc.*, 183, 1317, 1983.
57. Poelma, F. G., Peters, J. C., Mieog, W. H. W., and Zwart, P., Infektiose Peritonitis bei Karakal (Felis caracal), Erkrankungen der Zootier 13th Int. Symp., Helsinki, 1974, 145.
58. Porterfield, J. S., Immunological enhancement and the pathogenesis of dengue hemorrhagic fever, *J. Hyg. (Cambridge)*, 89, 355, 1982.
59. Potkay, B. A., Bacher, J. D., and Pitts, T. W., Feline infectious peritonitis in a closed breeding colony, *Lab. Anim. Sci.*, 24(2), 279, 1974.
60. Siddell, S., Wege, H., and Ter Meulen, V., The biology of coronaviruses, *J. Gen. Virol.*, 64, 761, 1983.
61. Slausen, D. O. and Finn, J. P., Meningo-encephalitis and panophthalmitis in feline infectious peritonitis, *J. Am. Vet. Med. Assoc.*, 160, 729, 1972.
62. Theobald, J., Felidae, in *Zoo and Wild Animal Medicine*, Fowler, M. E., Ed., W. B. Saunders, Philadelphia, 1978, 658.
63. Toma, B., Duret, C., Chappuis, G., and Pellerin, B., Echec de l'immunisation contre la peritonite infectieuse feline par infection de virus de la gastroenterite transmissible du Porc, *Rec. Med. Vet.*, 155, 799, 1979.
64. Tuch, K., Witte, K. H., and Wuller, H., Feststellung der Felinen Infecktiosen Peritonitis (FIP) bei Hauskatzen und Leoparden en deutschland, *Zentralbl. Vet. Med. B*, 21, 426, 1974.
65. Ward, J. M., Morphogenesis of a virus in cats with experimental feline infectious peritonitis, *Virology*, 41, 191, 1970.
66. Ward, J. M., Smith, R., and Schalm, O. W., Inclusions in neutrophils of cats with feline infectious peritonitis, *J. Am. Vet. Med. Assoc.*, 158, 348, 1971.
67. Weiss, R. C. and Scott, F. W., Pathogenesis of feline infectious peritonitis: nature and development of viremia, *Am. J. Vet. Res.*, 42, 382, 1981.
68. Weiss, R. C. and Scott, F. W., Antibody-mediated enhancement of disease in feline infectious peritonitis: comparison with dengue hemorrhagic fever, *Comp. Immunol. Microbiol. Infect. Dis.*, 4, 175, 1981.
69. Weiss, R. C. and Scott, F. W., Pathogenesis of feline infectious peritonitis: pathologic changes and immunofluorescene, *Am. J. Vet. Res.*, 42, 2036, 1981.
70. Wolfe, L. G. and Griesemer, R. A., Feline infectious peritonitis, *Pathol. Vet.*, 3, 255, 1966.
71. Wolfe, L. G. and Griesemer, R. A., Feline infectious peritonitis. Review of gross and histopathologic lesions, *J. Am. Vet. Med. Assoc.*, 158, 987, 1971.

72. Woods, R. D. and Pedersen, N.C., Cross-protection studies between feline infectious peritonitis and porcine transmissible gastroenteritis viruses, *Vet. Microbiol.*, 4, 11, 1979.
73. Woods, R. D., Cheville, N. F., and Gallagher, J. E., Lesions in the small intestine of newborn pigs inoculated with porcine, feline and canine coronaviruses, *Am. J. Vet. Res.*, 42, 1163, 1981.
74. Zook, B. C., King, N. W., Robinson, R. L., and McCombs, H. L., Ultrastructural evidence for a viral etiology of feline infectious peritonitis, *Pathol. Vet.*, 5, 91, 1968.
75. Zook, B. C., King, N. W., Robinson, R. L., and McCombs, H. L., Ultrastructural evidence for a viral etiology of feline infectious peritonitis, *Pathol. Vet.*, 5, 91, 1968.

Chapter 8

CANINE DISTEMPER VIRUS*

Steven Krakowka, Michael K. Axthelm, and Gayle C. Johnson

TABLE OF CONTENTS

* This work was supported by grant No. 2R01-14821, NIH, PHS.

I. INTRODUCTION

Canine distemper virus (CDV) infection is a pantropic endemic and epidemic viral disease principally of the Canidae and their close relatives. It is world-wide in its distribution and manifests itself as an acute contagious disease with clinical signs of respiratory, gastrointestinal, and/or nervous manifestations.

Although the viral etiology was first demonstrated in 1905 by Carré and colleagues, significant advances in our understanding of the biology of this virus in its relevant host cell systems were not made until the virus was adapted to growth first in embryonated eggs and later in tissue culture. This advance permitted the development of serologic methods for monitoring the course and nature of the disease and, of course, permitted the subsequent development of killed and attenuated vaccine products.

Modern molecular biology and virology techniques have demonstrated the close relationship between CDV and other members of the *Morbillivirus* group, namely measles virus (MV) and rinderpest virus. Thus, in addition to its obvious importance within the veterinary community, the study of CDV infection in recent years is promoted via the linkage to MV and its subsequent value to comparative medicine and biology.

A number of excellent reviews describing canine distemper virus[1,2] or measles virus[3] are available for reference and study. Further, virtually every textbook on veterinary medicine addresses various aspects of the CDV-related disease. For these reasons a comprehensive yet concise review of all CDV literature is not feasible. Rather, the views and opinions of one laboratory's experiences with CDV infection in dogs are given here.

II. PHYSICAL PROPERTIES OF CANINE DISTEMPER VIRUS

Canine distemper virus is an enveloped virus. It belongs in the family Paramyxoviridae within the genus *Morbillivirus*. Included in this genus are MV and rinderpest virus.[1,3,4] The virus is a relatively large paramyxovirus and is composed of a pleomorphic envelope of host cell origin that surrounds an internal nucleic acid core. The virion varies from 100 to 300 nm in diameter. It is a single negative stranded RNA virus. The virus has been banded in cesium chloride and potassium tartrate. The mean buoyant density of the virion in these substances is 1.23 to 1.233 g/mℓ. Infectious CDV bands between 32 and 48% on sucrose gradients. The antigenic or polypeptide structure of the virus is given schematically in Figure 1. Two internal or core proteins are recognized.[3,5-7] The major core polypeptide is designated as the NP, or nucleocapsid protein. It is closely associated with viral RNA. A second internal or core protein is associated with a phosphorylase enzyme. It is designated by the letter P. Three major envelope glycoproteins are observed. The M protein (or matrix protein) is the smallest of the major proteins. The two major envelope glycoproteins are the so-called H protein (or hemagglutinating-equivalent protein) and the F (or fusion) glycoprotein. The fusion glycoprotein is rapidly degraded into smaller subunits designated F_1 and F_2, respectively. The H protein is responsible for viral adsorption to target cells. The fusion glycoprotein is responsible for cell-to-cell fusion. This is also referred to as "fusion factor". As far as is known, all distemper virus isolates contain these common viral polypeptides. MV, the *Morbillivirus* closely related to CDV, has been studied extensively. It, too, contains the same viral polypeptides.[3] There are several differences, however. MV will hemagglutinate certain primate species' erythrocytes, whereas CDV does not hemagglutinate any species' erythrocytes. Although these viruses are extensively cross-reactive, there is a difference in the immune response to the H protein in primate vs. the canine species.

Canine origin antiCDV antibody will precipitate only CDV-H protein. In contrast,

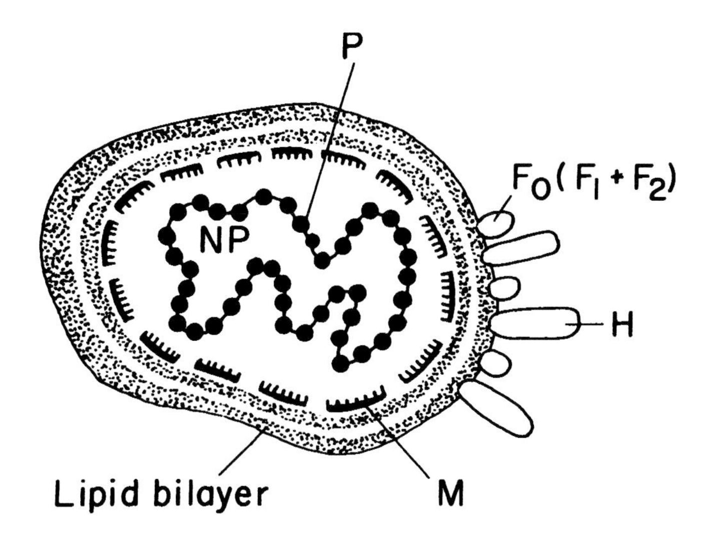

FIGURE 1. Schematic diagram of the canine distemper virus virion. NP = nucleoprotein, P = phosphorylase protein, M = matrix protein, H = hemagglutinin-equivalent protein, F = fusion glycoprotein.

human antiMV antibody will precipitate both CDV-H and MV-H polypeptides. The failure of canine origin antibody to bind to MV-H is the basis of the heterotypic vaccination procedure used in young animals possessing passively acquired antiCDV antibody.

Canine distemper virus, being an enveloped virus, is quite labile. Agents such as heat, UV light, and lipid solvents are all known to readily inactivate the viral infectivity. The virus is stable to lyophilization; however, it is known that some loss of titer, usually about 1 log (base 10), is lost during the lyophilization process. Virtually all of the commonly employed chemodisinfectant substances will inactivate the virus. For practical purposes, this means that decontamination of environment is quite easy.

III. CULTIVATION OF THE VIRUS AND IN VITRO HOST CELL RANGE

Canine distemper virus is a somewhat difficult virus to propagate in vitro. It is difficult to isolate from affected animals with any regularity, and it is exceedingly difficult to propagate the virus in its virulent form in vitro. However, once adapted to tissue culture, the virus can be readily transmitted to other cells in culture and can be readily propagated thereafter. The most reliable method for in vitro growth of virulent CDV is by use of primary canine pulmonary macrophage system. Appel and colleagues have shown that virulent virus will readily infect primary cultures of canine pulmonary macrophages.[8] Recently, similar findings were reported using ferret origin peritoneal macrophage cultures.[9]

A second cell culture type shown to be susceptible to virulent virus inoculation is the bovine fibroblast primary culture system reported by Metzler and colleagues.[10-12] He showed that inoculation of these cells with tissue suspensions containing virulent CDV resulted in the recovery of virulent virus as a persistent noncytolytic infection within these cells. Subsequent study has shown that virus from these bovine fibroblast cultures could be transferred to other continuous cell lines by co-cultivation methods.[13] A third culture system, primary monolayers of canine bladder epithelium, has also been used to isolate virulent CDV.[14]

An alternative method for initial isolation of canine distemper virus is to prepare

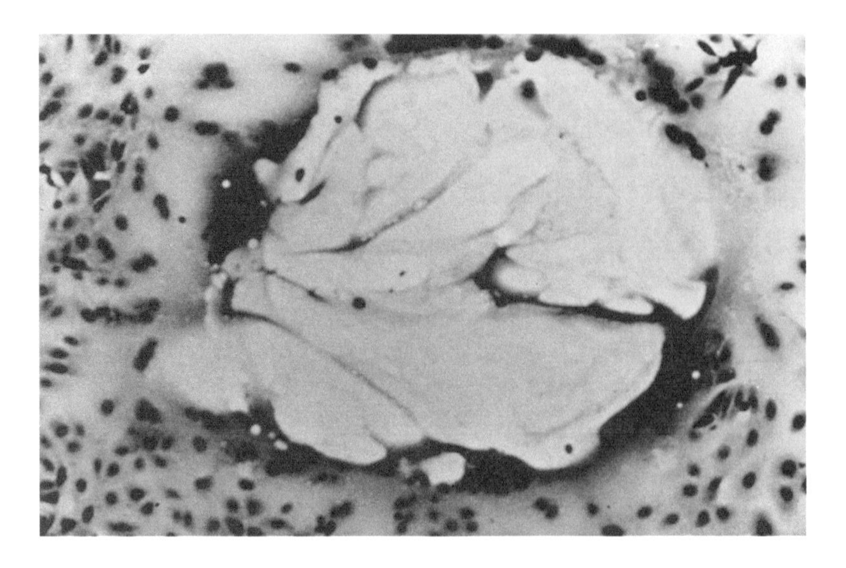

FIGURE 2. Multinucleated syncytial giant cell formation in a Vero cell monolayer induced by CDV.

primary cultures from infected tissues.[12,13] This explant technique is quite successful if tissues are explanted from animals in early or acute phases of disease. As time goes on, it becomes more and more difficult to recover virus from these cases, even though viral antigen and inclusion bodies may well be present in tissues. The reasons for this are unknown.

Avirulent or tissue-culture adapted canine distemper virus has been propagated in many different cell culture systems, including cells of avian, mustelid, dog, human, feline, and simian origin.[1] The easiest way to transfer infection from either the primary canine pulmonary macrophage or the explanted tissues to these cells is by performing co-cultivations of infected cultures with standard tissue culture cell lines.[13,14] For this a 1:1 ratio of cells is used, and cells are then handled as one would normally handle the primary cell culture lines. The cultures are screened for viral antigen by methods discussed below. Assuming that virus has been adapted to the cell line in question, the following series of general statements can be made. Virus readily absorbs to target cell monolayers; in fact, peak absorption probably occurs within 1 hr and is essentially complete by 4 hr. In our laboratory we routinely use 1-hr absorption. A number of investigators have published growth curves documenting both one-step and multistep replication of virus in tissue culture.[1] For most studies, investigators have shown that free infectious virus is released into the supernatant somewhere between 24 to 36 hr after inoculation.[13] Peak viral infectivity in supernatants is generally observed between 3 to 5 days after infection. In standard lytic infections, infectivity in supernatants always exceeds that associated with the cells. Canine distemper virus will produce cytopathic effects (cpe) when inoculated onto tissue culture cells. There are several distinct features of this in vitro virus-host cell interreaction. The most obvious cpe is the formation of multinucleated syncytial giant cells in culture (Figure 2). Fusion is mediated between adjacent infected cells by the envelope membrane associated F protein. Coincident with this is the appearance of both intracytoplasmic and intranuclear eosinophilic inclusion bodies (Figure 3). It should be noted that these inclusion bodies can best be seen when culture monolayers are stained with hematoxylin and eosin. Accompanying syncytia formation is host cell strand formation and eventual lysis.

Viral cytopathology is more obvious and more dramatic in young, actively growing cell cultures. The extent and duration of viral cpe depends upon the state or composi-

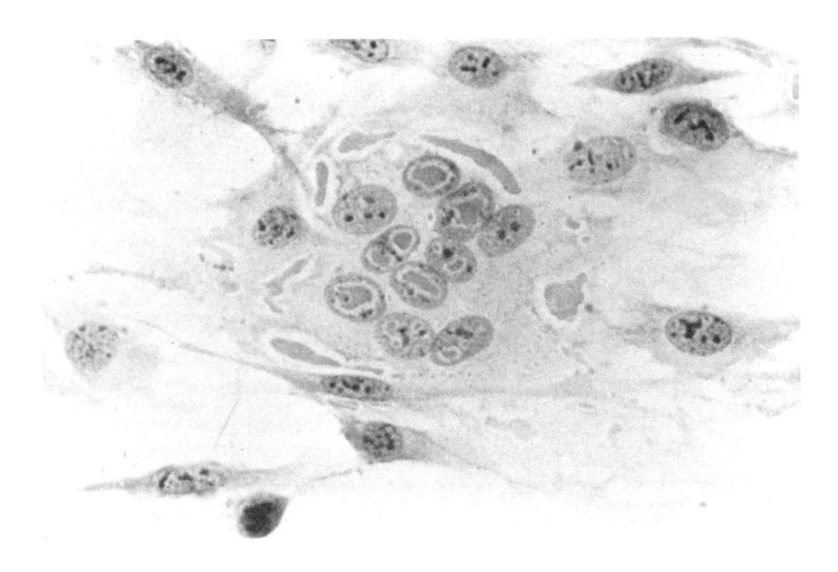

FIGURE 3. Multinucleated syncytial giant cells of explanted glial cell origin illustrating intracytoplasmic and intranuclear viral inclusion bodies.

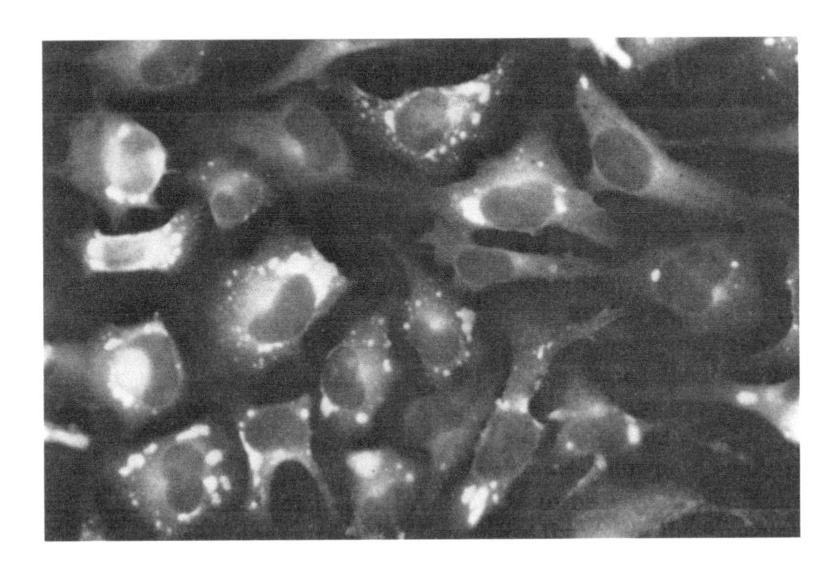

FIGURE 4. Persistent noncytolytic CDV infection in mink lung cell monolayers. Cytoplasmic viral antigen is stained by indirect immunofluorescence methods.

tion of the original viral inoculum. For example, stock inocula have been shown to contain more than one plaque type,[15,96] and this distinction has been made on the size of the multinucleated syncytial giant cell produced. Thus, plaque variants, large and small, exist, and the ratios of the same can affect the overall cpe. In addition to this problem, there exists in stock inocula viral variants that replicate without an overt viral cpe. In vitro this is recognized as a persistent infection.[12,16-18,97] Persistent infections (Figure 4) will contain intracytoplasmic and in some instances, intranuclear inclusion bodies of viral material, but monolayers will lack other manifestations of viral cpe. The morphologic appearance and growth characteristics for these infected cells are virtually identical to those noted in uninfected control cultures. Although the mechanisms for generation and maintenance of this in vitro persistent infection have not been determined, it is likely that either defective interfering particles and/or viral mutants, possibly deficient in functional fusion glycoprotein, can best be postulated to explain this result.

Another variable affecting viral cpe is the source, origin, or type of viral strain. There are a number of different CDV strains.[1] A partial list includes Snyder Hill, Green, Onderstepoort, the Rockborn strain, the R252-CDV, and the Cornell A75-17 strains. Although no exhaustive comparison of these viruses has been made, investigators have noted that these strains may vary in the amount of virus released, presence or absence of large syncytia, etc.

Laboratory strains of CDV have been plaque purified in cell culture.[19] In our laboratory, plaque assays for CDV in Vero cells have been adapted for use with all isolates except R252-CDV.[98] The method employs a methylcellulose overlay (1.0%) and visual determination of plaque-forming units on monolayers subsequent to staining of the monolayer with neutral red. Repeated attempts to plaque purify R252-CDV have thus far been unsuccessful. For most routine work, viral titrations are accomplished by the $TCID_{50}$ method on Vero cells or other appropriate cell culture monolayers. Propagation of virus should ordinarily start with multiplicities of infection of between 0.1 and 0.001.

IV. NATURAL HOST RANGE OF CANINE DISTEMPER

Canine distemper virus is an infectious disease of the Carnivora. Within the order Carnivora are a number of families that have been shown to be susceptible to CDV. Prominent among them are the Canidae: dogs, dingos, foxes, etc. Two other families of great importance are the Procyonidae, including the raccoons, kinkajous, the lesser panda, the Mustelidae that include the species ferret, mink, skunk, badger, and the Hyaenidae. There is some controversy over whether cats or other members of the felid family can be affected with canine distemper. Part of this confusion stems from the designation of feline panleukopenia as feline distemper in the lay literature. Concerning the domestic cat, no adequately documented case of naturally acquired infection with CDV has been published. Experimentally, newborn 6 to 8-week-old and adult cats were infected with Snyder Hill strain CDV.[20] The cats experienced limited replication of the virus, but did not develop significant clinical disease nor was viral material shed into excretions and secretions. The authors concluded that the cat is unlikely to be affected with CDV under natural conditions. There are some very old and poorly documented reports of distemper virus infection in captive large cats.[1] Recently, however, an inclusion body encephalitis has been described in tigers.[21,22] Although distemper virus was not isolated from these animals, at least one animal showed a rising titer to CDV during infection.[22] Formalin-fixed brain tissue from both animals stained positively for CDV antigen (Figure 5), using an indirect immunofluorescence procedure.[97] The role of distemper virus, or closely related paramyxoviruses, in infectious diseases of large captive cats must be further investigated.

Of the commonly employed laboratory animal species, perhaps the most work has been done with murine-adapted CDV.[1,2] Adaption of virus to mice is best accomplished by serial inoculation of suckling mice with suspect viral suspensions.[23] Most investigators adapt CDV to mice by intracerebral passage in 1- to 2-day-old suckling animals. In most instances the pattern of resultant infection is characterized as a monophasic acute encephalopathy with mild meningitis and focal to multifocal areas of confluent zones of necrosis within the brain.[23] This phenomenon is age-dependent.[24] Adult mice seem to be quite resistant to replication of even murine-adapted virus. Weanling mice, however, show an intermediate pattern of development. Approximately 40% of weanling mice inoculated with murine passaged-CDV die acutely. The

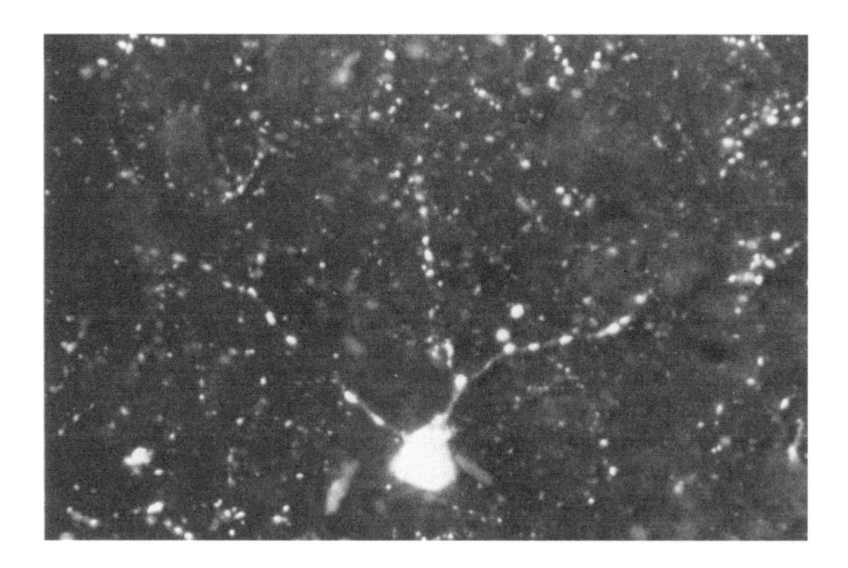

FIGURE 5. The demonstration of CDV antigens in the brain of a captive tiger affected with an inclusion body encephalitis.

remaining convalescent animals may live for long periods of time. The consequences of this infection are some neurologic deficits and at least two reports of an obesity syndrome in surviving affected animals.[25,26] It must be noted that there is also a mouse-strain dependent influence operable here. Resistant strains of mice, most notably the SJL strain, develop a slowly evolving encephalitis[25] rather than the acute fulminant disease noted in susceptible strains of mice.

Similar studies have been performed in hamsters;[27-29] however, an important additional observation has been made, namely that the neurovirulence potential of virus is somewhat dependent upon the plaque type of the inoculum. Previously reference was made to large and small plaque types of CDV. Recently investigators have inoculated these two different plaque variants into hamsters and have shown that the small plaque variant of tissue culture-adapted CDV is neurovirulent for the hamster, whereas the large plaque variant is not. This has a number of implications in not only the design and development of in vitro experiments, but also in terms of interpreting the mechanisms of viral encephalitis.

Finally, to complete the story of CDV infection in exotic (i.e., noncanid) species, the suggestion has been made that CDV by virtue of its similarity to MV may be infectious for primates, including man. It is known that primates are susceptible to experimental inoculation with virulent CDV[30] and that central nervous system (CNS) lesions produced mimic those noted with MV infection in either primates or man. Recently, it has been suggested that CDV, as a zoonotic infection, is involved in the etiology of multiple sclerosis (MS), a debilitating CNS demyelinating disease of unknown etiology.[31] A number of subsequent serological and epidemiologic studies have not supported this hypothesis,[32,33] nor has CDV ever been recovered from MS tissues. Nonetheless, reasonable caution should be exercised when handling this virus because there is a potential for replication in the primate species.

V. PATHOGENESIS OF INFECTION IN THE NATURAL HOST (DOGS AND/OR FERRETS)

The pathogenesis of acute infection of CDV in a fully susceptible host is probably identical regardless of species. There is general agreement that, under natural conditions, the most likely, obvious, and important route of infection to the host is by infective aerosol droplets. Thus transmission of the virus is facilitated by coughing and sneezing and close confinement in a warm, humid, closed environment. Initial studies with ferrets and aerosol exposure suggested that viral infection occurred first in the respiratory epithelium of the upper oronasal tract with subsequent spread to the deep pulmonary parenchyma.[1,2] Later studies, however, suggested that the epithelial components of the body are secondarily infected. There is little disagreement today that the initial events in the pathogenesis of distemper in either ferrets or dogs are similar and occur by the following scheme. Infective aerosols are inhaled into susceptible dogs. Tissue macrophages and monocytes located in or along respiratory epithelium and tonsils appear to be the first cell type to pick up and replicate CDV. Following a local burst of virus production in these sites, the virus is then spread by lymphatics and blood to distant lymphoreticular tissues. This is accomplished by viremia (Figure 6 A, B) and occurs anywhere from 2 to 4 days after initial infection. Viremia was formerly thought to be exclusively leukocyte-associated[1] and the initial phases of the disease restricted to blood origin monocytes. Subsequent study has shown that by at least 7 days after infection and probably sooner, significant quantities of free infectious virus appears in the plasma of CDV-infected dogs.[34] Nonetheless, between 8 and 9 days after infection, the virus will spread beyond lymphoreticular tissues to involve epithelial and mesenchymal cell elements.[35] This probably occurs in every animal regardless of the outcome of the disease process. It is at this stage of the process that specific host immune responses to viral antigens influence the outcome of disease.[1,2,35-37] At least one of three possible clinical syndromes can result from infection.[38] The acute fatal or fulminant form is characterized by unrestricted viral spread to virtually every tissue in the body. Virus can be found in every excretion and secretion and, by using IF methods or other antigen tracing techniques, antigen can be demonstrated in virtually every cell type within the body. For most of these animals, the most likely proximate cause of death, if uncomplicated by secondary bacterial infection, is fulminant fatal neurologic involvement.

A second group of animals exhibits clinically delayed progression of disease and modest convalescent immune responses. Clinical signs, if present, are subtle early in the disease and are a reflection of viral persistence within the CNS. Subsequent development of overt CNS disease is variable. The third group of animals exhibits essentially no overt clinical signs of disease and is recognized as convalescent, clinically normal animals.

The actual mechanism of viral spread from the portal of entry to distant sites within the body has been the subject of considerable study in recent years. As mentioned previously the spread of virus is thought to occur by a leukocyte-associated viremia. Virus can be detected as early as 2 to 3 days after infection in the blood of animals. Recent work from our laboratory has investigated in detail the mechanisms of spread of virus from the blood vascular system to extravascular tissue components, notably components of the brain.[39] Using very sensitive immunocytochemical viral antigen tracing techniques, we have shown the first component of the CNS that experiences viral infection is cerebrovascular endothelium and not the choroid plexus as previously thought. Infection of endothelium is likely achieved by contact with cell-free or platelet-IgG-virus complexes. As a result of this platelet-virus-endothelium or virus-endothelial interaction, the subadjacent tissues, whether it be brain or other tissues, acquire the infection. This occurs by one of two mechanisms. Either virus infected endothelial cells fuse with adjacent uninfected parenchymal cells and thereby transmit the infection, or infected endothelial cells contract thereby producing localized areas of enhanced vascular permeability. Subsequent to this increased vascular permeability, virus-infected leukocytes adhere to damaged endothelium or subendothelial collagen and

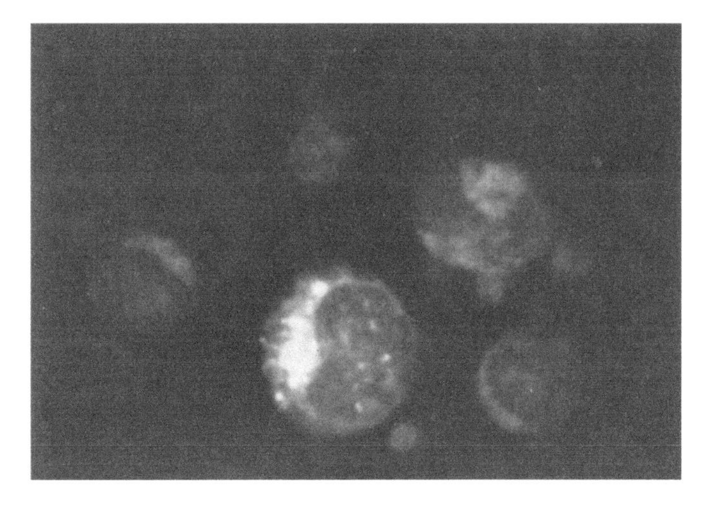

A

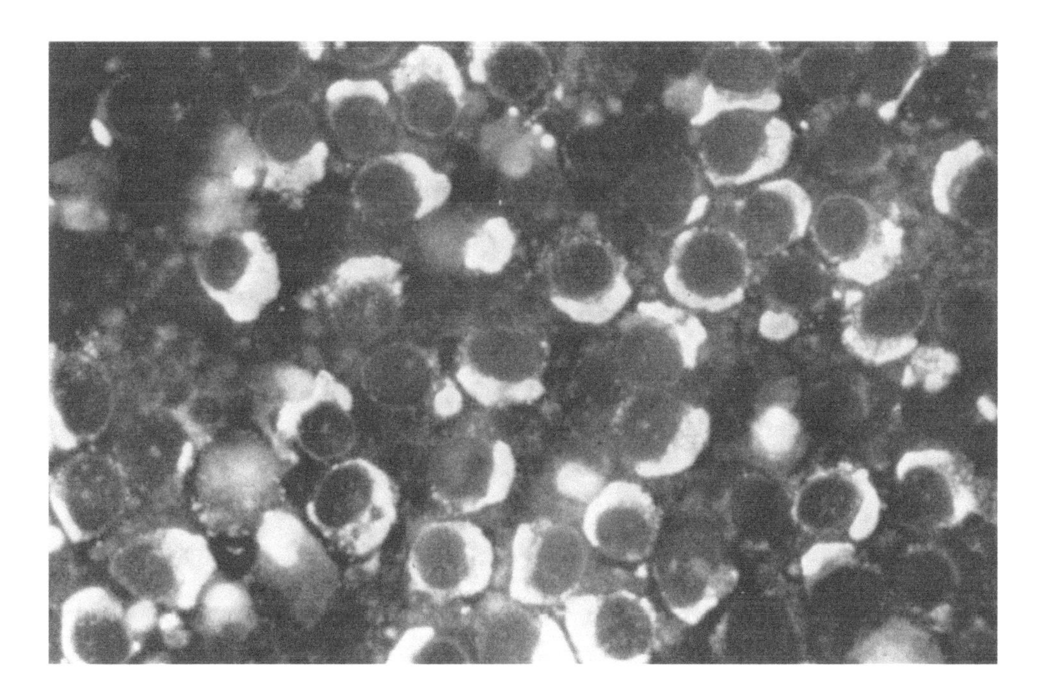

B

FIGURE 6. (A and B) Demonstration of CDV antigen by direct immunofluorescence in isolated peripheral blood mononuclear leukocytes in a dog with acute systemic distemper.

then emigrate into extravascular locations. This hypothesis is schematically illustrated in Figure 7. In any event, all animals regardless of the eventual state of their clinical immunity, do go through an extravascular phase of infection.

As mentioned previously, transmission from dog to dog is principally by infective areosols. Other routes of transmission are possible. Transplacental transmission of virus has been documented in this laboratory.[40] The sequence of events in in utero-infected puppies presumably occurs by the same pathways as outlined above.

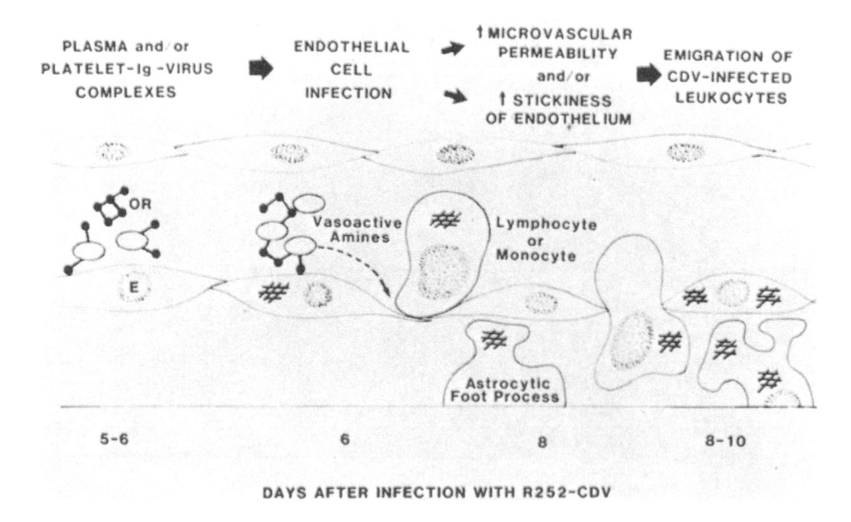

FIGURE 7. Proposed entry of virus into the brain of a CDV-infected dog.

VI. IMMUNE RESPONSES TO CANINE DISTEMPER VIRUS

Both humoral and cellular immune responses are thought to play roles in recovery of dogs or ferrets from CDV infection. In addition, there are nonspecific host defenses including intact mucosal surfaces, an active phagocytic system, high plane of nutrition, etc. that also contribute to resolution of the disease process. Fortunately, all laboratory and wild strains of CDV appear to share common envelope and core antigens and are thus extensively cross-reactive. This feature of CDV facilitates serological studies and simplifies the development of safe and effective vaccine products.

A. Humoral Immune Responses

Actively infected dogs destined to recover from CDV infection will demonstrate free circulating antiviral antibodies on or about PID 6 to 7.[35,36] Titers rise rapidly to high levels in early convalescence. A number of different systems have been used to measure antiviral antibody responses in dogs. For historical purposes, gel diffusion, complement fixation, and indirect immunofluorescence methods have been employed. By far the most accepted method for measuring antibody to distemper, is the virus neutralization test. In the early days, workers used CDV-susceptible animals for test purposes. This was obviously cumbersome and expensive. The adaptation of CDV to embryonated eggs facilitated the development of egg protection tests for CDV antibody.[1,2] For many years the ability of serum from convalescent CDV-infected animals to neutralize or reduce plaque formation on amniotic sacs of embryonated eggs was considered the serologic test of choice. With the subsequent implementation of tissue culture methodology, including the adaptation of the virus to growth in a number of different cell lines, virus neutralization capabilities then moved to the tissue culture laboratory. The Cornell group deserves much of the credit for the development of the microtiter virus neutralization assay.[1,8,41] This procedure has been standardized to test the ability of 100 μl of serum to neutralize 100 $TCID_{50}$[42] Onderstepoort CDV contained in 100 μl of media. After 1 hr incubation at 37°C, the mixture of virus and diluted antibody are then inoculated onto cell monolayers in microtiter plates and incubated at 37°C for 3 to 5 days. Plates are then fixed and stained and viral cpe is determined by light microscopy.

Recently, the concept of measuring distemper virus antibody using enzyme-linked immunosorbent assays (ELISA) has been introduced.[43,44] These test procedures have the advantages of being rapid, economical, reproducible, and can provide the investigator or the diagnostic laboratory with the ability to process large numbers of samples at one time. Further, with ELISA it is possible to determine class-specific immune responses to distemper virus antigen, and thus get an indication of time of exposure to CDV infection.[45]

It is also possible to determine the immune response to individual virion polypeptide antigens using a RIP-PAGE procedure.[7,46,47] For this, virus is radiolabeled with S^{35}-methionine. After harvest and solubilization, soluble nonaggregated labeled viral antigen is mixed with serum antibody. Antibody-bound radiolabeled antigen is separated from unbound antigen by use of solid phase immunoabsorption, usually with Cowan Strain *Staphylococcus aureus* protein A immunoabsorbent. Samples are then solubilized and electrophoresed on polyacrylamide gels. Autoradiographs are then made from the gels and the presence or absence of binding to each of the five major polypeptides can be determined by examining the autoradiographs. In this manner, the presence or absence of a humoral immune response to each of the virion polypeptides can be determined. Using this technique[46] it was shown that acute, fatally affected dogs produce no detectable antibodies to envelope glycoproteins and only trace amounts of antibody to core polypeptides. Further, it was shown that dogs at risk for developing delayed or chronic onset distemper virus-associated encephalitis did not produce adequate amounts of antibody to the envelope glycoproteins, whereas their immune response to core polypeptides was adequate.

In any event, regardless of the methods used to measure humoral immune responses, it is clear that an animal, once vaccinated or subclinically infected, recovers from the infection and develops a lifelong immunity. Of course the precise role of this antibody in protection of animals to subsequent exposure to CDV has not yet been determined in great detail. It is thought, however, that antibody serves to neutralize extracellular virus alone or with complement. This opsonizing function serves to prevent widespread reinfection within the immune host and thus serves as a mechanism of immediate protection to external reinfection.

B. Cellular Immunity

Several recent studies have examined the role of the virus-specific activated T-cell in recovery from and/or resistance to CDV infection. Essentially two types of assays have been used. The first series of studies demonstrated that lymphocytes from immune dogs (but not controls) were capable of inhibiting the formation of virus specific cpe, namely syncytia formation on a monolayer within a 24 hr incubation period.[37,48,49] This syncytia inhibition response was rapid, dependent upon viable effector cells, independent of circulating antibody, and was not histocompatibility or DLA (dog leukocyte antigen) restricted. Dogs expressing this syncytia inhibition response, when challenged with virulent CDV, demonstrate a rapid rise to high levels of activity in convalescence, and this response persisted for at least 45 days after rechallenge.[49]

The second type of assay used is a short-term cytotoxicity assay. For this, autologous and heterologous CDV-infected target cells labeled with 51-chromium, were incubated with effector T cells, and cells functioning in antibody directed cellular cytotoxicity (ADCC) reactions in various lymphocyte:target cell ratios.[50] Essentially the same results were obtained, namely that the cytotoxic T cell response following vaccination with a modified live product was shortlived (on the order of 1 to 3 weeks). Recall was rapid, and the cytotoxic response in dogs convalescing from virulent CDV infection was much more dramatic and much more prolonged. Both of these investigations concluded that cytotoxic T cells do play a role in recovery from CDV production. It is likely that cytotoxic T cells kill, lyse, or otherwise inactivate CDV-infected target cells in vivo, thereby decreasing the internal source of infection of virus.

C. Other Virus-Associated Immune Responses

It's important to mention the role of the canine natural killer cell (NK cell) in canine distemper virus infection. In both the studies cited above, preinoculation or preexposure low levels of syncytia-inhibition and cytotoxicity effects were noted. Subsequent study has revealed that this preexistent cellular immune response can be attributed to CDV-NK activity in the effector cell populations.[50,99] The canine NK natural killer cell is recognized as a cell of T cell lineage, i.e., it bears a Thy-1 surface antigen. It also bears an Fc receptor for IgG, but is negative for IgM receptors, complement receptors, and E-rosette markers.[99] Careful study has revealed that NK activity can be detected in preexposure samples as indicated, and the level or amount of NK dependent activity is somewhat age-dependent. Young or neonatal dogs lack, or are deficient in, NK activity. Adult levels in the dog are reached within 3 to 4 weeks of age. Although definitive in vivo data is lacking, it is presumed that this NK response, most likely modulated by virus-induced interferon production,[51] represents a third and integral part of host defenses against acute CDV infection.

VII. CLINICAL, IMMUNOLOGICAL, AND PATHOLOGICAL PHASES OF THE DISEASE

A. Epizootiology

In spite of the widespread use of safe and effective attenuated viral vaccines, canine distemper infection remains one of the most serious and important viral diseases of dogs. Not only is the disease itself fatal, but the virus, through its immunomodulating effects, can potentiate secondary lethal infection. On a population basis it is known that a seasonal prevalence to the disease occurs.[2] Most studies have shown that peak incidence occurs at two time periods of the year, namely late spring and again in early fall. This coincides with the successive introduction of crops of young, CDV-susceptible dogs into the canine population. While it has been shown that all ages of dogs are susceptible to infection if prior exposure cannot be documented, there is no doubt that the disease is much more severe in younger animals. Though less well-documented, it is likely that the disease as either epizootic or endemic forms is much more frequent in environments or geographical areas that experience high humidity and moist conditions, both of which tend to favor virus survival within the environment. In rural areas, it's important to consider the role of wildlife species, chiefly raccoons, skunks, foxes, and coyotes in maintenance of CDV within these populations between canine epizootics. Due to the inordinate sensitivity of wild species to CDV, most investigators would suggest that the infected dog is a greater danger to these species rather than wild populations serving as a reservoir source for infection to the dog population.

The question of the carrier state for canine distemper virus has not yet adequately been addressed. It is generally felt that an immune animal will not shed virus to its environment or to other CDV-susceptible animals. However, several lines of evidence suggest that this is an oversimplification. By the explant technique, CDV has been recovered from primary kidney cell cultures, derived from apparently healthy dogs.[52] On several occasions in our laboratory, gnotobiotic litters, derived from apparently healthy unconditioned pregnant bitches, have developed fatal CDV infection postnatally.[40] In the absence of postnatal exposure (as excluded via the isolation units), the source of infection was tentatively ascribed to the subclinical or carrier state in the bitches, whose expression was facilitated by either the stress of pregnancy or natural immune cyclicity or circadian rhythm phenomena.[53,54]

The third epidemiologic piece of evidence that tends to suggest that distemper virus can persist in its natural host long after the acute clinical episode is the rare development of old dog encephalitis (ODE) and related clinical syndromes in adult dogs.[55,56] While this aspect of CDV disease will be discussed in detail below, ODE as a clinico-

FIGURE 8. A young mixed-breed dog with acute systemic canine distemper virus infection.

pathologic entity is known to occur in adult dogs, often with complete vaccination and/or immunization histories. It is thought, though certainly not proven, that these dogs harbor virus for years prior to clinical expression.

The final piece of evidence relates to an interaction noted almost as a clinical anecdotal entity here at the Ohio State University. In recent years there has been an increase in the number of adult dogs presented to the clinic with CDV-associated chronic encephalomyelitis.[97] Many of these dogs have experienced or are experiencing concurrent canine parvovirus (CPV) infection. Expression of CDV seems to follow recovery from CPV-related enteric disease. Again, although there are many explanations for this phenomenon, the data suggest, particularly in light of evidence discussed below, that CPV reactivates latent canine distemper virus in these immune animals.

B. Clinical Features of CDV Disease

At the outset it must be stated that the clinical presentation of dogs infected with distemper is extremely variable. There exists currently no reliable set of clinical criteria that would exclude canine distemper virus from the etiologic diagnosis in an ill dog. For clarity, it's useful to describe the clinical presentation in one of two forms: an acute systemic form and a neurological form.

1. Clinical Features: Acute Systemic Form

Dogs (Figure 8) affected acutely with CDV show variable degrees of depression, anorexia, and fever. Their skin may be variably dehydrated, dry, roughened, and inelastic. A proportion of these animals will show photophobia, and some will have evidence of mucopurulent ocular-nasal discharge. Intermittent diarrhea is a common clinical sign. During this acute viremic phase of the disease, virus is shed in every excretion and secretion. As the disease progresses, pneumonia frequently due to secondary bacterial invaders, may develop. Clinicopathologic findings are unrewarding. Dogs in this stage of disease are moderately to severely lymphopenic. They frequently exhibit a coincidental leukocytosis and left shift, depending on the degree or amount of secondary infection. There are minor and nonspecific changes in serum protein as detected by electrophoresis. Of course, since the animals are viremic, a direct immunofluorescence test on leukocytes collected from these animals will demonstrate viral antigen. This confirms the diagnosis. It is important to emphasize that not every dog acutely affected with canine distemper will exhibit even these signs. Frequently the disease can be entirely subclinical.

2. Clinical Features: Nervous Form

There are essentially two clinical syndromes that present with CNS signs associated with CDV infection. For convenience, neurological forms can be further subdivided into those cases with CNS signs that occur and are accompanied by other systemic signs of CDV infection as indicated above, and those cases in which neurologic signs are unaccompanied by systemic signs of CDV infection.

Although acutely affected dogs can show virtually any combination of neurological signs, in its most common presentation, the dog presents in petit mal or grand mal seizures. These convulsive episodes occur over time and with increasing frequency. Neurological signs in these cases are quite complex and include disorders in cranial nerves, disorders attributable to meningitis, and signs attributable to diffuse cerebral disease, i.e., confusion and head pressing. A cerebellar form in which incoordination and instability may be seen, and a spinal cord form in which paralysis or paresis is a predominant presenting sign may also occur. Variable numbers of the dogs are photophobic and a few, in fact, are blind, a reflection of CDV-induced retinal and optic nerve damage. A characteristic neurologic sign is hyperkinesia or rhythmic tonic-clonic motor movements. This tic, or chorea, as it's called, is characteristic and may persist into apparent convalescence.

The second neurologic form of canine distemper is that which occurs with ODE, or occurs after subclinical infection and apparent recovery. Again, the CNS signs can be extremely varied in presentation and can be mistaken for brain tumor, head trauma, bacterial meningitis, hydrocephalus, spinal cord disc disease, etc. These cases are extremely difficult to attribute to CDV infection without necropsy and histopathologic examination for the following reasons. First of all, the dogs are long since past the viremic phase of the disease, so it is no longer possible to demonstrate viral antigen in leukocytes, excretions, or secretions. Second, since the presentation is so varied, other causes of neurologic disease in the dog must also be considered. Perhaps the best way to make the diagnosis of canine distemper in these cases is by examination of the cell-free and blood-free cerebral spinal fluid for the presence of antibodies to CDV.[35] Usually, but not invariably, these cases exhibit local antibody production within the CSF. The presence of antibody to CDV in this compartment is indicative of local active antibody synthesis and infers that there is a local active infection somewhere within the CNS.

C. Immunosuppression Associated with CDV Infection

Many of the signs of acute systemic CDV, e.g., diarrhea, fever, and malaise, are attributable to secondary and/or concurrent infection with various secondary bacterial, mycotic, and viral pathogens.[1,2] Many of the systemic manifestations of CDV can be controlled by appropriate antibiotic or supportive therapy, and the nonneurological signs are completely absent in CDV-infected gnotobiotic dogs.[57-59] Thus a major non-neural manifestation of CDV infection in dogs is CDV-associated immunodepression.[37,60] As mentioned previously, many of the signs of canine distemper virus infection are attributable to coincidental secondary infectious processes occurring in this debilitated animal. It has been known for many years that CDV infection is frequently complicated by bacterial and/or other viral diseases. On a clinical basis, veterinarians for years have spoken of the anergy of infection associated with this disease. Probably the chief and most important bacterial pathogens associated with disease in dogs are the pneumonic bacterial species including: *Bordetella bronchiseptica, Pasturella* spp., and of course, *Staphylococcus* and *Streptococcus* spp. These agents are responsible for the purulent conjunctivitis, rhinitis, and bronchopneumonia noted clinically in CDV-infected dogs. Mixed viral infections, chiefly of the respiratory type, are also common. In addition to canine adenovirus II infection, reovirus, canine parainfluenza virus, and presumably other viruses such as canine herpesvirus are all involved in this dual or multiple mixed infection problem.

As the means for successful chemotherapy have improved, mortalities attributable to secondary bacterial pneumonias, enteritis, etc. have declined dramatically. This is a mixed blessing in that a portion of these treated dogs eventually will develop acute or chronic neurological manifestations of disease. Nonetheless, in the acute phases of infection, treatment should certainly be directed toward supportive therapy, including fluids, warmth, reduced light, provision for and inclusion of adequate nutrition, and of course the prophylactic, judicious use of systemic broad spectrum antibiotics. None of these measures will materially affect the course of the viral infection, but will certainly limit the consequences of infection by the secondary invaders.

In examining this problem of secondary coincidental infection further, it's quite apparent that the major and most likely mechanism that would permit infection by these opportunists is via a direct or indirect viral effect upon the host immune system. This immunomodulating effect is a significant and important component of this disease. Mention has been made already of the lymphopenia,[58] noted as a hematological finding. In recent years, CDV-associated immunosuppression has been documented using a variety of different in vitro and in vivo assays.[37,60,61]

As a consequence of direct viral infection either in lymphoid cells and/or macrophages, lymphocytes from CDV-infected dogs are rendered incapable of generating effective in vitro and in vivo immune responses. The early virolytic effects of CDV upon the lymphoid system and macrophages also have the effect of suppressing established immune responses and/or normal host defenses. Immunosuppression is not simply due to a direct virolytic effect, as immunosuppression persists long after virus can no longer readily be demonstrated in lymphoreticular tissues.[60] Other explanations for this prolonged immunosuppression have been sought and can be summarized by the following. For months after infection, dogs exhibit delayed and diminished in vitro responses to plant lectins such as phytohemagglutinin and pokeweed mitogen.[60] Further, in quantitative immunization studies it has been shown that apparently normal convalescent dogs responded less well to an artificial immunogen than did their uninfected control litter mates.[37] Lymphocytes collected from immunosuppressed animals were shown to be capable of suppressing the in vitro proliferative responses to lectins in uninfected control animals.[61] This suppressor effect was tentatively attributable to a T cell suppressor lymphocyte subset, and the current thought is that as a consequence of CDV infection, the suppressor cell network is activated. The resultant suppression is, by definition, nonspecific and is capable then of suppressing *de novo* responses of lymphocytes to environmental and/or other immunogens. Paradoxically, however, the effect of CDV upon natural killer cell responses to various target cell systems including CDV-infected target cells appears not to be affected by acute viral disease.[99]

VIII. GROSS AND HISTOPATHOLOGIC LESIONS INDUCED BY CANINE DISTEMPER VIRUS

A. Gross Pathology

Relatively few specific changes indicative of CDV infection are noted upon gross examination of affected dogs. If the dog is young enough to have a thymus, i.e., less than 3 to 6 months of age, then one finding is thymic atrophy.[57,58] Normally, in a young animal, the thymus extends from the thoracic inlet to the base of the heart in the anterior mediastinum. In acute fatally affected dogs the thymus is reduced, often to microscopic remnant. The other virus-specific lesion observed is patchy or multifocal interstitial pneumonia.[1,2] This is recognized as tanned depressed areas, usually around main stem bronchi and hilar regions of the lung. In acute fatal cases, of course, secondary bacterial invaders can produce a suppurative bronchial pneumonia as well as a nonspecific pustular dermatitis. Mucopurulent nasal discharge and catarrhal to hemorrhagic enteritis may also be observed.

B. Microscopic Changes

Histopathologic changes are many and varied and have been described by a number of authors. For convenience it is useful to divide the histopathologic findings into two general categories: extraneural findings and neural findings. In cases of acute fatal systemic distemper, lymphoid tissues show systemic lymphoid depletion and accompanying reticulum cell hyperplasia.[37,58] There is severe reduction in the thickness of the cortex of lymph nodes that is caused by an absolute reduction in the number of lymphocytes. Additional lesions in acute cases include: focal to multifocal areas of lymphoid necrosis and subtle and occasional syncytial giant cell formation in the node. In addition, the subcapsular sinus areas are devoid of recirculating T lymphocytes. Underlying this lymph nodal depletion reaction is concomitant and subsequent reticulum cell hyperplasia. This lesion or change is most prominent in the medullary regions and can be viewed as a lymphoreticular response to injury. Lymph nodes affected with these changes are extensively infected with CDV as determined by immunofluorescence. If the dogs in question go on to develop the delayed or chronic forms of CDV infection, the lymph nodes and lymphoreticular tissues will repopulate, presumably from bone marrow-origin stem cell precursors. Consequently, the lymph nodes experience varying degrees of nodular to diffuse lymphoreticular hyperplasia. In addition, the thymus will repopulate with thymocytes. A residual CDV lesion seen in these dogs is the formation of germinal centers or B cell dependent zones within the medullary regions of the thymus.[58]

The lungs are another area noted for histopathologic occurrence of CDV-associated lesions. Pulmonary alveolar macrophages can be seen to be the chief resident cells carrying virus.[62] These cells can be found in alveoli frequently as indistinct syncytial or multinucleated giant cells. Interstitial, that is interalveolar, spaces are thickened chiefly by an influx of monocyte-macrophages and occasional lymphocytes. Of course, a bacterial pneumonia, if present, may supercede this viral lesion. Eosinophilic viral inclusion bodies can be found in the cytoplasm and nuclei of bronchiolar epithelium and within resident alveolar macrophages.

Within the gastrointestinal tract, about the only CDV-specific lesion is the appearance of CDV eosinophilic inclusion bodies in intestinal epithelium. These are best sought for in the glandular elements of the stomach, or in the deep crypt areas of the intestinal tract. These are usually focal in distribution and must be searched for diligently to demonstrate them. Within the urinary tract, eosinophilic inclusion bodies can be demonstrated in renal pelvic epithelium and in transitional epithelium of the bladder. The seminiferous tubules and epididymis of male dogs are consistently affected by CDV.

Recently the effects of CDV infection on the musculoskeletal system have been documented. Young animals fatally affected with CDV will show multifocal areas of mineralization and necrosis within the myocardium.[63] This lesion is superficially similar to that associated with CPV infection, but can be distinguished from CPV on the basis of lack of inflammation with CDV, and lack of CPV-associated intranuclear inclusion bodies. It has been known for many years that young dogs that survive acute systemic CDV infection will demonstrate pitting of enamel on permanent teeth. This has been shown to be due to virus induced focal destruction of the enamel forming organ of the developing tooth bud.[64] Recently it was shown that canine distemper virus can produce sclerotic lesions in the metaphyses of long bones in young dogs that in many ways resemble hypertrophic osteodystrophy.[100] This disease, a failure to remodel actively growing bone in these regions, is thought to be due to CDV-induced interference with osteoclastic resorption capabilities within the primary spongiosa. Immunocytochemical labeling has shown CDV antigen infection in osteoclasts, osteoblasts, vascular endothelium, and stromal cells.

IX. NEUROLOGICAL FORMS OF CDV INFECTION

The morphological expression of CDV infection within the brain and neuraxis of affected animals has received a great deal of attention in the last 10 years. It can safely be said that the lesions within the CNS vary greatly in nature and occurrence. References were made previously to the ability of CDV to mimic virtually every other neurologic disorder of the dog. Thus it is not surprising that lesions within the brain can be found anywhere within the neuraxis and can be focal, multifocal, or diffuse in nature. Classically, distemper virus lesions have been thought to be associated with the ventricular system of the brain, the choroid plexus, and the area postrema around the fourth ventricle and associated white matter tracts of the cerebellum.[1,2,65] In fact, virtually every standard veterinary histopathology textbook refers to these sites as predilection sites of viral infection and appearance of lesions. In our experience these sites, while frequently involved, are certainly not the first, and certainly not the only sites affected by CDV. In fact, using viral histochemical labeling techniques, one can state with a great deal of certainty that routine histologic screening of brain tissues for lesions greatly underestimates the amount, and the distribution of virus within the neuraxis.[39] Using viral antigen tracing procedures, it has been shown that there are really no predilection sites for viral involvement within the brain. Viral entry into the brain is essentially a random event. The cerebrum, being the most voluminous of the major subdivisions of the brain, thus can be expected to contain the most viral antigen.[39] This is, in fact, the case. In any event, it is useful when describing or categorizing the types of neurological lesions associated, to divide them into one of four or five different histopathologic syndromes.

A. Acute Encephalopathy

This lesion is a very common lesion in cases of acute fulminant, fatal, and overwhelming CDV infection in young animals.[59,66,67] It is a consequence of viral infection, chiefly of neuronal elements in the cerebrum and cerebellum. The short clinical course and fulminant nature of the disease virtually assures that there will not be dramatic and obvious neuropathologic changes.[68] Affected neurons, if they can be identified (Figure 9), are shrunken and pyknotic. They may or may not express small intracytoplasmic or intranuclear eosinophilic inclusion bodies. Lymphoplasmacytic cellular infiltrations are essentially absent in these cases. While most lesions contain viral antigen, some neuronal changes have been attributed to hypoxia secondary to seizure activity. This is particularly true of the hippocampus. The authors have no opinion as to which is the likely cause other than to state that experimentally, when seizuring is

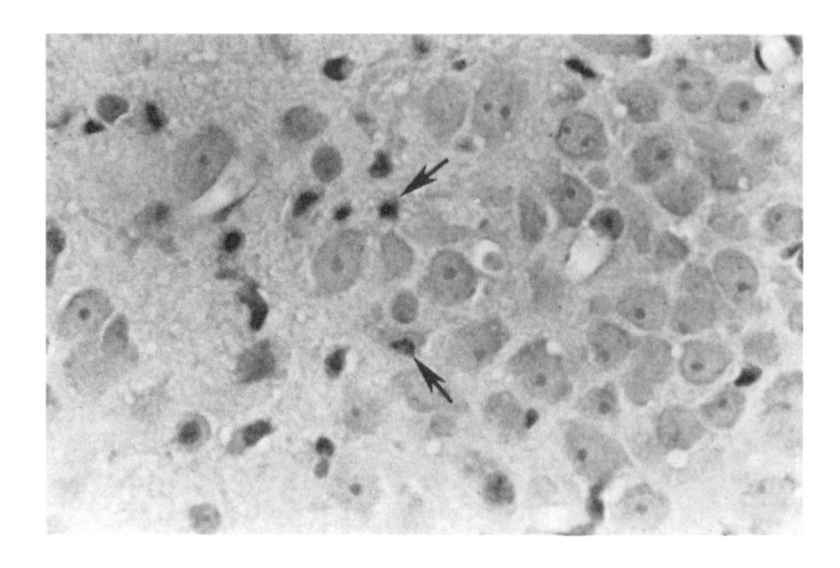

FIGURE 9. Acute lesionless encephalopathy: single cell neuronal necrosis due to CDV infection in a young dog.

controlled through the use of anticonvulsants such as dilantin or phenobarbital, neuronal degeneration and occasionally malacia still occur. Paradoxically, acute encephalopathy can be enhanced experimentally by the administration of high titers of immune serum to affected animals.[67] The reasons for this effect are not known.

B. Acute Encephalitis

This form tends to occur in slightly older dogs and, as the name implies, is characterized in part by the presence of an inflammatory response. Histologically, this form of fatal CDV infection is manifested by more obvious neuronal involvement than occurs in acute, fatal encephalopathy above. In addition, reactive astrocytic changes, consisting of multifocal astrocytic hypertrophy and hyperplasia are also parts of the lesion.[38,65,69] Immunohistochemically, viral antigen can be demonstrated in neurons and astrocytes within the CNS, and intranuclear and intracytoplasmic inclusion bodies may be found in these cells by light microscopy (Figure 10). As before, the lesion is multifocal in nature and is accompanied by subtle or slight perivascular cuffing responses.[70,71] These infiltrating inflammatory cells are chiefly lymphocytes and cells of monocyte-macrophage lineage and many of these cells contain cytoplasmic viral antigen.[70,72] Viral antigen can be demonstrated in cells of the choroid plexus (Figure 11), and in ependymal cells (Figure 12) lining the ventricular system. In these cases, white matter tracts subadjacent to areas of cortical or ventricular involvement show some evidence of vacuolar degeneration. These white matter changes are best interpreted as Wallerian type degeneration.

C. Subacute to Chronic Encephalitis with Degeneration and Demyelination

This histopathologic form occurs largely in weanling and adult dogs and reflects long-term interaction between viral infection within the CNS and host immune responses directed toward virus and/or brain antigens. There is extensive multifocal lymphoplasmatic and macrophage infiltration within the meninges, around parenchymal blood vessels, and within neural parenchyma. Astrocytic hypertrophy and hyperplasia are dramatic.[75,78] Reactive vascular changes are frequently a striking part of the parenchymal lesion and consist of segmental hypertrophy and hyperplasia. Multifocal to confluent zones of demyelination and malacia are conspicuous changes in some of the

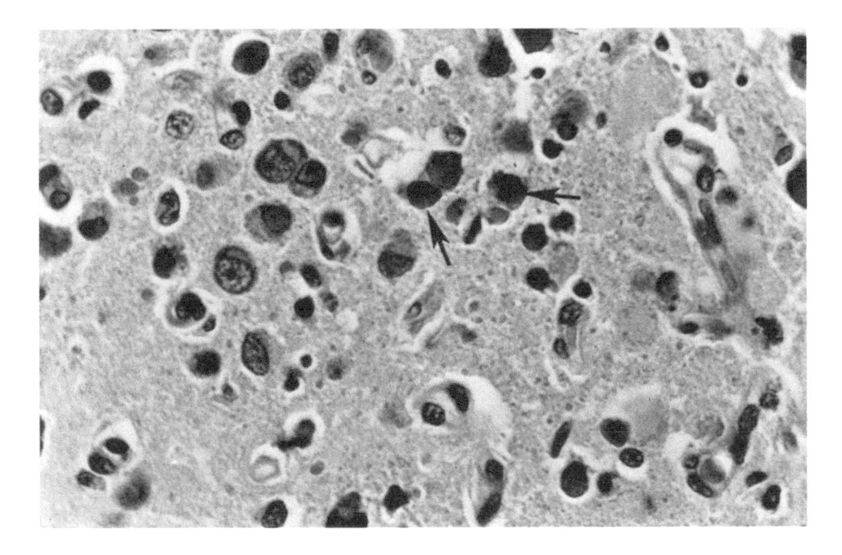

FIGURE 10. Virus-specific inclusion bodies in a dog with acute CDV-associated encephalitis.

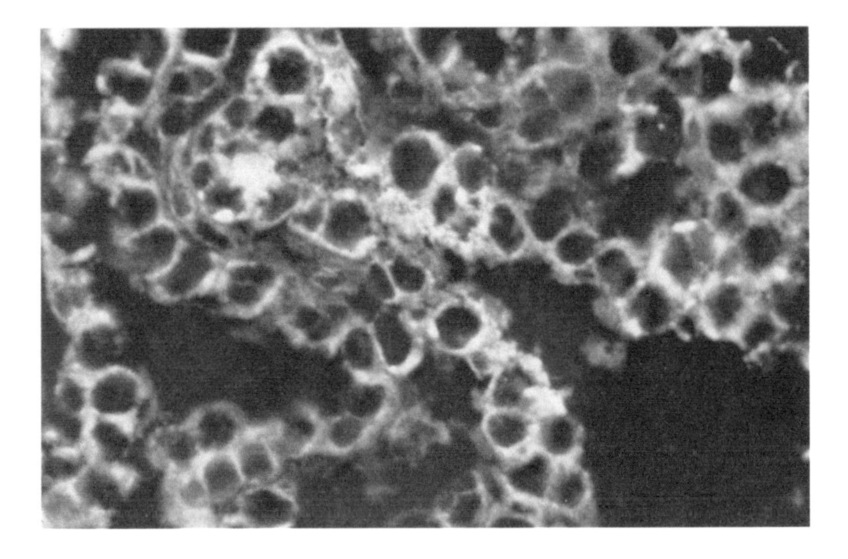

FIGURE 11. Viral infection of the choroid plexus as determined by immunofluorescence in a young dog with acute CDV-associated encephalitis.

cases and are associated with areas of intense leukocytic infiltrations.[38,73-77] They are recognized as areas of myelin pallor by routine staining procedures and can be best appreciated with a myelin-specific stain such as luxol fast blue. Within areas of demyelination, mononuclear phagocytes can be seen actively phagocytizing myelin lipid (Figure 13). By immunohistochemistry, viral antigen can always be demonstrated within and subjacent to lesions either as structurally distinct inclusion bodies or diffuse cytoplasmic antigen. In contrast to the two forms above, viral antigen is found predominantly within glial cells and macrophages. Neuronal infection is a minor component.[77] The reason for the change in viral distribution in this form of the disease is unknown. Glial preference may possibly reflect an age-associated loss of neuronal susceptibility

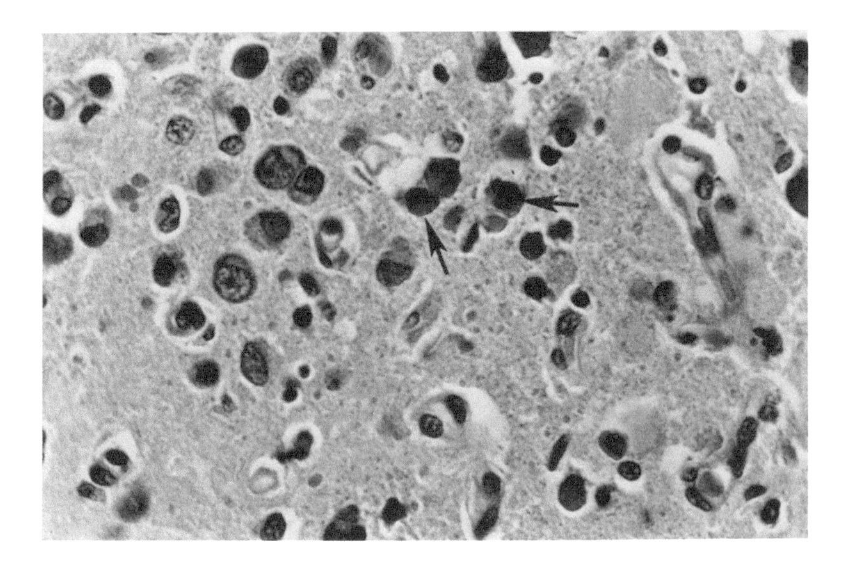

FIGURE 12. Cytoplasmic viral inclusion body in an ependymal cell lining in central canal of the spinal cord.

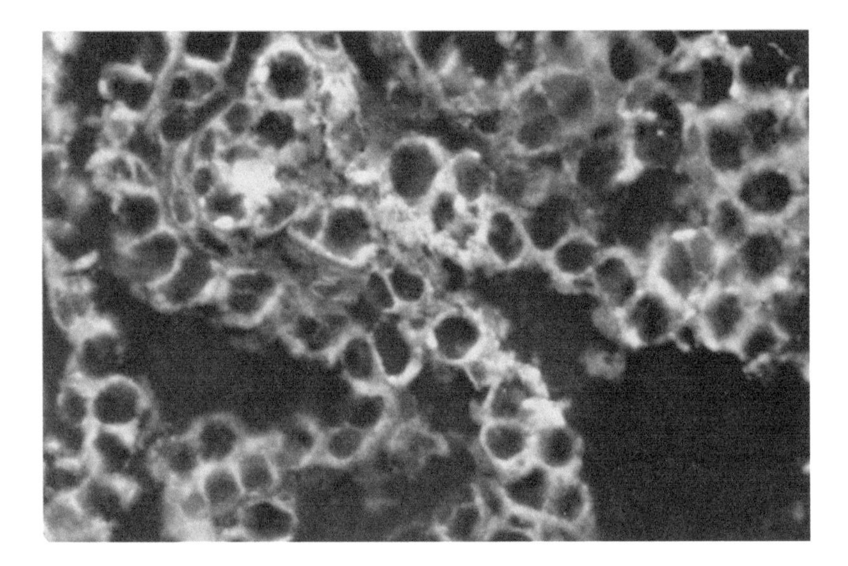

FIGURE 13. Infiltrating mononuclear phagocytes in an area of complex degeneration and demyelination associated with CDV infection. Serine esterase stain with methyl green counterstain.

to viral infection. Others have shown that both IgG and IgM containing lymphoid cells are a prominent component of inflammatory lesions.[79] Further, there are extracellular deposits of IgG and IgM.[79,80]

The CNS demyelinating component (Figure 14) has intrigued many investigators, chiefly due to the similarities between demyelination noted with CDV and that seen in human demyelinating diseases, notably multiple sclerosis and other MV-related disease syndromes. The cellular mechanisms of demyelination have not been worked out in any great detail in CDV. Primary infection of the oligodendrocytes, the glial cell responsible for production and maintenance of CNS myelin has been suggested as a

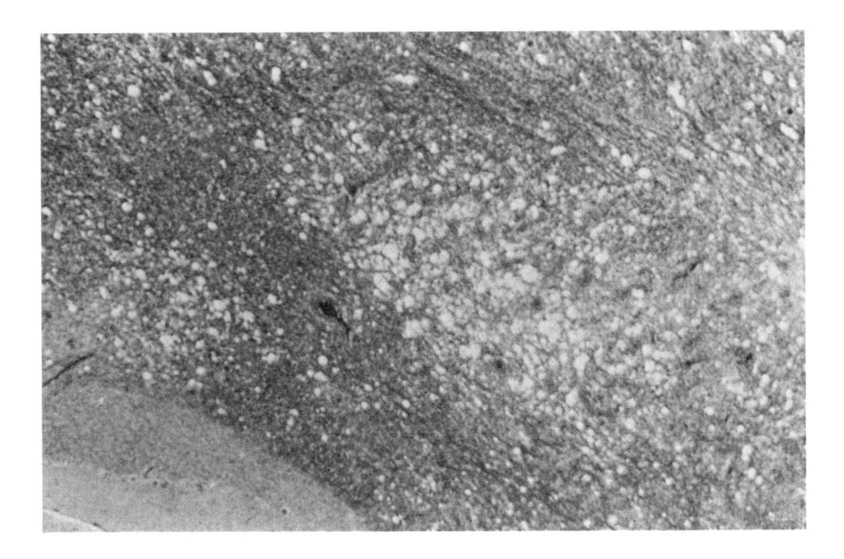

FIGURE 14. Canine distemper virus-associated white matter demyelination. Luxol fast blue-PAS stain.

mechanism of demyelination in this disease. It is known that CDV will infect the oligodendrocyte in vivo; however, most authors feel that primary infection of the oligodendrocyte is of minor importance to demyelination.[75]

Essentially two pathways for demyelination have been proposed: a humoral and a cellular pathway. The humoral pathway envisions the generation of antimyelin[81] or antimyelin basic protein antibodies[80] that occurs during the course of CDV infection. This autoimmune response presumably in conjunction with serum complement or acting as intermediate in antibody-directed cellular cytotoxicity reactions, then directs an effector-cell response to sequestered CNS antigens. The resultant lesion produced is demyelination. Some of the features of naturally occurring demyelinating diseases demyelination have been produced by injection of antisera to myelin into white matter of laboratory animals.[82] Both antimyelin and antibasic protein antibody have been demonstrated in serum and, in the latter,[80] in cerebrospinal fluid of CDV-affected dogs.

The second or cellular scenario implicates the blood origin macrophage-monocyte[38,71] and possibly the astrocyte,[75] as the major effector cell in demyelination. It is thought that, by analogy to allergic encephalomyelitis, T cells sensitized to CNS antigens direct or effect macrophage activation and/or astrocytic activation and instruct these cells to actively phagocytize structurally normal CNS myelin, thereby resulting in a demyelinating lesion.

Very little can be said in an attempt to resolve these mechanisms of demyelination at the present time. Humoral demyelinating factors[65] have been demonstrated in serum from CDV-infected dogs. In addition, work from this laboratory and elsewhere has directly implicated the macrophage as the primary cell responsible for myelin destruction. Also, it is known that the astrocyte, once activated, can engage in myelin phagocytosis. There is no reason not to suspect the astrocyte as also having a role in genesis of the lesion.

D. Old Dog Encephalitis (ODE), Delayed Onset Encephalitis, and Adult Forms of CDV Infection within the Brain

A rare and histologically distinct form of CDV infection has been recognized for many years. This lesion has been called, variously, Dawson's encephalitis of dogs,

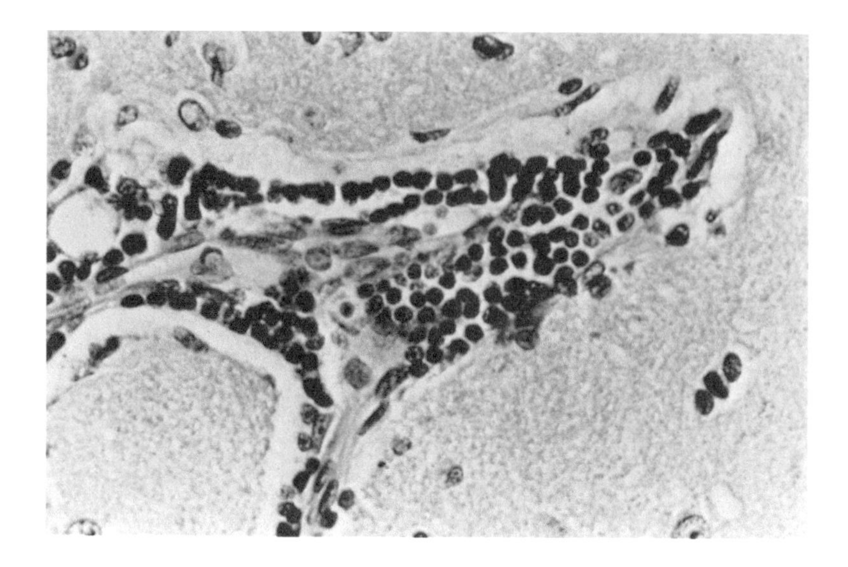

FIGURE 15. Perivascular lymphoplasmacytic and monocytic cellular infiltrations into the brain of a dog affected with old dog encephalitis.

FIGURE 16. The demonstration of viral antigen by indirect immunofluorescence within the central nervous system in a case of old dog encephalitis.

multiple sclerosis of dogs, subacute sclerosing encephalitis of dogs, or simply old dog encephalitis (ODE). The connection of this lesion to canine distemper virus was made definitively in 1971 by Gorham and colleagues.[55,56] The essential clinicopathologic and histopathologic features of ODE are as follows. It is a diffuse, cerebral disease, clinically presented as cerebrocortical or cerebral derangement in adult, middle-aged dogs. The lesion (Figures 15, 16) within the brain consists of extensive, chiefly neuronal infection by CDV attended by extensive nonsuppurative meningitis and perivascular cuffing and glial activation.[55] Demyelination is not a prominent feature of the disease although some areas of poliomalacia have been observed. Although similar lesions may occur in the cerebellum and spinal cord, lesions appear primarily in the cerebral hemispheres. All attempts to infect animals by direct inoculation of ODE brain material, have so far been unsuccessful. Virus can be recovered from these animals with some difficulty and only by explantation of affected brain.[83] Due to the rarity of this disease and the extensive clinical work-up needed to exclude other neurologic problems of the dog, very few cases have been made available for virologic study. Rather, the diagnosis of ODE is usually made on the basis of its unique histologic appearance. Nonetheless, once virus has been isolated, it has been shown to be CDV by immunofluorescence and polypeptide tryptic map digest studies. Using the latter technique, several ODE isolates have been compared to laboratory strains of CDV. Minor differences were noted in polypeptide composition.[83] Importantly, however, is the fact that ODE-CDV isolates possess all major virion polypeptides, including the M or matrix protein.[47] This is an important distinction because an analogous human disease syndrome, namely subacute sclerosing panencephalitis (SSPE), an MV related disease of humans, is noted for its lack of production or synthesis of the matrix protein.[47,84] Thus, ODE, while an excellent model of chronic paramyxoviral neuronal disease, is not a perfect replica of human SSPE. The reasons for persistence of ODE virus within the CNS of these animals has not been determined. For example, it is not known if ODE represents host infection by a unique mutant of wild-type CDV that has a predilection to infect and persist in neurons, or is the result of mutation of an established infection as suggested for visna in sheep. This dilemma has been partially resolved by work recently completed in our laboratory. An animal, infected with our R252-CDV for over 3 years and kept isolated in a gnotobiotic isolation unit throughout that time interval, developed a clinical syndrome that was clinically and histologically compatible with old dog encephalitis.[39] Direct animal transmission studies were not successful; however, brain explant cultures established from this case eventually resulted in recovery of infectious viral progeny. Histologically, the lesions were extensive and consisted chiefly of multifocal areas of lymphoplasmacytic cellular infiltration, patchy neuronal involvement, active and chronic perivascular demyelination, consistent with repeated bouts of damage to the brain. This latter is not at all unlike that seen in multiple sclerosis, in which astrocytic scars and subacute and chronic zones of demyelination and inflammation can all coexist within the same brain.[85] Further studies with this ODE virus derived from R252-CDV stock are continuing.

E. Virus-Associated Hydrocephalus

Most investigators believe that hydrocephalus in dogs is chiefly a genetic or inherited defect in development. Recently, however, canine parainfluenza virus was shown to induce hydrocephalus when inoculated into neonatal or weanling gnotobiotic dogs.[86-88] Subsequently the suggestion was made that canine distemper virus presumably due to its ability to replicate in ependymal lining cells of the ventricular system, was also capable of producing viral hydrocephalus.[89] At the present time, documentation of this effect of CDV within the brain of dogs, either under natural or experimental conditions, is lacking.

F. Neuropathological Findings in Noncanid Species

Surprisingly enough, relatively little work has been done regarding the details of histopathologic findings of CDV infection within the CNS of noncanid species. Best documented are the lesions observed with infection of ferrets and related mustelids. In virtually all cases the extreme susceptibility of this species to CDV infection has resulted in the production of acute fatal encephalopathy-like lesions.[1,2] There is virtually nothing in the way of perivascular cuffing, and inclusion bodies are rare and difficult to demonstrate.

X. TREATMENT AND PREVENTION OF CDV INFECTION

Today, fortunately, there exists a number of safe and efficacious, modified-live viral vaccine products. Virtually every veterinary biological supply house offers a range of products with CDV included in them. The virus, consisting chiefly of laboratory adapted and attenuated Rockborn or Lederle strains of CDV, is propagated today in canine origin continuous cell lines. One inoculation into a CDV-susceptible clinically healthy, normal dog, will provide effective and probably lifelong immunity. Thus, under typical clinical circumstances, immunity can be assured by delivering the product according to manufacturer's instructions to the animal or animals in question. There are, however, a number of different exceptions to this generality that should be mentioned. First, it has been known for many years that it is difficult to vaccinate young animals for canine distemper virus. This is due chiefly to the acquisition of colostrum origin, maternal, or passive antibodies.[1,2] In a typical situation, these antibodies will decline to zero within 10 to 12 weeks of age. However, since within a litter the amount of colostrum ingested is variable, as is the size of the dog, etc. there exists a window of vulnerability to acquiring CDV infection in the postnatal period. It is important to stress that vaccination of a passively immune dog will not result in protection. To circumvent this difficulty, serum samples can be tested for viral antibody using the Cornell Nomograph System.[1] Once the antibody level has dropped to baseline values the animals can be safely and efficaciously vaccinated. This is an excellent and effective method for use in expensive and closely watched animals. It is not particularly practical for routine use.

Another avenue around this difficulty is to immunize young animals, dogs less than 10 weeks of age, with a so-called heterotypic vaccine product. The heterotypic vaccine consists of modified-live MV in addition to CDV.[1] Previously reference was made to unidirectional cross-reactivity of MV-H protein and the CDV-H protein. This failure of passive antiCDV antibody to bind and inactivate MV via its H protein antigenic determinant, then serves as the basis of heterotypic resistance. Thus, dogs possessing high levels of colostrum-origin CDV antibody can be vaccinated with modified-live MV and thus make an immune response to the MV-H antigen. This procedure thus renders the dog actively immune during the time of greatest concern, the immediate postnatal period. The only disadvantage of this procedure is that, if heterotypic vaccination is used in subsequent generations of puppies, the bitch, having been previously vaccinated with MV as a puppy, will transmit MV-H antibodies to her puppies in colostrum, thereby, in effect circumventing or eliminating any advantage that MV vaccination has for her progeny.

Another precaution to be cognizant of is the use of modified-live products in exotic animal species.[90] Sadly, very valuable and very rare animals, including species such as the lesser panda[91] and black-footed ferret[92] have been killed with modified-live (MLV)-CDV vaccine. Even species within the Canidae family may show heightened susceptibility to CDV. For example, MLV-CDV was recently shown to be dangerous for use in grey foxes, but not in red foxes.[93] We think the standard rule of thumb when vaccinating exotics is to use inactivated CDV products, multiple doses at, at least yearly, intervals.[90] If the animal is extremely valuable, serum can be checked for antibody activity following multiple immunizations with inactivated products. It is important to stress, also, that sick or debilitated animals may not respond adequately to MLV vaccination, or that the possible suppressive effects of MLV-CDV may potentiate other disease.[90,94] Under clinical conditions, of course, the most likely reason for a sick and debilitated dog is likely to be preexistent canine distemper virus infection. Vaccinating animals with active CDV-associated disease will not provide any measure of protection. Even worse, however, it will provide both the veterinarian and the client with the false assumption that the resultant mortality is due to the MLV vaccine product and not antecedent CDV infection. Although in utero transmission of MLV-CDV has not yet been documented, first time vaccination of pregnant dogs with MLV-vaccine cannot be recommended.

The final precaution relates to the clinical interaction of canine parvovirus infection with canine distemper virus infection. This is a controversial topic and one in which no definitive answers are yet available. On a clinical basis, we and others have noticed that canine parvovirus infection potentiates the lethal effects of either concurrent CDV infection or CDV fatalities associated with vaccination with an ordinarily safe modified-live product. Again, this is a clinical association and has not been subjected to experimental verification although it is known that dogs actively infected with virulent canine parvovirus are at risk for developing fatal CDV-vaccine associated neurologic disease.[95]

The obvious question is, what is the relationship between vaccination of a dog with an attenuated CPV product in combination with an MLV-CDV? Again, the definitive data is lacking on this subject, although it is known that not all attenuated CPV-vaccines are completely innocuous.[94] At this time, it is the authors' personal belief and recommendation that the vaccination for canine distemper virus and canine parvovirus be separated by at least several weeks. This, then, by definition excludes the haphazard use of the combination MLV vaccine products so popular today. We would argue that canine parvovirus, either in its attenuated or inactivated form should be administered first, followed 2 weeks later by an MLV-CDV series. Our laboratory and others are actively seeking way to circumvent this synergistic difficulty.

REFERENCES

1. Appel, M. J. G. and Gillespie, J. J., Canine distemper virus, *Virol. Monogr.*, 11, 1, 1972.
2. Gorham, J. R., Canine distemper, in *Adv. Vet. Sci.*, 6, 288, 1960.
3. Fraser, K. B. and Martin, S. J., *Measles Virus and Its Biology*, Academic Press, New York, 1978, 193.
4. Imagawa, D. T., Relationship among measles, canine distemper and Rinderpest viruses, *Prog. Med. Virol.*, 10, 160, 1968.
5. Waterson, A. P., Rott, R., and Ruckle-Enders, G., The components of measles virus and their relationship to Rinderpest and distemper, *Z. Naturforsch.*, 18B, 377, 1963.
6. Waters, D. J. and Bussell, R. H., Polypeptide composition of measles and canine distemper virus, *Virology*, 55, 554, 1973.
7. Hall, W. W., Lamb, R. A., and Choppin, P. W., The polypeptides of canine distemper virus synthesis in infected cells and relatedness to the polypeptides of other morbilliviruses, *Virology*, 100, 433, 1980.
8. Appel, M. J. G. and Jones, O. R., Use of alveolar macrophages for cultivation of canine distemper virus, *Proc. Soc. Exp. Biol. New York*, 126, 571, 1967.
9. Whetstone, C. A., Bunn, T. B., and Gourlay, J. A., Canine distemper virus titration in ferret peritoneal macrophages, *Cornell Vet.*, 71, 144, 1981.
10. Metzler, A. E., Higgins, R. J., Krakowka, S., and Koestner, A., Characterization of bovine cells supporting *in vitro* growth of both virulent and attenuated canine distemper virus, *Am. J. Vet. Res.*, 42, 1257, 1981a.
11. Metzler, A. E., Higgins, R. J., and Krakowka, S., Virulence of tissue culture-propagated canine distemper virus, *Infect. Immunol.*, 29, 940, 1980.
12. Metzler, A. E., Higgins, R. J., Krakowka, S., and Koestner, A., Persistent *in vitro* interaction of virulent and attenuated canine distemper virus with bovine cells, *Arch. Virol.*, 66, 329, 1981b.
13. Confer, A. W., Kahn, D. E., Koestner, A., and Krakowka, S., Biological properties of a canine distemper virus isolate associated with demyelinating encephalomyelitis, *Infect. Immunol.*, 11, 343, 1975.

14. Bui, H. D., Tobler, L. H., Van Pelt, L. F., Howard, E. B., and Imagawa, D. I., Canine bladder epithelial cells in culture: susceptibility to canine distemper and measles viruses, *Am. J. Vet. Res.*, 43, 1268, 1982.
15. Jackwood, D. and Rice, J., personal communication, 1984.
16. ter Meulen, V. and Carter, M. J., Morbillivirus persistent infections in animals and man, in *Virus Persistence*, Mahy, W. J., Minson, A. C., and Darby, G. K., Eds., Cambridge University Press, New York, 1982, 97.
17. ter Muelen, V. and Martin, S. J., Genesis and maintenance of a persistent infection by canine distemper virus, *J. Gen. Virol.*, 32, 431, 1976.
18. Narang, H., Ultrastructural study of long-term canine distemper virus infection in tissue culture cells, *Infect. Immunol.*, 36, 310, 1982.
19. Ho, C. K. and Babiuk, L. A., A new plaque system for canine distemper: characteristics of the green strain of canine distemper virus, *Can. J. Microbiol.*, 25, 680, 1979.
20. Appel, M., Sheffy, B. E., Percy, D. H., and Gaskin, J. M., Canine distemper virus in domesticated cats and pigs, *Am. J. Vet. Res.*, 35, 803, 1974.
21. Gould, D. H. and Fenner, W. R., Paramyxovirus-like nucleocapsids associated with encephalitis in a captive Siberian tiger, *J. Am. Vet. Med. Assoc.*, 183, 1319, 1983.
22. Blythe, L. L., Schmitz, J. A., Roelke, M., and Skinner, S., Chronic encephalomyelitis caused by canine distemper virus in a Bengal tiger, *J. Am. Vet. Med. Assoc.*, 183, 1159, 1983.
23. Gilden, D. H., Wellish, M., Rorke, L. B., and Wroblewska, Z., Canine distemper virus infection of weanling mice: pathogenesis of CNS disease, *J. Neurol. Sci.*, 52, 327, 1981.
24. Lyons, M. J., Faust, I. M., Hemmores, R. B., Buskirk, D. R., Hirsch, J., and Zabriski, J. B., A virally induced obesity syndrome in mice, *Science*, 216, 82, 1982.
25. Bernard, A., Weld, L. F., and Tripier, M. F., Canine distemper infection in mice: characterization of a neuroadapted virus strain and its long-term evolution in the mouse, *J. Gen. Virol.*, 65, 1511, 1982.
26. Lyons, M. J., Hall, W. W., Petito, C., Cam, V., and Zabriski, J. B., Induction of chronic neurologic disease in mice with canine distemper virus, *Neurology*, 30, 92, 1980.
27. Massanari, R. M., Paterson, P. Y., and Lipton, H. L., Potentiation of experimental allergic encephalomyelitis in hamsters with persistent encephalitis due to measles virus, *J. Infect. Dis.*, 139, 297, 1979.
28. Cosby, S. L., Lyons, C., Fitzgerald, S. P., Martin, S. J., Presdee, S., and Allen, I. V., The isolation of large and small plaque canine distemper viruses which differ in their neurovirulence for hamsters, *J. Gen. Virol.*, 52, 345, 1981.
29. Cosby, S. L., Morrison, J., Rima, B. K., and Martin, S. J., An immunologic study of infection of hamsters with large and small plaque canine distemper viruses, *Arch. Virol.*, 76, 201, 1983.
30. Yamanouchi, K., Yoshikawa, Y., Sato, L. A., Katow, S., Kobune, F., Kobune, K., Uchida, N., and Shishido, A., Encephalomyelitis induced by canine distemper virus in non-human primates, *Jpn. J. Med. Sci. Biol.*, 30, 241, 1977.
31. Cook, S. and Dowling, P., The possible association between house pets and multiple sclerosis, *Lancet*, i, 980, 1977.
32. Appel, M. J., Glickman, L. J., Raine, C. S., and Toutellotte, W. W., Canine viruses and multiple sclerosis, *Neurology (New York)*, 31, 944, 1981.
33. Barridge, M. J., Multiple sclerosis, house pets and canine distemper: critical review of recent reports, *J. Am. Vet. Med. Assoc.*, 173, 1439, 1978.
34. Krakowka, S., Higgins, R. J., and Metzler, A. E., Plasma phase viremia in canine distemper virus infection, *Am. J. Vet. Res.*, 41, 144, 1980d.
35. Appel, M. J. G., Pathogenesis of canine distemper, *Am. J. Vet. Res.*, 30, 1167, 1969.
36. Krakowka, S., Olsen, R. G., Confer, A. W., Koestner, A., and McCullough, B., Serologic response to canine distemper viral antigens in gnotobiotic dogs infected with R252-canine distemper virus, *J. Infect. Dis.*, 132, 384, 1975b.
37. Krakowka, S., Higgins, R. J., and Koestner, A., Canine distemper virus: a review of structural and functional modulations in lymphoid tissues, *Am. J. Vet. Res.*, 41, 284, 1980c.
38. McCullough, B., Krakowka, S., and Koestner, A., Experimental canine distemper virus-induced demyelination, *Lab. Invest.*, 31, 216, 1974b.
39. Axthelm, M. K., Studies of Canine Distemper Virus Entry into the Central Nervous System, Doctoral dissertation, Ohio State University, Columbus, 1984.
40. Krakowka, S., Hoover, E. A., Koestner, A., and Ketring, K., Experimental and naturally occurring transplacental transmission of canine distemper virus, *Am. J. Vet. Res.*, 38, 912, 1977.
41. Appel, M. and Robson, D. S., A microneutralization test for canine distemper virus, *Am. J. Vet. Res.*, 34, 1459, 1973.
42. Reed, L. J. and Muench, H., A simple method of determining fifty percent end points, *Am. J. Hyg.*, 27, 493, 1938.

43. Noon, K. F., Rogul, M., Binn, L. N., and Appel, M., Enzyme-linked immunosorbent assay for evaluation of antibody to canine distemper virus, *Am. J. Vet. Res.*, 41, 605, 1980.
44. Bernard, S. L., Shen, D. T., and Gorham, J. R., Antigen requirements and specificity of enzyme-linked immunosorbent assay for detection of canine IgG against canine distemper viral antigens, *Am. J. Vet. Res.*, 43, 2266, 1982.
45. Winters, K. G., Mathes, L. E., Krakowka, S., and Olsen, R. G., Immunoglobulin class response to canine distemper virus in gnotobiotic dogs, *Vet. Immunol. Immunopathol.*, 5, 209, 1983.
46. Miele, J. A. and Krakowka, S., Antibody responses to virion polypeptides in gnotobiotic dogs infected with canine distemper virus, *Infect. Immunol.*, 41, 869, 1983.
47. Hall, W. W., Imagawa, D. L., and Choppin, P. W., Immunological evidence for the synthesis of all canine distemper virus polypeptides in chronic neurological diseases in dogs. Chronic distemper and old dog encephalitis differ from SSPE in man, *Virology*, 98, 283, 1979.
48. Krakowka, S., Wallace, A., and Koestner, A., Syncytia inhibition by immune lymphocytes: an *in vitro* test for immunity to canine distemper, *J. Clin. Microbiol.*, 7, 292, 1978c.
49. Krakowka, S. and Wallace, A. L., Lymphocyte-associated immune responses to canine distemper and measles viruses in distemper-infected gnotobiotic dogs, *Am. J. Vet. Res.*, 40, 669, 1979.
50. Shek, W. P., Schultz, R. D., and Appel, M. J. G., Natural and immune cytolysis of canine distemper virus-infected target cells, *Infect. Immunol.*, 28, 724, 1980.
51. Tsai, S. C., Summers, B. A., and Appel, M. J. G., Interferon in cerebrospinal fluid. A marker for direct persistence in canine distemper encephalomyelitis, *Arch. Virol.*, 72, 257, 1982.
52. Smith, O. K., Dunlap, R. C., Thiel, J. F., Newman, J. T., and Palmer, A. E., Isolation of viruses from primary dog cell cultures and the occurrence of viral antibody in donor animals, *Proc. Soc. Exp. Biol. Med.*, 138, 560, 1970.
53. Shifrine, M., Taylor, N., Rosenblatt, L. S., and Wilson, F., Seasonal variation in cell-mediated immunity of clinically normal dogs, *Exp. Hematol.*, 8, 318, 1980.
54. Shifrine, M., Garsd, A., Christiansen, J. A., and Rosenblatt, L. S., Photoperiod and cell-mediated immunity of clinically normal dogs, *J. Interdisciplinary Cycle Res.*, 13, 177, 1982.
55. Lincoln, S. D., Gorham, J. R., Ott, R. L., and Hegreberg, G. A., Etiologic studies of old dog encephalitis. I. Demonstration of canine distemper viral antigen in the brain in two cases, *Vet. Pathol.*, 8, 1, 1971.
56. Lincoln, S. D., Gorham, J. R., Davis, W. C., and Ott, R. L., Studies of old dog encephalitis. II. Electron microscopic and immunohistologic findings, *Vet. Pathol.*, 10, 124, 1973.
57. Gibson, J. P., Griesemer, R. A., and Koestner, A., Experimental distemper in the gnotobiotic dog, *Pathol. Vet.*, 2, 1, 1965.
58. McCullough, B., Krakowka, S., and Koestner, A., Experimental canine distemper virus-induced lymphoid depletion, *Am. J. Pathol.*, 74, 155, 1974a.
59. McCullough, B., Krakowka, S., Koestner, A., and Shadduck, J., Demyelinating activity of canine distemper virus isolates in gnotobiotic dogs, *J. Infect. Dis.*, 130, 344, 1974c.
60. Krakowka, S., Cockerell, G., and Koestner, A., Effects of canine distemper virus on lymphoid function *in vitro* and *in vivo*, *Infect. Immunol.*, 11, 1069, 1975a.
61. Krakowka, S., Mechanisms of *in vitro* immunosuppression in canine distemper virus infection, *J. Clin. Lab. Immunol.*, 8, 187, 1982a.
62. Miry, C., Ducatelle, R., Thoonen, H., and Hoorens, J., Immunoperoxidase study of canine distemper virus pneumonia, *Res. Vet. Sci.*, 34, 145, 1983.
63. Higgins, R. J., Krakowka, S., Metzler, A. E., and Koestner, A., Canine distemper virus-associated cardiac necrosis in the dog, *Vet. Pathol.*, 18, 472, 1981.
64. Dubielzig, R. R., Higgins, R. J., and Krakowka, S., Lesions of the enamel organ caused by canine distemper virus, *Vet. Pathol.*, 18, 684, 1981.
65. Koestner, A., Animal model of human disease. Animal Model. Distemper-associated demyelinating encephalomyelitis, *Am. J. Pathol.*, 78, 361, 1975.
66. Krakowka, S. and Koestner, A., Age-related susceptibility to canine distemper virus infection in gnotobiotic dogs, *J. Infect. Dis.*, 134, 629, 1976.
67. Krakowka, S., Mador, R., and Koestner, A., Canine distemper virus-associated encephalitis: modification by passive antibody administration, *Acta Neuropathol.*, 43, 235, 1978d.
68. Higgins, R. J., Krakowka, S., Metzler, A. E., and Koestner, A., Experimental canine distemper encephalomyelitis in neonatal gnotobiotic dogs: a sequential ultrastructural study, *Acta Neuropathol. (Berlin)*, 57, 287, 1982a.
69. Wright, N. G., Cornwell, H. J. C., Thompson, H., and Lauder, I. M., Canine distemper: current concepts in laboratory and clinical diagnosis, *Vet. Rec.*, 94, 86, 1974.
70. Summers, B. A., Greisen, H. A., and Appel, M. J. G., Possible initiation of viral encephalomyelitis in dogs by migrating lymphocytes infected with distemper virus, *Lancet*, i, 187, 1978.
71. Summers, B. A., Greisen, H. A., and Appel, M. J. G., Early events in canine distemper demyelinating encephalomyelitis, *Acta Neuropathol.*, 46, 1, 1979.

72. Higgins, R. J., Krakowka, S., Metzler, A. E., and Koestner, A., Primary demyelination in experimental canine distemper virus-induced encephalomyelitis in gnotobiotic dogs: sequential immunologic and morphologic findings, *Acta Neuropathol. (Berlin),* 58, 1, 1982b.
73. Lisiak, J. A. and Vandevelde, M., Polioencephalomalacia associated with canine distemper virus infection, *Vet. Pathol.,* 16, 650, 1979.
74. Innes. J. F. M. and Saunders, L. Z., Viral and rickettsial encephalomyelitis; C. Canine distemper, in *Comparative Neuropathology,* Academic Press, New York, 1962, 373.
75. Raine, C. S., On the development of CNS lesions in natural canine distemper encephalomyelitis, *J. Neurol. Sci.,* 30, 13, 1976.
76. Wisniewski, H., Raine, C. S., and Kay, W. J., Observations on viral demyelinating encephalomyelitis. Canine distemper, *Lab. Invest.,* 26, 589, 1972a.
77. Vandevelde, M. and Kristensen, B., Observations on the distribution of canine distemper virus in the central nervous system of dogs with demyelinating encephalomyelitis, *Acta Neuropathol. (Berlin),* 40, 233, 1977.
78. Vandevelde, M., Bichsel, P., Cerruti-Sola, S., Steck, A., Kristensen, F., and Higgins, R. J., Glial proteins in canine distemper virus-induced demyelination, *Acta Neuropathol. (Berlin),* 59, 269, 1983a.
79. Vandevelde, M., Fankhauser, R., Kristensen, F., and Kristensen, B., Immunoglobulins in demyelinating lesions in canine distemper encephalitis. An immunohistologic study, *Acta Neuropathol. (Berlin),* 54, 31, 1981.
80. Vandevelde, M., Kristensen, F., Kristensen, B., Steck, A. J., and Kihm, U., Immunological and pathological findings in demyelinating encephalitis associated with canine distemper virus infection, *Acta Neuropathol. (Berlin),* 56, 1, 1982b.
81. Krakowka, G. S., McCullough, B., Koestner, A., and Olsen, R. G., Myelin-specific auto-antibodies associated with central nervous system demyelination in canine distemper infection, *Infect. Immunol.,* 8, 819, 1973.
82. Williams, R. M., Krakowka, S., and Koestner, A., *In vivo* demyelination by anti-myelin antibodies, *Acta Neuropathol. (Berlin),* 50, 1, 1980.
83. Imagawa, D. T., Howard, E. B., Van Pelt, L. F., Ryan, C. P., Bui, H. D., and Shapshak, P., Isolation of canine distemper virus from dogs with chronic neurological diseases, *Proc. Soc. Exp. Biol. Med.,* 164, 355, 1980.
84. Johnson, R. T., *Viral Infections of the Nervous System,* Raven Press, New York, 1982.
85. Raine, C. S., The etiology and pathogenesis of multiple sclerosis. Recent developments, *Pathobiol. Ann.,* 7, 347, 1977.
86. Baumgartner, W. K., Metzler, A. E., Krakowka, S., and Koestner, A., *In vitro* identification and characterization of a virus isolated from a dog with neurological dysfunction, *Infect. Immunol.,* 31, 1177, 1981.
87. Baumgartner, W. K., Krakowka, S., and Koestner, A., Acute encephalitis and hydrocephalus in dogs caused by a canine parainfluenza virus, *Vet. Pathol.,* 19, 79, 1982a.
88. Baumgartner, W. K., Krakowka, S., and Koestner, A., Ultrastructural evaluation of acute encephalitis and hydrocephalus in dogs due to canine parainfluenza virus, *Vet. Pathol.,* 19, 305, 1982b.
89. Summers, B. A. and Appel, M. J. G., Virus-induced hydrocephalus in the dog, *Vet. Pathol.,* 20, 513, 1983.
90. Montali, R. J., Bartz, C. R., Teare, J. A., Allen, J. T., Appel, M. J., and Bush, M., Clinical trials with canine distemper vaccines in exotic carnivores, *J. Am. Vet. Med. Assoc.,* 183, 1163, 1983.
91. Bush, M., Montali, R. J., Brownstein, D., James, A. E., and Appel, M. J. G., Vaccine-induced canine distemper in a lesser panda, *J. Am. Vet. Med. Assoc.,* 169, 959, 1976.
92. Pearson, G. L., Vaccine-induced canine distemper in black footed ferrets, *J. Am. Vet. Med. Assoc.,* 170, 103, 1977.
93. Halbrooks, R. D., Swango, L. J., Schnurrenberger, P. R., Mitchell, F. E., and Hill, E. P., Response of gray foxes to modified live-virus canine distemper vaccines, *J. Am. Vet. Med. Assoc.,* 179, 1170, 1981.
94. Kauffman, C. A., Berman, A. G., and O'Connor, O. P., Distemper virus infection in ferrets: an animal model of measles-induced immunosuppression, *Clin. Exp. Immunol.,* 47, 617, 1982.
95. Krakowka, S., Olsen, R. G., Axthelm, M. K., Rice, J., and Winters, K., Canine parvovirus infection potentiates canine distemper encephalitis due to modified-live virus vaccine, *J. Am. Vet. Res. Assoc.,* 180, 137, 1982c.
96. Krakowka, S., unpublished observation.
97. Axthelm, M. K., unpublished, 1984.
98. Krakowka, S., unpublished, 1984.
99. Ringler, S., unpublished, 1984.
100. Boyce, R., unpublished, 1984.

Chapter 9

RABIES

H Fred Clark and Bellur S. Prabhakar

TABLE OF CONTENTS

I. INTRODUCTION

The pathology and the natural history of rabies have been intensively studied for more than 100 years by investigators on several continents.[1] Until, perhaps, the past 2 decades, such studies have been largely limited to observations of pathogenesis and the induction of a protective immune response in the whole mammalian organism. Rabies virus resisted early efforts at efficient propagation in cell culture until the breakthrough successes of Kissling,[2] Fernandes et al.,[3] and others[4] in the early 1960s. Adaptation of rabies virus to cell culture has now facilitated the analysis of the molecular and antigenic composition of the virion to a degree of refinement equal to that of many animal viruses that were propagated much earlier in cell culture.

The enormous mass of rabies research and the plethora of fascinating findings have certainly enhanced our understanding of the disease. Yet few of the ultimate questions about rabies have been answered. Rather, each advance in our comprehension of the disease has tended to expose new mysteries that again defy ready resolution. A list of examples is at least a partial catalog of the problems currently comprising the challenge to a complete comprehension of rabies.

The association of rabies with infection of the brain was demonstrated unequivocally by Pasteur more than 100 years ago.[5] The association of rabies with characteristic intraneuronal inclusion bodies within the central nervous system (CNS) was established by Negri[6] at the beginning of this century. However, in the preponderance of natural cases of rabies, the degree of tissue destruction and inflammatory response in the brain is so minimal that the reason for death of the host cannot be explained on this basis. Similarly, the extraordinarily high mortality rate (approximately 100%) of natural cases of rabies remains unexplained.

It has been convincingly demonstrated in experimental studies that the transit of rabies virus from peripheral exposure sites to the CNS occurs through peripheral nerves at a rather rapid and consistent rate.[7] The structural form of the virion or its component(s) that travels through the peripheral neurons remains unknown. Likewise, the site and manner of its apparent sequestration in nonnervous tissue prior to entry to the nervous system in the common cases with very prolonged incubation periods remain undefined.

Pasteur, again, described a century ago a postexposure vaccine regimen for protection against rabies that continues in effect (with assorted modifications) to this day.[8] Nevertheless, few critical analyses of "Pasteur treatment" have succeeded in unequivocally confirming its efficacy.[9] Present highly purified vaccines of cell culture origin appear to be highly (but not infallibly) effective for postexposure use in man. The mechanism of protection induced by postexposure immunoprophylaxis remains to be explained. Also, efforts are still in progress to establish the spectrum of antigenic determinants required in a vaccine expected to protect against all feral strains of rabies virus.

The epidemiology of rabies is characterized by the existence of different but characteristic patterns of transmission, usually by a single predominant animal host in a given region[10] or occasionally by a succession of predominant hosts over a period of time (such as that observed in the changing cycles of importance of fox, skunk, and raccoon rabies in the U.S.). The ecological and/or virological explanation for the selective infection of certain mammalian species and the sparing of others in most regions of enzootic infection remains a potentially fertile field for exploration.

These problems have yielded very reluctantly to resolution by traditional approaches. The rabies virion is now being thoroughly characterized at the molecular level. The immune responses to rabies, both cellular and humoral, are being characterized by numerous investigators in newly exquisite detail. It is to be hoped that application of modern techniques of molecular biology, cell biology, and immunology to the study of rabies and in particular, rabies virus-cell interactions will yield new insights that contribute to the explanation of the more global issues of rabies virus pathogenesis and epidemiology. It will be the purpose of this chapter to present "the state of the art" of characterization of the rabies virion, the disease, and the immune response, not as definitive explanations, but as a progress report on an always dynamic process of multidisciplinary investigation.

II. THE CLINICAL DISEASE

The clinical picture of rabies is primarily an admixture of depressed signs typical of viral encephalitis with signs of agitation ("derangement") that are not invariable in their occurrence, but are particularly characteristic of rabies. Signs of peripheral paresthesia and localized paralysis also may occur.

The course of clinical disease is extremely variable. Such factors as host species, virus substrain, and site of bite exposure undoubtedly contribute to this variation. Unexplained extreme variation in the individual host response has also frequently been observed even in experimental studies in which apparently uniform groups of animals have been given identical inoculations with the same virus preparation.

Naturally occurring rabies has been most often described in man and in the dog since antiquity, with the most extremely violent types of behavior attracting the most attention. However, in one modern analysis of 49 confirmed cases of human rabies, it was noted that "it was extremely difficult for the clinician to make the diagnosis of rabies", and even the reviewing pathologist suspected an alternative diagnosis in 12 cases.[11]

In man the typical course of clinical signs and symptoms has been divided into five stages by Hattwick and Gregg.[12] The incubation period, which may vary from about 2 weeks to rarely a year or more, is characterized by total absence of virus-specific symptoms. The onset of clinical disease is characterized by a brief (a few days only) period of nonspecific neurologic symptoms of anxiety, irritability, or depression preceded in about 50% of cases by neuritic pain at the site of bite exposure. The third "acute neurologic stage" is characterized in most, but not all cases, by both behavioral aberrations such as hallucinations, disorientation, and fits of "furious" hyperactive behavior, and by paralysis. During this phase, fear of painful pharyngeal spasms often leads to the classically observed hydrophobia. Paralysis rapidly evolves as the predominant symptom leading to death by failure of essential respiratory functions, or to coma. The fifth stage described by Hattwick and Gregg is recovery. This phenomenon, although recently recorded in a child intensively treated with supportive therapy,[13] is extremely rare in man (serologic surveys indicate a fairly common occurrence of survival in several species of wild animals — see Section III. Epidemiology). When vital functions are maintained therapeutically, rabid individuals usually die after a prolonged course of illness because of failure of a variety of body systems, the role of which is probably of secondary importance in acute rabies not subject to therapeutic intervention.[12,14]

In the dog the incubation period varies as in man, after which a brief prodromal stage characterized by changes in behavior may be noted. The ensuing acute neurologic stage normally first attracts attention. Increasing irritability, excitability, and pica are noted.[15] Viciousness, aimless wandering, and drooling caused by paralysis of the muscles of the pharynx and the muscles of deglutition are observed. Convulsions may occur. Paralysis gradually predominates, leading to coma and death.

In the major wild animal reservoir species, the fox, skunk, and raccoon, the same types of vicious neurologic signs may develop, with the paralytic signs attracting less attention. In all vector species the frequent concurrence of excessive biting behavior and salivary shedding of virus obviously facilitates transmission. In wild animals other forms of behavioral derangement are equally important. Animals that are normally timid lose all fear of man; normally nocturnal animals go abroad throughout the day. A tendency to wander aimlessly has been shown to cause foxes to travel far beyond their normal territories, enabling a single animal to significantly extend the limits of an infection region.[16] However, the development of vicious behavior is far from invariable; a review of the literature by Parker[17] indicated that only approximately half of naturally infected populations of rabid foxes, skunks, raccoons, or dogs exhibited aggressive behavior.

Rabid bats are frequently alleged to show no increase in vicious behavior over their normal unpleasant dispositions. Baer[18] concluded that "furious" behavior is not characteristic of rabid vampire, frugivorous, or colonial insectivorous bats, but is a common feature of rabies in the solitary species of insectivorous bats.

Common laboratory animals very rarely exhibit any signs of rabies other than those of depression, encephalitis, and paralysis, regardless of the strain of virus, route of inoculation, etc. Laboratory animals (especially mice) also seem to differ from man and most species of domestic and wild animals in exhibiting a rather frequent occurrence of recovery from rabies. "Survival with sequellae",[19] especially with paralysis of one or more limbs or with persistent encephalitic signs is commonly induced in mice inoculated with rabies by a variety of virus strains and experimental regimens.[19-22] Yet we are unaware of any report of a wild or domestic animal demonstrated to exhibit nervous system sequellae after a naturally occurring rabies infection.

Johnson[23,24] has attempted to correlate the appearance of the typical signs of rabies with the coursing of the viral infection through the CNS. It was suggested that the selective susceptibility to rabies infection of neurons in the brainstem, hippocampus, septal nuclei, and limbic cortex coupled with a far lesser degree of susceptibility of neurons in the neocortex may be correlated with the clinical signs of alertness and aggressiveness necessary to sustain the infectious cycle. A comparison of this localization of rabies virus in the CNS in species and individual animals alternatively exhibiting "furious" and "dumb" clinical disease would obviously be of great interest.

III. EPIDEMIOLOGY

The basic event associated with the most common means of rabies transmission — the bite of a rabid animal — has been recognized since antiquity. Indeed, in few infectious diseases can the acquisition of infection be traced so unequivocally to a single readily identified event.

We are concerned about the existence of rabies in a given geographic area primarily because of the occasional incidental exposure of man or domestic animals. Yet the patterns of feral animal transmission ultimately responsible for such exposures are varied and complex, and not usually simplistically explained solely by the nature of the potentially infectable mammalian fauna. It may be assumed that a thorough understanding of the diversity of animal transmission patterns will require increased knowledge of virus strains and their host species-specific pathogenic potential as well as the more traditional mammalian population studies. Limited but tantalizing experimental studies of the variability of rabies pathogenesis in different mammalian vector hosts suggest that pathogenic processes vary in ways that may significantly affect the potential for further dissemination of the disease. Thus, a brief discussion of epidemiology seems particularly appropriate to an account of the "biology" of rabies.

Given the enormous variety of mammalian fauna and the world-wide distribution of rabies virus, a surprisingly limited number of fundamental virus transmission and animal reservoir cycles have been identified. The relative importance of different transmission cycles changes apparently because of natural phenomena and directed human interference, but the known wildlife cycles seem to be relatively stable over periods of at least decades. We are aware of no important new mammalian reservoir pools that have been identified in the last quarter century.

Historically, and presently in much of the developing world, rabies in dogs causes by far the greatest number of infections of man. In regions with large populations of unvaccinated dogs it was assumed, probably correctly, that dog-to-dog bite transmission provided a mechanism capable of perpetuating the infection over prolonged periods. The dog obviously is capable of inflicting severe bites capable of delivering rabid saliva deep into susceptible muscle tissues. The social habits of even stray dogs bring them into very frequent contact with each other and with man.

Unfortunately, there is a paucity of quantitative data on the susceptibility of the dog and the efficiency with which canine rabies virus is shed from the saliva. It appears that the dog is less susceptible to rabies than are several species of wild animals.[15] In studies of dogs experimentally inoculated with street rabies isolates, as many as 50% of the infected animals have died without ever excreting detectable levels of virus in the saliva.[25,26] Virus titers were usually $\leq 10^{3.0}$ mouse lethal doses per milliliter of saliva.

Control of canine rabies by reduction of feral dog populations and/or effective vaccination of pet dogs has invariably led to amelioration of the problem of human exposure. Control of this domestic animal reservoir has also revealed a variety of wild animal reservoir systems capable of maintaining enzootic infection in many regions. In the fortunate island countries of Australia, Japan, and Great Britain, the elimination of dog rabies apparently led to the complete extinction of the disease. In many other regions, virus has persisted well established in wildlife. According to Murphy,[10] "rabies has never been eliminated from a complex, multiple host ecosystem involving wildlife."

The wildlife rabies cycles identified are diverse but finite in number. Among the better characterized systems are:

Fox rabies — In either the arctic, red, or gray fox, rabies is widespread over the northern portion of all continents in the Northern Hemisphere.[16] Because of occasional long incubation periods and great distances traveled by foxes, the species seems capable of maintaining the infection alone with spill-over into other domestic and wild species being incidental to entrenchment of the infection. The current European rabies outbreak emerging westward from Poland in the 1940s appears to be based fundamentally on fox infection. Intensive study of this outbreak including computer modeling methods has led to the suggestion that there is an optimal level of transmission efficiency well below 100%, since perfect transmission efficacy would totally eliminate susceptible animals, and therefore the disease.[27] Arctic rabies seems to be maintained primarily by the arctic fox, but in temperate North America the fox shares reservoir potential with other mammals (see below).

Mongoose rabies — In the introduced mongoose population in several Caribbean islands, rabies seems to represent a stable essentially single host reservoir. Studies of many hundreds of clinically normal mongoose on the island of Grenada indicated an antibody incidence consistently greater than 20%. It was suggested that many of these animals must recover from rabies and that rigorous mongoose control programs might be counterproductive because of potential removal of immune animals that may modulate rabies transmission cycles.[28]

Vampire bat rabies — Represents one of the few reservoir systems not primarily recognized because of danger to man. Although an occasional cause of human disease, rabies of vampire bat origin has acquired notoriety primarily because of devastating losses induced in cattle in Latin America. In a single year the estimate of economic loss to the cattle industry has been as high as 500,000 cattle valued at $50,000,000 (1967 dollars).[29] It was suggested that the vampire bat may possess a unique capacity to secrete virus into the saliva over prolonged periods of nonsymptomatic infection.[30]

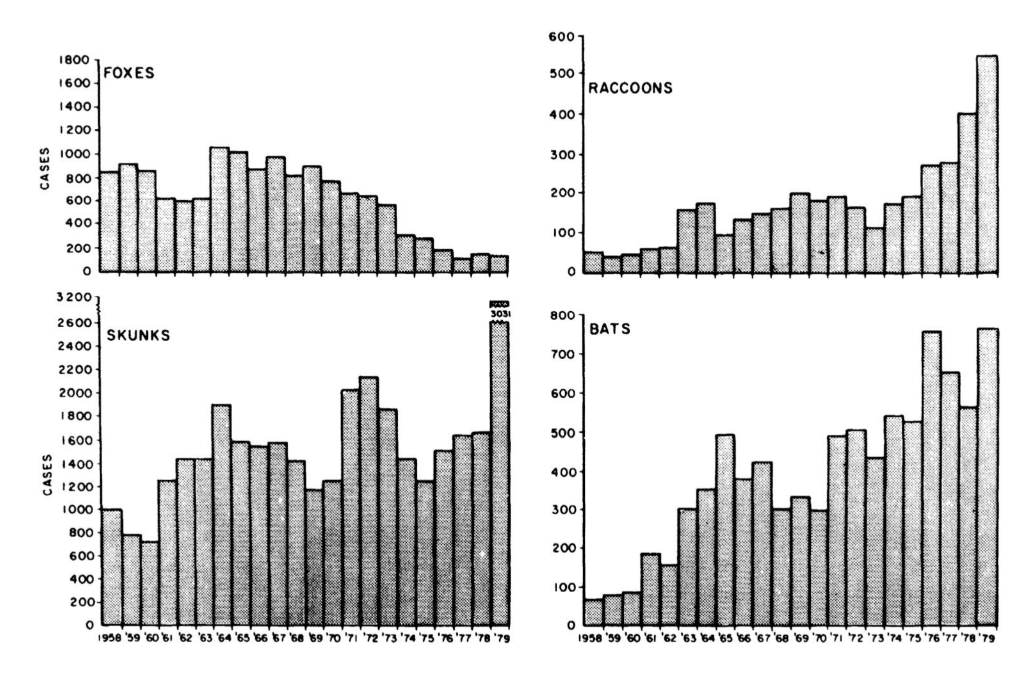

FIGURE 1. Wildlife rabies in the U.S. 1958 — 1979. (From Rabies Surveillance, Annual Summary 1979, U.S. Department of Health and Human Services, Washington, D.C., 1979.)

That cattle frequently survive vampire bat-origin rabies infection was suggested by the finding of rabies-specific antibodies in clinically normal cattle in affected regions at an incidence as high as 27%.[18]

Rodent rabies — Has been assumed to be very rare in most enzootically infected regions; many public health laboratories in the U.S. refuse to examine rodents suspected of rabies infection. However, isolation of rabies from several small rodents, primarily European voles (genus *Microtus*), has been reported in Czechoslovakia and southern Germany.[31,32] These viruses were reported to be antigenically typical of rabies, but also to be of extremely low pathogenicity for laboratory mice. The significance of voles as a reservoir host in Europe remains undetermined, but the claimed potential of this species to maintain rabies is of particular interest because voles may attain enormous population densities.

Multiple animal reservoir rabies — Although multiple host rabies reservoir cycles may exist in the rich fauna of infected African and tropical Asian areas of the developing world, the characterization of such cycles is at best primitive. It is fortuitous that the only well-monitored multihost reservoir system has evolved in the U.S. where resources are available for its continued characterization.

A brief recounting of patterns of rabies infection in the U.S. in the past 30 years will illustrate its evolution and its complexity. In 1953 there were reported 7334 cases of domestic animal rabies (5688 in dogs) and only 1479 cases of wild animal rabies, very predominantly in foxes (1033 foxes, 319 skunks, 38 raccoons, 8 bats, 81 other wild animals).[17] By 1958 the number of dog rabies cases had declined to 1643 and was exceeded by the number of fox and skunk rabies cases (845 foxes and 1005 rabid skunks). The number of dog rabies cases continued to decline until it stabilized at approximately 200 cases per year in recent years.[33]

The pattern of wildlife rabies has evolved in the manner illustrated in Figure 1. During the years when numerous cases of both fox and skunk rabies were observed, the cases were not indiscriminately mixed: fox rabies tended to predominate in Appalachia and the northeastern states while skunk rabies predominated in the central U.S., this despite the fact that both species coexisted in most areas. The subsequent dramatic decline in fox rabies and increase in skunk rabies was not associated primarily with displacement of one cycle with another in fox rabies areas. Rather, skunk rabies increased in the original skunk rabies region as fox rabies declined or disappeared in previous fox rabies areas. Fox rabies persisted primarily in areas also affected with skunk rabies, possibly as continued spill-over from that reservoir.

In very recent years, the ubiquitous raccoon has emerged from a prolonged role as the predominant reservoir host only in Florida and Georgia[34] to become a significant rabies vector in a range rapidly progressing northward along the East coast and reaching Pennsylvania in 1983.[35] Raccoons present a special concern because of their adaptibility to survival in very close proximity to human habitations, even in densely populated suburban areas.

Superimposed on these diverse patterns of terrestrial mammal rabies is the emergence of extraordinarily widely disseminated rabies in insectivorous bats. The number of rabid bats identified in the U.S. increased from 8 in 1953 to 858 in 1981. Even more significantly, rabid bats were found in 46 states in 1981.[36] Bat rabies is not correlated geographically with any particular terrestrial mammal reservoir; rabid bats have often been found in areas completely free of rabies in other wild mammals.

Enormous interest has attended the demonstration that colonial freetail bats may produce rabies virus in such quantity that the air in infested caves may cause respiratory infection of exposed mammals, an observation contrary to centuries of rabies dogma.[37] That such colonial bats efficiently infect each other (apparently with frequent survival) is indicated by studies showing a very high incidence of rabies antibody.[18]

The basis for the apparent stability of the infection cycle in solitary bats is not known. However, some surveys have revealed a much higher rate of infection in solitary than in colonial bats.[18,38] Experimental studies suggest that bats are capable of infecting various terrestrial animals by biting, but are rather inefficient.[39,40] It is a paradox that bats in the U.S. maintain a stable widely disseminated pattern of infection in their own populations while apparently exerting little effect on the fundamental cycles of rabies transmission in other wild or domestic terrestrial mammals.

Since the relative incidence of rabies in different animal species is not randomly distributed, experimental studies of comparative pathogenesis have been pursued in an effort to explain epidemiologic observations. Despite the long-established twin dogmas both: (1) that rabies is a monospecific virus consisting of a single strain, and (2) that all mammals are susceptible, preliminary experimental studies have re-examined the role of both the virus isolate and the host animal in determining the outcome of infection.

The most illuminating of such studies are found in the classical reports of fox and skunk rabies isolates inoculated into foxes and skunks reported by Sikes[41] and by Parker and Wilsnack.[42] Sikes, using a fox rabies virus isolate demonstrated that foxes secrete much less rabies virus than do skunks, but may also be infected by doses of virus almost 100-fold less than those required to infect skunks. Thus, foxes secrete enough salivary virus to infect foxes, but not skunks. Skunks secrete salivary virus in high concentrations capable of infecting either skunks or foxes. However, foxes are so exquisitely susceptible to rabies that they may succumb to encephalitis from high doses, such as those found in skunk saliva, before their infection has disseminated into their own salivary glands. Parker and Wilsnack, using a rabies virus isolated from skunks, found skunks and foxes to be equally susceptible, but repeated the observations: (1) that skunks secrete virus in higher titers, and (2) that foxes inoculated with virus of high titer die without themselves secreting salivary rabies virus. These elegant observations seem to explain the possibility that skunks might remain free of rabies despite exposure to rabid foxes, as well as the converse situation.

The most numerous studies of rabies pathogenesis in wild species have been conducted in bats.[18,43] Interest in bats was first attracted by the results of experimental studies performed in Trinidad[30] and Brazil[44] in the 1930s that indicated that bats could act as asymptomatic carriers and potential transmitters of rabies virus over periods of many months. Similar studies were not repeated for over 40 years; in a single modern study, vampire bats inoculated with a vampire bat isolate either i.m. or s.c. were observed to sicken and die following periods of presymptomatic salivary virus secretion that did not exceed 8 days.[45] The validity of the early South American studies must be considered questionable until they can be repeated by modern methods.[46]

Experimental studies of North American insectivorous bats have indicated that they also tend to follow a disease course that is basically similar to that of other mammals: they suffer fatal encephalitis after varying incubation periods with periods of salivary shedding of virus that only slightly exceed those commonly observed in terrestrial mammals.[47] Sulkin et al.[48] demonstrated that progress of rabies infection in insectivorous bats may be delayed during hibernation, allowing such animals to potentially act as over-wintering hosts of rabies. Rabies virus seemed to be often localized in brown fat during the preclinical phase of the infection.[49]

Constantine demonstrated that rabies virus isolated from colonial insectivorous bats had a normal potential to cause typical disease in a variety of wild animal species inoculated artificially by parenteral routes.[50,51] In his experimental bat bite studies, however, infected bats efficiently infected only the highly susceptible fox and coyote.[39] It is unfortunate that relatively few studies of bat-to-bat bite transmission have been reported, given the frequent compartmentalization of the bat rabies transmission cycle within the mammalian wildlife ecosystem.

IV. THE VIRUS

Rabies is classified as a rhabdovirus of the genus *Lyssa*. Members of the family Rhabdoviridae have a characteristic bullet-shaped morphology and possess a genome consisting of a single molecule of negative strand RNA. Although animal rhabdoviruses are found in extremely diverse vertebrate (mammal, bird, fish and reptile) and invertebrate hosts, rabies virus has been called "by far the most important rhabdovirus infecting vertebrates".[52] Rabies virus is unique among rhabdoviruses in its very broad world-wide geographic distribution, in its predilection to cause encephalitis in mammalian hosts (a tropism shared with antigenically related lyssaviruses), and in its status as a significant public health risk to man.

By a unique nomenclature, rabies virus strains are categorized as either "fixed" or "street".[53] Fixed strains have been adapted to a laboratory animal by serial intracerebral (IC) passage to a point such that fatal encephalitis ensues after a very few days (e.g., 5 to 10),[54] and further passage does not lead to further reduction in the incubation period ("the virus is fixed"). Characteristically, fixed virus infects very efficiently when administered by the IC route and very inefficiently when administered by a peripheral route in comparison with feral or "street" rabies virus. Fixed virus differs from street rabies in growing to very high infectious titers in the brain (about 10^6 infectious doses per gram compared to 10^3 for street virus)[55] and in *not* causing infection of salivary glands. Virtually all rabies strains that have been adapted to growth in cell culture have been fixed strains.[4,56] Therefore, the reader must be aware that all descriptions of the physicochemical properties of rabies virions and essentially all studies of virus replicative processes in cell culture deal with viruses whose biological phenotypes differs in many respects from that involved in the etiology of naturally occurring infections.

A. Physicochemical Properties

Rabies is a bullet-shaped virus 75 to 80 nm in diameter and approximately 180 nm in length. Particles examined by negative contrast electron microscopy[57,58] reveal an outer layer of surface spikes about 8 nm in length arranged in honeycomb symmetry separated by intervals of 4 to 5 nm. These spikes represent the hemagglutinin. Beneath the spikes is a unit lipoprotein membrane envelope derived from cell membrane. Within the membrane the helically coiled nucleocapsid is visualized as cross-striations with an interval of 4.5 nm. A central electron-lucent core or apical channel is approximately 40 nm in diameter. Particles that are complete (infectious) have been reported to vary somewhat in length; also virions of some fixed strains appear more conical than cylindrical in outline. However, these variations in gross appearance have never been correlated with identifiable differences in the composition of the molecular components of the virion.

Molecular structure — As rabies virus does not efficiently suppress host cell macromolecular synthesis[59] nor apparently impinge directly on the host cell genome, its mechanism for inducing cytopathology remains obscure. A thorough understanding of cytopathic effects (and possibly their therapeutic amelioration by specific antiviral drugs) will depend upon the further definition of the virus molecular constitution and replication strategy. The gross chemical makeup of the virus is approximately 74% protein, 22% lipid, 3% carbohydrate, and 1% RNA.[60] The RNA consists of one single-stranded molecule of mol. wt. ca. 4.0×10^6 daltons complexed with nucleoprotein to form the helical nucleocapsid that is approximately 1.0 μm in length and 100Å in diameter.[61] There are four major structural proteins. In addition to the nucleoprotein of the virus core, two membrane proteins and a glycoprotein make up the viral envelope, with the glycoprotein comprising the surface spikes (hemagglutinin). A very large (L) protein is present in few copies and presumably represents the virus transcriptase.[62]

Glycoprotein — The glycoprotein is of primary biological importance as it is responsible for eliciting protective virus-neutralizing antibody[63,64] and as the hemagglutinin is presumably also the prime determinant governing the ability of the virus to attach to receptor sites on susceptible cells. The glycoprotein is currently receiving special attention from molecular biologists hoping to precisely define the antigenic domain(s) of the molecule capable of eliciting a protective immune response. Such knowledge should contribute significantly to the understanding of both strain variation and the mechanism of specific virus neutralization by antibody and, most attractively, to the possibility of developing a strategy for producing a rabies subunit vaccine by modern means of genetic engineering.

The isolated rabies glycoprotein has been shown to be an effective protective immunogen, especially when prepared under conditions that favor the formation of large aggregates of the 80,000 mol. wt. monomers.[65,66] Such aggregates are more potent immunogens than the glycoprotein as presented on the native virion itself.[64] Four separate sites on the glycoprotein have been demonstrated to react with virus-neutralizing monoclonal antibodies.[67] Three distinct peptide fragments produced by CNBr cleavage of the glycoprotein were shown to elicit virus-neutralizing antibodies.[68] However, none of the CNBr cleavage products were bound by neutralizing monoclonal antibodies, indicating an important role for secondary and tertiary structure in establishing the antigenic configuration of the entire molecule. Reagan and Wunner[69] have recently demonstrated that purified glycoprotein preparations are not capable of blocking specific rabies receptor sites on cultured cells.

The tertiary structure and manner of presentation of the glycoprotein are apparently also critical in the process of hemagglutination, as in early studies hemagglutinating capacity of rabies was found to be associated only with intact virus particles.[70] Subsequently, however, as in the case of the protective antigenic moiety, it was demonstrated that surface glycoproteins could be removed from the rabies virion in such a manner as to maintain their hemagglutinating potential.[71]

Intense interest attends the possibility of cloning the rabies virus glycoprotein gene and obtaining expression of the corresponding antigenic polypeptide in a prokaryotic vector. Anilionis et al.[72] successfully cloned a cDNA to the strain ERA rabies glycoprotein mRNA into a pBR322 plasmid in *E. coli*. From the base sequence of the cloned cDNA they deduced a structure for the complete polypeptide that is comprised of 524 amino acids. A C-terminal hydrophilic sequence of 44 amino acids was considered to be potentially involved in binding to the internal virion M_2 and N proteins whereas the adjacent 22 amino acid highly hydrophobic sequence was presumed to represent the transmembrane region of the molecule.

By pursuing a similar strategy, the gene for the glycoprotein of strain CVS rabies virus has also been cloned in plasmid pBR322 by Yelverton et al.[73] Apparent expression of two protein gene products has been demonstrated by specific reaction with rabies antiserum. The ability of these nonglycosylated polypeptides to elicit virus neutralizing antibodies has not yet been determined.

N protein — The other (nonglycoprotein) structural polypeptides of the virion have received less attention. The nucleoprotein (N protein) is bound to the RNA genome to form the nucleocapsid, representing 96% of the mass of the nucleocapsid.[61] The N protein does not elicit virus-neutralizing antibody following inoculation into animals.[63] Antibodies to N protein of rabies do cross-react with the N proteins of all other viruses of the genus *Lyssavirus* in a variety of in vitro assays; therefore, N protein is considered the group-specific antigen of the genus.[74] Although these and other studies suggest that the structure of the N protein is selectively conserved during evolution, studies with monoclonal antibodies have demonstrated epitopic variation in the N proteins of rabies strains that vary in neutralization antigen composition and in virulence.[75-77]

M_1 and M_2 proteins — The membrane proteins (M_1 and M_2) have been little characterized, although they have been shown to be incapable of eliciting virus neutralizing antibodies.[64] Indeed the propriety of their designation both as membrane proteins and the original assumptions regarding their physical configuration within the virion have been questioned. The protein traditionally designated M_1 has been shown to be selectively associated with the N and L proteins, whereas the M_2 protein associates with the G protein. This has led to the conclusion that M_2 is indeed a membrane protein, but M_1 is a nucleocapsid protein analogous to the core protein NS of the prototype vesicular stomatitis rhabdovirus.[78-80] The latter conclusion suggests that the molecular arrangement of rabies virus is much more similar to that of other rhabdoviruses than was believed previously.

L protein — Early observations of rabies virion-associated polypeptides also indicated that rabies differed from vesicular stomatitis virus in lacking the L polypeptide,[81] shown to be necessary for RNA transcription in other rhabdoviruses.[82] However, Madore and England were successful in demonstrating this large (~190,000 mol. wt.) polypeptide in infected cells,[62] and it was subsequently identified in both purified rabies virions and in isolated nucleocapsids.[79] The difficulty in characterizing this protein may be attributed to the very small amounts found in each virion; estimates range (possibly varying with virus strain) from 17 to 150 copies per virion.[62,83]

In vesicular stomatitis virus it has been shown that although the L and NS proteins are less firmly bound to the virion RNA than is the N protein,[84] they do form a subviral complex with infectious potential.[85] Rabies nucleocapsids isolated from infected cytoplasmic extracts under conditions allowing recovery of only the RNA:N protein complex were shown to be noninfectious.[61] Demonstration of a more complex rabies component with bound L protein and M_1 (NS analogue) with infectious potential would have great pathogenic significance, as a nucleoprotein complex has often been postulated as the minimal rabies subviral unit capable of transmitting infection through intraneuronal pathways in the infected animal.

Structural lipid and carbohydrate — These components have received less attention as they are neither apparently important immunogens nor are they encoded by the rabies virus genome. That viral lipid is essential for maintenance of infectivity was indicated by experiments showing inactivation of rabies virus by ether performed more than 50 years ago.[86] Removal of a portion of the phospholipids from purified rabies virions has been shown to cause dissolution of typical bullet-like morphology with retention of infectivity.[87] Blough et al.[88] compared the lipid composition of the attenuated Flury HEP rabies virus and of the membranes of host BHK tissue culture cells. They concluded that since viral lipid differed from that of both plasma membrane and endoplasmic reticulum, considerable amounts of virus must maturate *de novo* without using either cell membrane system as template. These observations may not be applicable to other virus-host systems in which electron microscopic observation has indicated varying patterns of virion maturation sites *(vide infra).*

Although a considerable amount of carbohydrate is presumed to be present in glycolipids and mucopolysaccharides of the virus membrane,[89] only the carbohydrate moiety of the rabies virion glycoprotein has been characterized. It has been shown that each glycoprotein molecule of rabies virus contains three carbohydrate side chains. These are composed of mannose, fucose, galactose, glycosamine, and sialic acid, with the concentration of the latter component varying according to virus growth conditions.[90]

B. Replication of Rabies Virus

The manner by which pathogenic rabies viruses induce cytopathology or cell physiologic dysfunction is not known. A number of processes involved in rabies viral replication may be specifically deleterious to the host, but none has been proved to be cytopathic. Furthermore, the replication of rabies virus has been studied only in a fragmentary fashion. Most of the presumed replicative pathway has been deduced from studies of VSV, but numerous critical steps in VSV replication also remain unexplained.[91] Hence replication will be discussed only very briefly here with evidence of apparent specific cytopathic mechanisms described under Section VI. F., Cell-Virus Interaction In Vitro.

Rabies virus attaches to apparently rhabdovirus-specific receptor sites on the host cell membrane.[69] Electron microscopic studies of fixed rabies virus interacting with BHK cells in monolayer indicate that attachment occurs preferentially by alignment of the flat basal end of the virus particle with the cell plasma membrane.[92] Penetration may then occur either by fusion of the viral and host cell membranes allowing release of the nucleocapsid complex into the cytoplasm or by phagocytosis of the intact virion. In the latter case, fusion of the virion membrane with the membrane of the phagocytic vacuole has been shown to occur, thereby allowing release of virus core components into cytoplasm.

Within the cytoplasm, transcription of the rabies RNA (negative strand) into complimentary "plus strand" RNA proceeds under the influence of the virion-associated RNA polymerase (transcriptase). The virus-bound transcriptase has been demonstrated in purified virions by several investigators,[93,94] but exhibits considerably lower efficiency than the transcriptase of VSV and several other more rapidly replicating rhabdoviruses.

The transcription product has been identified as five separate size classes of polyadenylated message recoverable from infected cell ribosomes and complementary to virion RNA.[95] The mRNA molecules are translated to produce the five distinct rabies structural proteins detectable in infected cells and in purified virions. Each of the proteins G, N, M_1, and M_2, and L has been shown to be distinct by two-dimensional peptide map analysis, indicating that none of those polypeptides represents a processed cleavage product of a common precursor; rather, each is a separate monocistronic gene product.[96] The identified gene products represent the preponderance of the coding capacity of the virus genome. Nonstructural proteins encoded by the rabies genome have not been detected.

Many of the processes following translation are incompletely understood or are deduced from studies of VSV.[84] The newly synthesized G and membrane M_2 proteins are rapidly inserted into either host cell plasma or internal membranes in which they displace the preexisting host cell membrane proteins. These proteins are rarely found free in excess in the cytoplasm; the synthesis of M_2 protein may be the rate-limiting process for replication of complete virions. The N, M_1 (NS), and L polypeptides are released into the cytoplasm where particularly the N protein may accumulate in excess as nucleocapsid, comprising the primary component of the aggregates visualized as matrix by electron microscopy and as cytoplasmic inclusions (including Negri bodies) by light microscopy.

The N protein is bound very firmly to viral RNA to form the nucleocapsid. Only that nucleocapsid comprising full length negative strand rabies viral RNA can contribute to the formation of infectious virions. The replication of this negative strand RNA is poorly understood. It appears later in the course of infection than the transcribed mRNA molecules and is never found in quantity free in the cytoplasm. For lack of an alternative explanation, it has been assumed that the L protein transcriptase may also act as replicase to direct synthesis of the negative strand genome. Such replicase action would occur after prior binding of the NS protein to a full length virus plus strand complementary RNA molecule also in the form of nucleocapsid.[91]

The means of final maturation of the virion is not understood. The genome-containing nucleocapsid-NS protein-L protein complex must make contact with a cell membrane containing the appropriately inserted G and M (M_2) proteins to which it presumably attaches at a specific molecular recognition site. The nucleocapsid complex then buds through the now viral membrane (by unknown means) to produce the mature virion. The preferred sites of membrane maturation (external or internal) and the proportion of nucleocapsid enveloped as virus as compared to that accumulated as matrix both vary considerably with virus strain, host cell type, and the time of observation postinfection.[92,97-100]

C. Defective Interfering (DI) Particles

Rabies virion components may occasionally be assembled into shortened (truncated = "T") particles consisting of a normal complement of virion polypeptides, but an abbreviated deletion mutant RNA genome. If these particles can be shown to be (1) noninfectious, (2) capable of replicating in the presence of complete infectious "standard" virus, and (3) capable of interfering with replication of standard rabies virus, they fulfill the criteria for being identified as defective interfering (DI) viruses.[101]

Although a role for DI rabies virions in pathogenesis has often been postulated, the DI rabies particle populations actually characterized have been largely derived from a limited number of fixed virus strains propagated in cell culture. DI particles were identified in BHK/21 cell cultures supporting chronic infection with either of the attenuated fixed rabies strains Flury HEP[102] or ERA.[103] It was postulated that the interfering capacity demonstrated for either DI particle may have spared enough cells to permit indefinite maintenance of the viable infected cell lines. The ERA defective particle effectively protected mice against challenge with ERA standard virus when a mixture containing excess DI particles was administered by the intracerebral route, but not when given by a peripheral (intraplantar) route.[103]

DI particles were first detected among the progeny of cell cultures acutely infected with Flury HEP rabies virus by Crick and Brown.[104] Subsequently DI virions were detected in the progeny of acute infections of cultured cells with virus of strains CVS and ERA as well by Clark and colleagues.[105] The latter authors showed that such particles derived from different strains differed in both size (which is derivative of the size of the DI virion RNA molecule) and interfering efficiency in cell culture. DI particle-mediated interference was particularly marked in neuroblastoma cells. Strain ERA DI particles of two different sizes effective in interfering with rabies virus replication in cell culture were extremely inefficient in protecting mice against fatal rabies encephalitis. Protection was demonstrated only when mice were given a dose of 5×10^8 DI particles mixed with less than 50 mouse lethal doses of standard virus.

The propensity of a given virus stock to form DI particles can be eliminated by serial passage at low multiplicity of infection (MOI). ERA rabies virus freed of DI particles by serial passage at low MOI subsequently regenerated a variety of DI particle sizes identical to that of the parental virus stock during subsequent high MOI passages, indicating that the spectrum of DI particles produced is strain-specific.[106]

Flury HEP rabies virus passaged in neuroblastoma cells reacquires virulence for mice (see below). A DI particle-free preparation of Flury HEP virus was passaged serially in neuroblastoma cells and shown to simultaneously acquire both the propensity to generate DI particles and increased mouse virulence, suggesting a minimal role for DI particles in attenuation.[106]

Unfortunately it has been difficult to demonstrate DI particles in rabies-infected brain tissue because of difficulty in purifying virions from such tissue. Attempts to demonstrate the generation of rabies DI particles *de novo* in virus serially passaged in newborn mouse brain have led to conflicting results.[106,107] At present, the attractive hypothesis that DI particle formation might explain in vivo rabies phenomena such as frequent low yields of virus from brain[55] and frequently prolonged incubation periods remains an open question for further investigation.

D. Genetics and Virus Variation

Until recent years rabies was considered a "single strain" virus because of the extensive antigenic cross-reactivity of known strains. Genetic studies of rabies with classical methods involving characterization of temperature-sensitive (ts) mutants were unrewarding because complementation between such mutants has not been demonstrable.[108] Nevertheless, there exists a rich variety of stable phenotypes of rabies virus differing in pathogenicity. These offer a potentially powerful tool for correlating virus structure with the pathological effects of rabies virus.

The dichotomy of fixed and street rabies strains was described by Pasteur a century ago.[53] More than 35 years ago, Habel and his co-workers examined many strains of fixed virus, all derived from the original Pasteur isolate and all passaged in an essentially similar manner in animal brains, and concluded that, "Substrains of Pasteur fixed rabies virus are not identical in their antigenicity, incubation period, duration of symptoms, pathology, or in their ability to overcome rabies immunity in immunized mice".[109,110] Variation in street viruses is also readily detected. Baer et al.[21] compared the mouse-pathogenic potential of street virus contained in salivary glands of 11 rabid animals (foxes, skunks, and bobcats) identified in the state of Arizona. The isolates exhibited remarkable differences in pathogenicity as characterized either by length of incubation period or by "invasiveness" — the relative efficiency of infection by footpad inoculation as compared to the intracerebral route of infection.

Especially useful for the study of factors leading to rabies virus-induced death are the variety of live vaccine strains of fixed rabies attenuated to varying degrees by serial passage in chick embryos or in cell culture.[54] The most attenuated of these, Flury HEP, and Kelev viruses, show, respectively, little[111] or no[112] capacity to cause fatal encephalitis in intracerebrally inoculated adult mice.

With the development of efficient plaque assays for fixed rabies viruses,[113] attempts were made to correlate such phenotypic virus characters as small plaque size and replication restriction at elevated incubation temperatures with attenuation. However, these strain markers were found to vary independently from mouse-virulence in a comparison of fixed strains CVS, PM, ERA, and Flury HEP.[114] A similar lack of correlation of small plaque or ts phenotypes with reduced mouse virulence was subsequently shown for the rabies-related lyssaviruses Mokola and Lagos bat virus.[115]

The molecular structural composition of rabies virus strains CVS, PM, ERA, and Flury HEP has been compared by Dietzschold et al.[83] Tryptic peptide analysis of the structural polypeptides revealed marked differences only in the G protein. The proteins of PM and CVS appeared more closely related to each other than to those of the less pathogenic ERA and Flury HEP strains. The PM and CVS strains were also reported to contain the highest number of L protein molecules per virion. It is not possible at present to determine whether these observed molecular differences are implicated in phenotypic differences in strain virulence or merely reflect the common origin of the PM and CVS virulent strains from original Pasteur strain virus.

Attenuation of rabies vaccine viruses by prolonged passage in chick embryos or cell culture presumably leads to accumulation of multiple mutations, many of which may not be involved in attenuation. Hence, possible attenuation by presumably point-mutation of the rabies virus genome by chemical mutagenesis has been investigated. Clark and Koprowski[22] identified one of five ts mutants of strain CVS virus as inducing an aberrant death pattern in intracerebrally inoculated mice: most mice survived infection with undiluted virus, but fatal encephalitis occurred frequently with lower doses. The same virus caused disease following a normal dose-response curve when inoculated via the intraplantar route, but the affected mice differed from those similarly infected with CVS virus by frequently surviving indefinitely with residual paralysis. Subsequent further analysis of the "mutant" virus causing this unusual pathogenic pattern revealed that the virus was in fact a relatively stable mixture of ts and revertant viruses.[116] Neither the ts virus nor the revertant virus alone caused the observed aberrant death patterns.

Aubert et al.[117] studied the response of mice to intramuscular inoculation with three different ts mutants of CVS virus isolated by Bussereau and Flamand.[118] Each of these viruses was representative of a different mutant group characterized according to the stage of replication blocked by nonpermissive temperature. Two of the three mutant viruses required increased doses (in pfu) to induce lethal disease; two of three also exhibited reduced titers of virus in mice with fatal encephalitis when compared with mice infected with wild-type CVS virus. Virus recovered from fatally infected mouse brains was demonstrated to retain the ts phenotype. The mutant that was least pathogenic was that which exhibited a ts lesion at the earliest stage in the replication cycle, belonging to a mutant group that exhibited no secondary transcription or replication at nonpermissive temperature. Lack of pathogenicity of this mutant may also have been caused by a particularly low nonpermissive temperature.

Factors favoring pathogenicity of rabies (and other) viruses have frequently been studied by comparison of wild-type virus with virus attenuated by very prolonged passage in an atypical host system. An alternative method of studying factors affecting the pathogenicity rabies virus became available when it was discovered that attenuated rabies viruses of strains Flury HEP and Kelev *increase* in virulence when passaged in either mouse C1300 neuroblastoma cells or in one of several human neuroblastoma cell lines.[119] Such reversion had previously been demonstrated only by passage in vivo in newborn mouse brain.[120] Reversion to virulence in C1300 cells occurred after very few (one to five) passages. The rapidity of this phenotypic change suggested that identification of the genotypic change associated with increase in virulence might be facilitated in this system since its rapidity would limit the accumulation of possibly concomitant irrelevant gene mutations.

When randomly selected attenuated Flury HEP virus clones were serially passaged in C1300 cells, reversion to virulence appeared after a variable number of passages. This suggested that these neuroblastoma cells selectively favored replication of randomly occurring back-mutations to virulence.[121] However, other experiments indicated that the mutation(s) favored by cultivation in neuroblastoma cells may not represent all of the genomic factors involved in virulence. Thus, certain attenuated clones derived from ts mutants of CVS rabies virus did not increase in virulence following passage in C1300 cells. Similarly, the low degree of virulence associated with strain ERA rabies virus and certain strains of Mokola, Lagos bat, Kotankan and Obodhiang lyssaviruses was not markedly enhanced by passage in these cells.[121]

Antigenic variation — Rabies was long considered to be a virus with a single antigenic serotype; for this reason antigenic composition historically attracted little attention in the preparation of rabies vaccines.[9] However, antigenic differences in fixed rabies strains were suggested by the results of mouse cross-protection tests performed by Habel and co-workers[109,110] and more recently by Crick et al.[122]

Wiktor and Clark demonstrated antigenic differences in four strains of cell culture-adapted rabies virus compared by means of plaque-neutralization and neutralization kinetic tests.[123] The existence of antigenic variation in both fixed and feral strains of virus has recently been unequivocally demonstrated by several investigators using monoclonal antibodies to rabies virus.

E. Monoclonal Antibodies

Since the advent of hybridoma technology in 1975,[124] our understanding of the structure of viruses has increased considerably. Monoclonal antibodies are highly specific probes that can be used to delineate antigenic determinants on proteins, including viral proteins. Generation of hybridomas synthesizing monoclonal antibodies against rabies virus[76] has resulted in a better understanding of the structure and function of various rabies virus polypeptides. Initially, all of the antibodies directed against viral glycoproteins (GP) were shown to neutralize the virus,[125] whereas, antinucleocapsid (NC) protein antibodies were nonneutralizing.[126] This clear distinction allowed the recognition of viral glycoproteins as essential components for both neutralization of the virus and induction of neutralizing antibodies. Although antiNC antibodies were unable to differentiate between ERA, CVS, Flury-LEP, and Kelev strains of fixed rabies viruses, antiGP antibodies readily distinguished these and other strains from one another. These studies also provided a basis for distinguishing and grouping rabies and rabies-related viruses. The most striking observation was the antigenic diversity detected in comparisons of the various strains of rabies that were previously thought to be antigenically homogeneous. Monoclonal antibodies were subsequently used to identify the virus strain in impression smears of brains from different species of animals. These experiments revealed the antigenic composition of street rabies strains is related to both the host species and geographic source of a given isolate.[127]

Monoclonal antibodies have been used to generate a number of antigenic variants in the laboratory. Plaque-purified parental virus, when grown in the presence of a monoclonal antibody, will give rise to progeny virus that no longer reacts with the selecting antibody. Based on these results, it was shown that, for a given antigenic determinant, frequency of variation in rabies virus can be as high as 10^{-4} (i.e., 1 in 10,000 particles will be antigenically different). The possible relevance of these antigenic variants to rabies immunoprophylaxis became apparent when it was shown that immunization of animals with these variants failed to confer complete protection against a challenge with the parental virus.[128] The relevance of this observation was confirmed by the studies of Schneider and Meyer.[129] These investigators showed that vaccination of mice with Pitman Moore (PM) strain of rabies (vaccine strain used for human vaccine production) virus fails to protect against a challenge with certain natural isolates. These observations are very important considering that vaccines in use for humans are all derived from a limited number of the original fixed strains isolated many years ago.

Antigenic characterization of a number of virulent strains of fixed rabies virus using a panel of 19 antiGP antibodies and a panel of 19 antiNC antibodies revealed that, in general, there are a larger number of GP determinants expressed on the virulent strains. Upon passage in chronically infected cultures, these strains lose their virulence with concomitant change in a number of determinants on both NC and GP.[77] Similarly, when avirulent viruses are passaged in neuroblastoma cells, they become virulent and also exhibit changes in antigenic determinants expressed. These studies showed that laboratory-induced changes in the biological properties of rabies virus may be accompanied by changes in the antigenic phenotype. These observations were further supported by the work of Coulon et al.,[130] who showed, conversely, that the biological phenotype can be altered by selection of antigenic variants with monoclonal antibodies. These investigators showed that avirulent mutants of the CVS strain (which is highly virulent) can be readily isolated from virus populations reacted with neutralizing monoclonal antibodies.

Recently, Dietzschold et al.[131] have provided new insights at the molecular level on how antigenic alterations influence viral pathogenesis. Avirulent variants of CVS strain were selected in the presence of neutralizing monoclonal antibody. The amino acid sequence data obtained for glycoprotein of both the parental virulent virus and the avirulent mutant have shown that a single amino acid substitution in position 333 was responsible for the alteration in the biological property. Virus with arginine in position 333 of the glycoprotein was virulent, but substitution of this amino acid with either isoleucine or glutamine rendered the virus avirulent.

Antigenic variants in what were once thought to be homogeneous virus populations may be one of the explanations for some vaccine failures. Deaths due to rabies virus infection in previously immunized animals and individuals are thought to be due to failure of vaccine to elicit an adequate immune response. However, such deaths may very well be caused by antigenic variants that arise in nature and differ from the vaccine strain.[129]

V. PATHOLOGY OF STREET RABIES

A. Central Nervous System (CNS)

The histopathological features of rabies encephalitis caused by street virus are essentially similar in wild, domestic, and laboratory animals infected naturally or artificially and in man. The features of this encephalitis have been thoroughly reviewed by Perl,[132] who acknowledged that most of the significant pathologic changes had been described by the end of the 19th century! Experimental studies on virus transit and the dynamic aspects of rabies pathogenesis have attracted much more attention in recent decades; these studies have been ably reviewed by Murphy,[10,55] Schneider,[133] and others.

Despite the very common occurrence of irritation of presumably virus-specific origin at the wound site in man and in some species of animals, no specific gross or histological pathology at this location has been described. The only grossly visible changes described in the rabid brain are cerebral edema and meningeal congestion.[55,132]

The microscopic changes in animals and in man (when the disease is not prolonged by immune intervention) are primarily remarkable for their failure to indicate disease of a severity likely to result in the nearly universally observed fatal outcome. Neither do the histological changes reflect the very widespread dissemination of virus infection indicated by fluorescent antibody studies of the brain.

A perivascular infiltration of inflammatory cells (predominantly mononuclear) is commonly observed, most prominently in the brain stem (especially pons and medulla), the grey matter of the spinal cord, and in certain spinal ganglia. Degenerative changes in the spinal ganglia, particularly the gasserian, have been described to be characteristic enough to allow specific diagnosis of rabies.[134,135] Neuronophagia occurs and many predominantly microglial cells may accumulate around degenerating neurons in the parenchyma of the brain, constituting what have been called Babes' nodules. Although often found in rabies encephalitis, similar lesions also occur in other encephalitides.[132] Since the sum of these lesions does not suggest a critical degree of cell killing, the challenge of modern rabies research is to identify a presumed fatal neurophysiologic lesion.

Analysis of CNS histopathology was instrumental in the proper diagnosis as rabies of the epidemic of vampire bat-transmitted fatal "ascending myelitis" that occurred in Trinidad in 1929 and 1930. Hurst and Pawan[136] noted lesions compatible with paralytic rabies in that "the spinal cord bears the brunt of the infection" with much nerve cell destruction and microglial infiltration. These authors claimed that rabies damage in the cord affected anterior and posterior cords equally, thus allowing rabies to be distinguished from poliomyelitis.

Dupont and Earle[11] studied a series of 49 human brains of persons dying of rabies on four continents collected over a period of more than 20 years. Although a typical pathologic picture could be ascertained, pathological examination alone did not lead to a diagnosis of rabies in 12 cases. In six of these, the only histopathological change was "mild congestion".

Rabies-characteristic human cases displayed primarily changes in the brain stem and cord. Perivascular cuffing was seen in 94% of cases and neuronophagia in 57%. The order of frequency of lesions in specific regions of the CNS in this series was first the medulla, followed by pons, cord, thalamus, cerebral cortex, basal ganglia, cerebellum, hippocampus, and peripheral ganglia. This distribution of lesions is in approximate agreement with virus distribution in CNS tissues of three human rabies cases studied by Leach and Johnson.[137] Highest rabies virus titers were consistently recovered from the thalamus and cervical, thoracic and lumbar cord, and the pons and medulla. However, virus was also consistently recovered from most regions of the cerebral hemispheres and cerebellum. Virus was not found in cerebral spinal fluid.

The intensity of the inflammatory lesions of street rabies encephalitis may be affected by the species of host, the virus strain, and by the course of the disease. Thus, it has been reported that inflammatory changes may be more severe in the dog than in pigs or herbivores.[132] Very severe inflammatory and necrotic changes have been associated with rabies virus isolated from Mexican freetail bats (see Figure 2), with Flury LEP rabies vaccine virus that caused sporadic rabies in vaccinated dogs (Figures 3 and 4), and with the rabies serogroup Mokola and Lagos bat viruses.[138] Intense encephalitic lesions are also associated with prolonged clinical disease. Histopathology associated with "chronic progressive rabies" that persisted for 136 weeks after inoculation of a cat both i.m. and i.p. with a bat rabies isolate[139] is illustrated in Figure 5.

B. The Negri Body

Given the frequent paucity of histopathologic lesions in rabies and the fact that such lesions often do not allow one to distinguish rabies from other encephalitides, great interest has focused upon the characteristic cytoplasmic inclusion body of rabies first described by Negri in 1903[6] (see Figures 6 and 7). Negri bodies are round or oval neuronal cytoplasmic inclusions that vary in diameter from about 1 to 30 μm. The idealized Negri body has a basophilic "inner body" surrounded by a more acidophilic peripheral region.

The size and incidence of Negri bodies may vary with virus strain and host animal

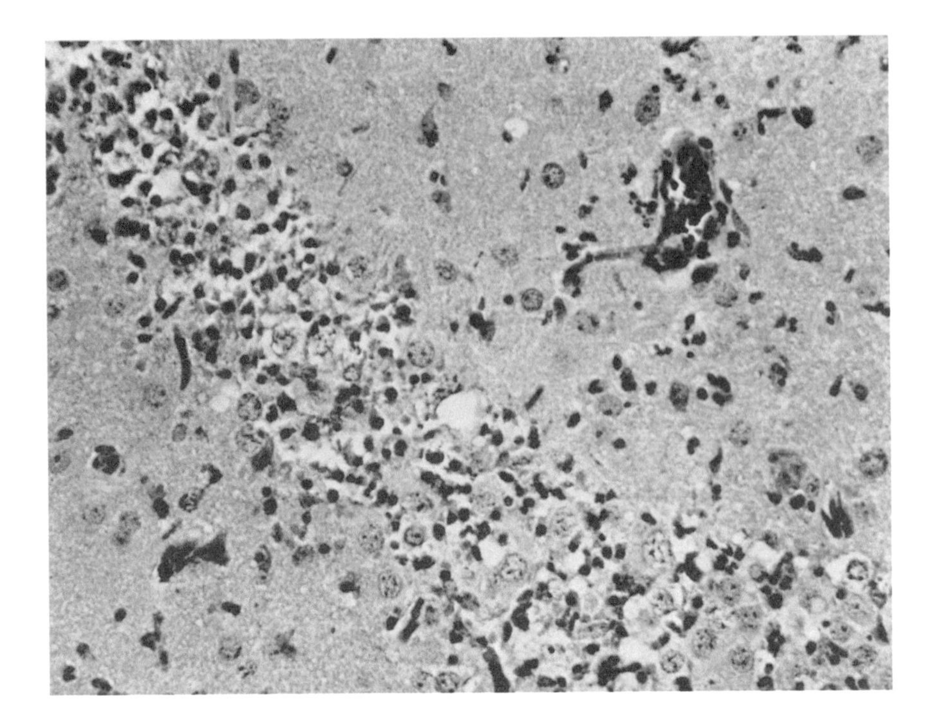

FIGURE 2. Severe rabies encephalitis: Mexican freetail bat isolate in mouse brain 5 days postinoculation; necrosis and edema. (Photo courtesy of F. A. Murphy.)

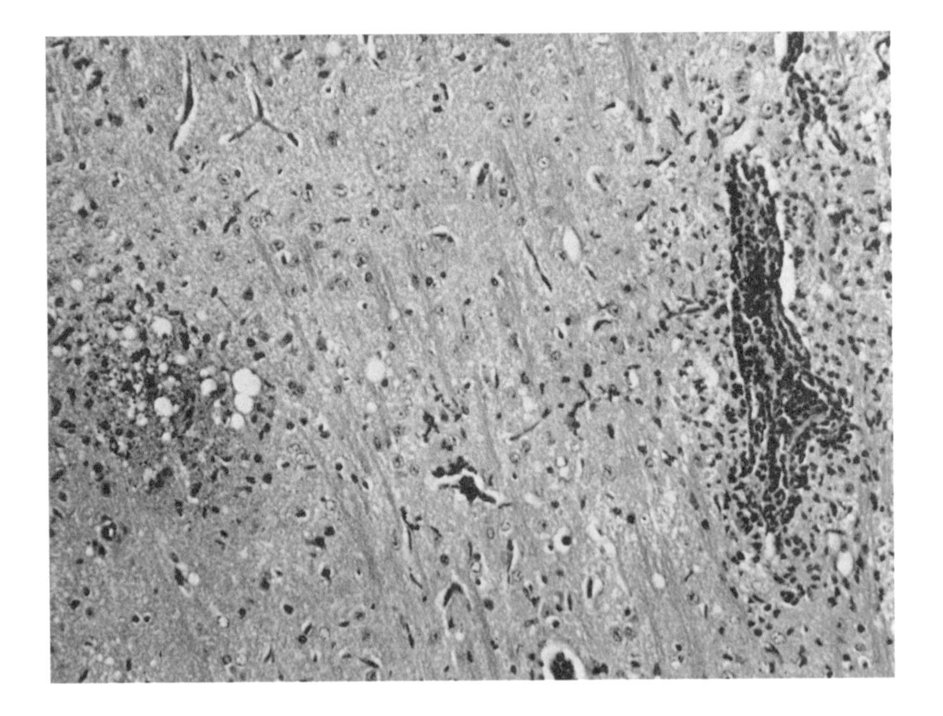

FIGURE 3. Severe rabies encephalitis: brain of mouse inoculated with Flury LEP rabies virus recovered from a dog with fatal encephalitis caused by LEP vaccine. Florid inflammatory cell infiltration. (Photo courtesy of F. A. Murphy.)

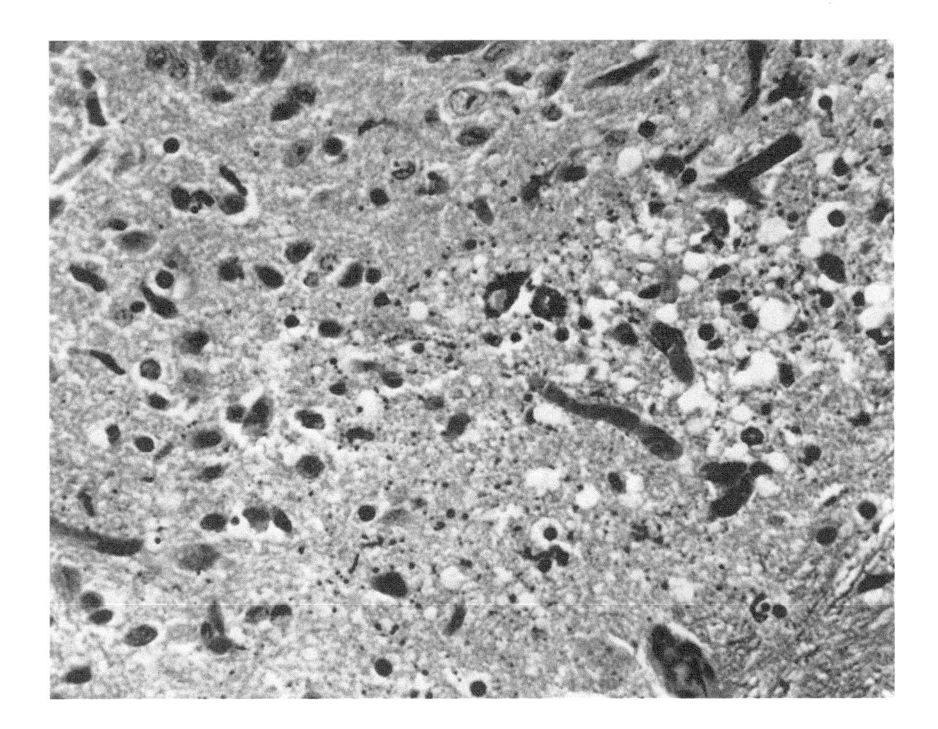

FIGURE 4. Severe rabies encephalitis: same mouse brain depicted in Figure 3; necrosis and edema. (Photo courtesy of F. A. Murphy.)

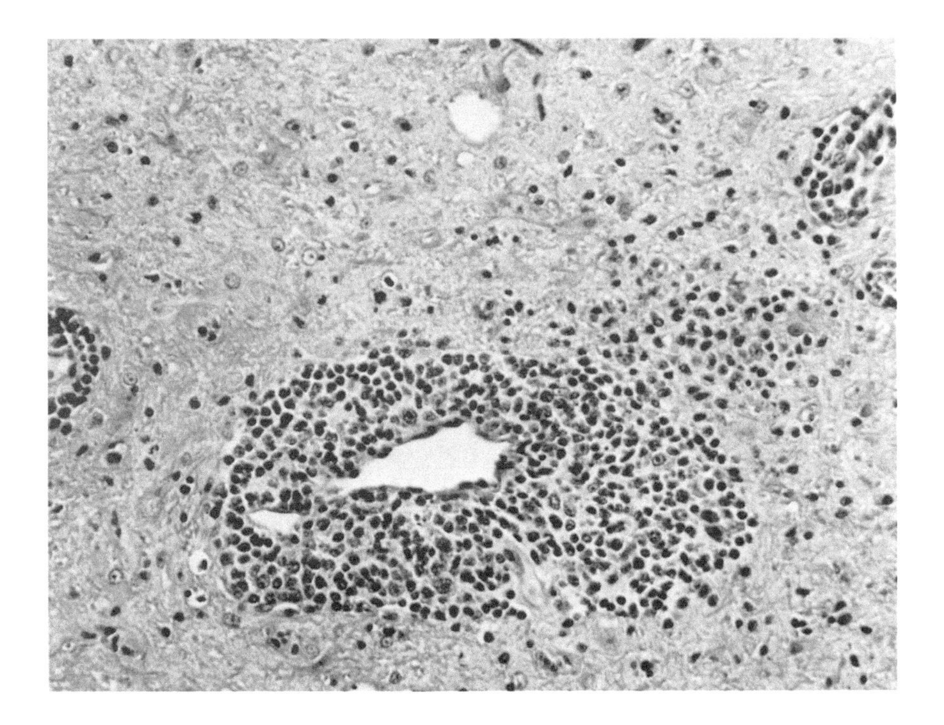

FIGURE 5. Severe chronic rabies encephalitis: brain of cat chronically infected with rabies for over 2 years.[139] Severe inflammatory cell infiltrate. (Photo courtesy of F. A. Murphy)

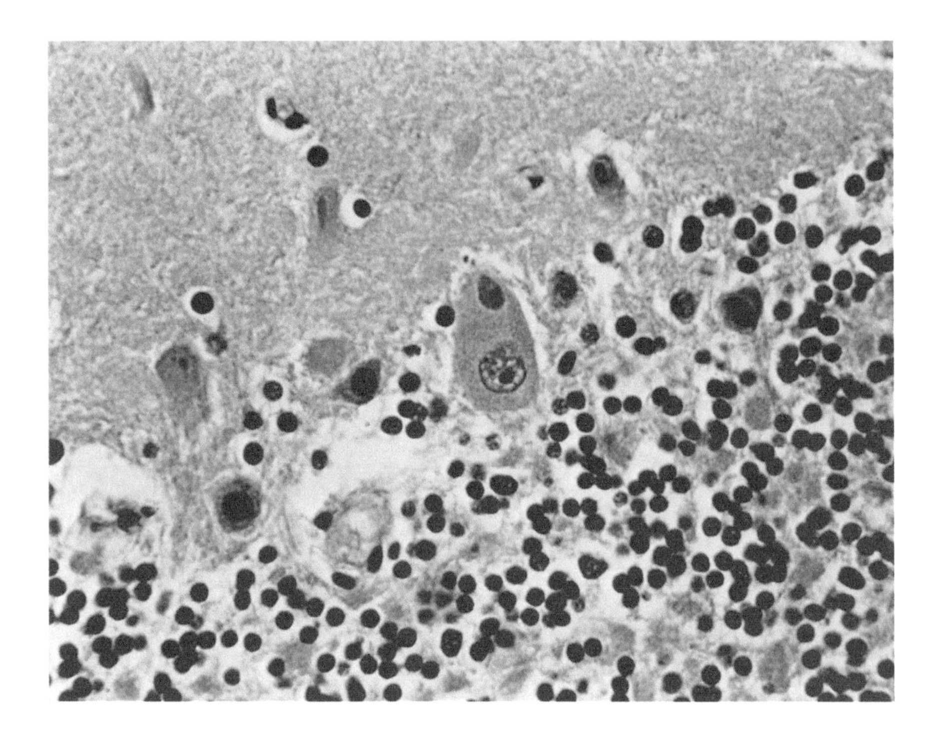

FIGURE 6. Negri body; street rabies virus in mouse brain. (Photo courtesy of F. A. Murphy.)

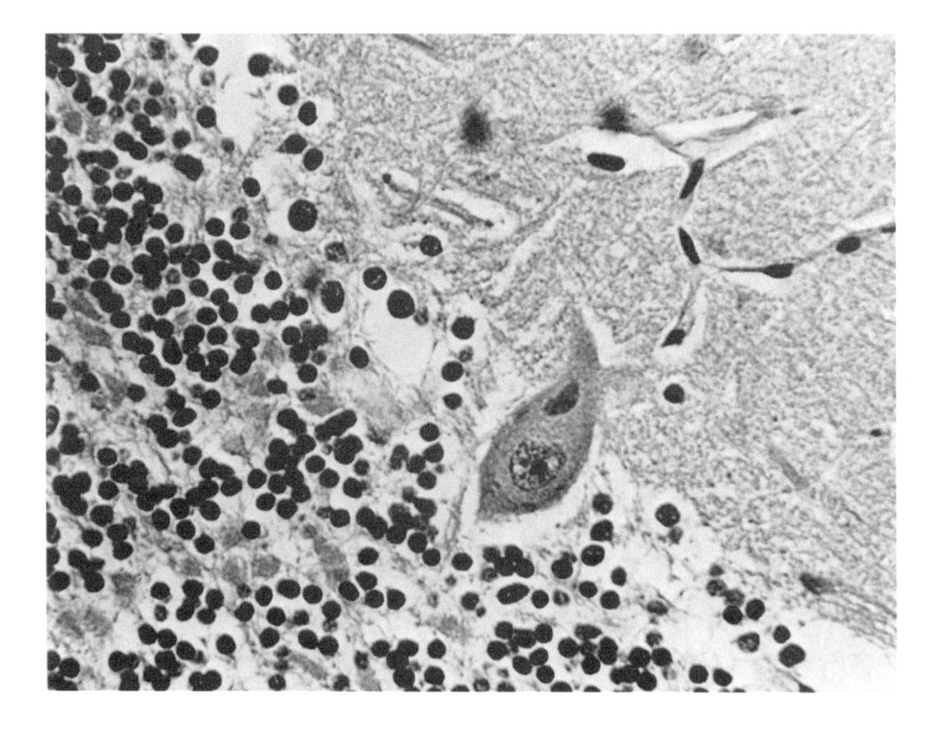

FIGURE 7. Negri body; inclusion in Purkinje cell of rabies virus-infected human brain. (Photo courtesy of F. A. Murphy.)

species.[140] Development of the largest Negri bodies tends to occur in the largest neurons, particularly in the ganglion of Ammon's horn noted by Negri and in the Purkinje cells of the cerebellum (see Figure 7). Inclusions that appear homogeneous in composition have been distinguished and named lyssa bodies by certain authors.[141] The latter may be confused with naturally occurring inclusions in some species of animals.

With the advent of the rabies fluorescent antibody technique, the Negri body was clearly shown to be composed of rabies viral antigen.[142] The exact composition of the Negri body was first convincingly demonstrated by Miyamato and Matsumoto[143] by careful light and electron microscopic observations of alternate thick and thin sections of street rabies virus-infected Ammon's horn tissue. The predominant eosinophilic ground substance of the Negri body was shown to be matrix of viral ribonucleoprotein strands. Small Negri bodies contained only this matrix and contained no basophilic inner body. They thus corresponded to lyssa bodies, indicating that distinction between the two types of inclusions is unnecessary. Matsumoto[144] found the inner bodies to be composed of virions accompanied by certain cell constituents and concluded from this observation that the Negri body is the site of virus replication. Murphy[57] has alternatively suggested that the inner body is composed of various cell organelles apparently randomly trapped in the matrix. It is not totally clear why the incidence of Negri bodies is so much less than the incidence of viral nucleocapsid accumulations observed by electron microscopy or by fluorescent antibody staining. However, it appears that the Negri body forms only in cells in which the normal cell architecture is largely spared.

Despite the diagnostic usefulness of the Negri body, its presence in brains of naturally infected animals is clearly not invariable; various studies have indicated its absence from 25% or more of rabid brains.[132] Fortunately rabies infection of brains not exhibiting Negri bodies can be readily confirmed by the mouse inoculation test or, in recent years, by the more rapid and reliable fluorescent antibody test.

C. Pathologic Change in Nonnervous Tissue

The absence of documented virus-specific pathologic change in the tissues involved at the site of the inoculation wound has been previously noted. Furthermore, despite the common widespread centrifugal postencephalitic dissemination of rabies virus in the body, only minimal documentation of associated peripheral pathologic changes exists.

It has been suggested that, given an adequately intensive search, rabies may occasionally be found in almost any organ of the body.[55,133] For example, rabies virus was isolated from extraneural tissues of 15 of 17 naturally occurring cases of human rabies observed in Cali, Columbia.[145] In a study of six naturally rabies-infected skunks from Kansas, rabies antigen was detected in each of 33 different tissues examined.[146]

Because of its role in virus transmission, the salivary gland is obviously of most critical importance, and is in fact the most often detected infected extraneural organ. However, the fact that salivary gland infection does not occur in the absence of CNS infection was convincingly demonstrated by Wanderler et al.[147] in a study of over 2000 rabid foxes, badgers, and stone martins. The histopathological appearance of the rabid salivary gland has received little attention, although Johnson[23] has reported that degeneration of acinar epithelial cells and mononuclear cell infiltration of interstitial tissue may occur.

Infection of olfactory mucosa of naturally infected bats was described by Constantine et al.[148] who suggested that resulting infected respiratory secretions might contribute to the airborne infections observed in certain bat caves. Winkler et al.[149] suggested that rabies-infected urine possibly contaminated by infected kidney and bladder epithelial cells of rabid foxes was the most likely source of an apparently airborne rabies outbreak in an experimental animal colony.

The successful use of fluorescent antibody examination of facial skin biopsies for diagnosis for rabies in man[150] and wild animals[151] is apparently based primarily upon infection of peripheral cranial nerve fibers. However, diagnosis of rabies by examination of corneal impressions is based on infection of corneal epithelium.[152]

At least two human cases of rabies have been induced by surgical transplantation of infected corneas.[152]

The only extraneural organ in which rabies virus infection has been frequently associated with both frank histopathological change and clinical signs is the human heart.[12] In a study of 11 cases of human rabies occurring in Abidjan, Ivory Coast, histopathological lesions were observed in both the myocardial parenchyma and the cardiac interstitial tissue of every patient.[153] It is noteworthy that in such a very widely disseminated infection, this is the only site outside of the CNS in which the infection has commonly been associated with an inflammatory cell response.

VI. EXPERIMENTAL PATHOGENESIS

The course of rabies infection in experimentally infected animals has been the subject of an enormous literature well summarized in a number of excellent reviews.[55,133,154,155] In this report we will attempt to concentrate only on those aspects of pathogenesis that remain speculative and hence are attractive areas for continued study.

A. Initiation of Infection

Despite intensive study, many critical questions remain unanswered concerning the events that intervene between introduction of rabies at a peripheral bite (or experimental inoculation) site and the subsequent coursing of infection through peripheral nerves into the CNS. The efficiency with which systemic infection is initiated is profoundly affected by the strain of the virus (street is much more efficient than fixed), the site of inoculation (proximity to the brain favors), the age of the host (young animals are more susceptible), and the species of host (see Section III). In none of these cases has the precise molecular or cellular characteristics of the virus or host that mediate such differences been identified. More exact characterization of these factors is indispensable to the development of better informed strategies for planning immuno- or chemotherapeutic intervention in the course of the infection.

The precise anatomical sites of entry of rabies virus into the peripheral nervous system have been determined largely through a series of elegant immunofluorescent and electron microscopic studies by Murphy et al.[10,55,156,157] In studies of street virus-inoculated newborn hamsters, sensory nerve endings in skeletal muscle proprioceptors were the most important sites of virus entry into nerves, with motor nerve endings of less importance. Watson et al.[158] presented evidence that cholinesterase-positive sites representing motor nerve end plates in skeletal muscles of adult mice may also serve as virus entry sites. Sensory nerve endings of the olfactory nerves in the nares and corresponding sensory nerve endings in the epithelium of the skin have been implicated as sites of virus entry for less typical infections initiated at these anatomical sites.

The potential for infecting laboratory animals and various species of wildlife with rabies via the gastrointestinal tract has been repeatedly demonstrated and is of particular interest to public health workers advocating this approach to immunization of wildlife.[159] Rabies infection has been initiated by feeding virus[160] or virus-containing baits,[161] by esophophagea intubation,[162] and by direct installation into duodenum.[159] Virus recovery studies suggested that fed virus might initiate infection through buccal or lingual mucosa, lung, or intestine,[163] but specific nerve end organs involved in the gastrointestinal tract have not been identified.

It is possible that administration of virulent virus naturally or artificially into traumatized muscle tissue may sometimes fortuitously cause virions to be directly apposed to susceptible nerve end organs, initiating neural infection without delay. Such an event would explain the observation of Johnson[23] that newborn mice inoculated in the footpad with highly mouse-virulent CVS rabies virus could not be protected from fatal encephalitis by amputations performed later than 2 hr after inoculation. However, at the other extreme, Baer and Cleary[164] protected a majority of adult mice inoculated in the footpad with a street rabies isolate when proximal amputation was performed as late as 18 days after inoculation. These authors concluded that the wound site was the point of virus sequestration during naturally occurring lengthy incubation periods, despite the fact that virus cannot be recovered from such sites. The mode in which virus may persist in peripheral wound sites is uncertain. The possibility that replication may occur therein extraneurally has been tested in numerous studies with diverse and contradictory results.

Johnson[24] reviewed the extensive literature on the subject published prior to 1971 and noted that although persistence of rabies virus (fixed or street) in inoculation sites of laboratory animals had been detected for up to 6 days, an increase in virus titer had never been observed. Similarly, no evidence of immunofluorescent antigen at the inoculation site had been reported. Subsequently, Schneider and Schoop[32] reported that although no evidence of multiplication of "classical" rabies virus could be detected in inoculation sites in mice, evidence of replication provided by immunofluorescent antibody studies could be found in striated muscle at sites inoculated with the Lagos bat virus (rabies serogroup) or Nigerian horse sickness virus (probably true rabies virus).

In an extensive series of experiments utilizing newborn hamsters inoculated with a variety of fixed and street rabies strains, Murphy and colleagues[156,157] have since provided definitive evidence of wound site replication of rabies virus in this host system. Virus replication was proved by sequential tissue titration as well as by fluorescent antibody and electron microscopic studies. Electron microscopy revealed initial productive infection of myocytes followed by involvement of neuromuscular and neurotendinal spindles before virus could be initially detected in peripheral nerves.

These observations were supported by the findings of Charlton and Casey[165] utilizing 6- to 10-month-old striped skunks infected i.m. with street rabies virus. Virus infection of muscle fibers near the inoculation site was detected by immunofluorescence prior to infection of peripheral nerve. However, the authors concluded from their own and other published observations that "infection of skeletal extrafusal muscle cells is not extensive" and occurs only in close proximity to the inoculation site.

In conclusion, it appears that in some rabies virus-host systems, replication of virus in myocytes may occur and enhance the probability of eventual entry of virus into the nervous system. Striated muscle cells may also provide a site for sequestration of virus during prolonged incubation periods. However, in the light of repeated failure to isolate virus from wound tissues during such intervals it would seem necessary to postulate persistence of a subviral particle retaining infectious potential. The structural constitution of such a hypothesized particle remains unknown. Also unknown is the requisite physiologic mechanism(s) to eventually trigger the initiation of infectious rabies virion replication leading to eventual neurologic infection.

B. Transit of Virus to the Central Nervous System

Extensive experimental studies of the means of neural transit of rabies virus from peripheral tissue exposure sites to the CNS have been thoroughly reviewed by Baer[155] and Murphy.[55]

Although transit of virus via peripheral nerve was long suspected on the basis of observations of the pathogenesis of the disease, the first conclusive experimental evidence was provided by the classic studies of Dean et al.[7] A variety of species of laboratory animals as well as foxes and dogs were subjected to sciatic and/or saphenous neurectomy prior to intraplantar inoculation of fixed or street rabies virus; significant sparing was associated with neurectomy in each experiment. These observations confirmed the implication of peripheral nerve in virus transit suggested by many earlier investigations,[155] and have been subsequently supported by other studies of rabies sparing by amputation of limbs inoculated distally.[23,164]

Baer et al.[166] found that neither removal of the perinuclear tissues from sciatic nerve nor mechanical injury leading to demyelination and allegedly total axonal degeneration prevented progression of rabies virus from a footpad inoculation site. However, subsequent studies have led to a consensus that rabies virus infectious potential does migrate presumably passively, in the axoplasm. Murphy et al.[156] and Jenson et al.[167] both demonstrated rabies virions and matrix in low concentration in the axoplasm of peripheral nerves. Furthermore, sparing of peripherally rabies-inoculated animals has been demonstrated following treatment with the axoplasmic flow inhibitors vinblastine[168] and colchicine.[169] The rate of virus transit of 3 mm/hr suggested by the amputation experiments of Dean et al.[7] corresponds to physiologic estimates of the rate of retrograde axoplasmic flow in normal peripheral nerves.[170]

Using rats infected in the footpad, Baer et al.[166] were able to demonstrate rabies antigen in sciatic nerve 48 hr after inoculation. This was prior to a time when infectious virus could first be isolated from the excised nerves. The latter observation is supported by reports of the difficulties encountered by numerous investigators in demonstrating immunofluorescent rabies antigen in peripheral nerve. As in the case of quiescent infection of wound sites during infections with prolonged incubation periods, it appears that a subinfectious virion component first traverses the peripheral nerve. This subviral component remains to be characterized.

There is no convincing evidence that hematogenous spread of rabies virus plays a significant role in the pathogenesis of experimental or naturally occurring rabies.

C. Infection of the Central Nervous System

In the centripetal progression of rabies virus infection, it is clear that the spinal ganglia represent the first site of readily observed neuropathologic effects and apparently the first definitive site of replication of infectious virus as well.[155] The necessity of virus replication at this site for continued central spread of infection is unknown;[24] Murphy has concluded from his experimental studies that it is probable that the virus genome can progress through the ganglion "directly to the CNS synapses in the cord".[55]

Once the cord is infected, virus reaches the brain very rapidly, often within a few hours,[55] and the outcome of the infection is irreversible. Virus replication may continue in the cord causing histopathological changes of varying severity that may be associated with posterior paresis.[133]

Descriptions of the evolution of rabies infection within the brain vary because of variation in all of the components of the model systems utilized, particularly virus strain, animal host, and route of inoculation, as well as the method of study. Careful sequential studies have shown that the site of earliest appearance of rabies antigen in the brain can be correlated with the anatomic site of inoculation in peripherally infected animals. Thus, as early as 1925, Goodpasture[141] reported localization of specific rabies inclusion bodies in the fifth motor nucleus when rabbits were inoculated into the masseter muscle. Murphy et al.[171] have shown that intranasal infection of hamsters led to initial localization of brain infection (indicated by immunofluorescent antigen) in the olfactory bulb. In the same study, oral administration of rabies caused irregular patterns of early brain antigen distribution, apparently associated with variable sites of infection along the gastrointestinal tract. In the majority of experimental studies designed to mimic natural disease, infection has been initiated in a peripheral limb and virus reaches the brain via the spinal cord.

Regardless of the route of infection, virus upon reaching the brain spreads very rapidly. Johnson[23,24] described early preferential infection of limbic system as a correlate to the clinical signs of furious rabies symptoms. Murphy et al.[171] described initial extension of infection from the cord to selected areas of brainstem and midbrain of hamsters, followed by very rapid dissemination to neocortex and "virtually every neuron" of the brain. This eventual very widespread involvement is similar in peripherally and intracerebrally inoculated animals.[133] With occasionally noted minor exceptions involving astrocytes and glia, the infection is restricted to neurons without involving supporting tissues.

As in the case of naturally occurring infections, the histopathological lesions observed in experimental rabies infections do little to explain the severity of the clinical disease. Considerable variation in the severity of meningoencephalitis can be associated with different virus strains and host species in model systems that are nevertheless equally uniformly fatal in outcome.

A prolonged course of clinical encephalitis has been associated with increased inflammatory lesions.[55] On the other hand, fixed virus strains that kill most rapidly have been generally observed to cause more severe inflammation in the brain than the more slowly acting street strains. Miyamoto and Matsumoto[172] compared the changes in Ammon's horn neurons of mice infected with a street or a fixed strain of rabies: selectively severe neuronal degenerative changes caused by the fixed virus were purported to prevent the development of large mature viral cytoplasmic inclusion bodies (Negri bodies) that were characteristic of the street virus infected hippocampal neurons that appeared otherwise undamaged. Karasszon[173] described marked perivascular infiltration, gliosis, neuronal degeneration, and neuronophagia widely distributed in the brains of hamsters infected with the Hōgyes strain of fixed virus. An especially severe involvement of the reticular formation of the brain stem was thought sufficient to explain the fatal course of the disease.

It is evident, however, that as in the case of typical street strains, fixed rabies virus may also sometimes kill without inducing remarkable histopathologic changes in the brain. Johnson[23] stated that the most striking observation in mice infected peripherally or intracerebrally with fixed strain CVS virus was the lack of inflammatory reaction and necrosis in the brain. Murphy[171] noted only "mild to moderate" inflammatory changes in suckling hamsters infected with either CVS or street rabies virus. Conversely, relatively severe inflammatory changes in brain may be characteristic of street virus infection in dogs[174] and bears.[175] Severe inflammation and necrosis characteristically induced by a street rabies strain isolated from a free-tailed bat *(Tadarida)* is illustrated in Figure 2. Very severe histopathological changes have also been characteristic of canine rabies deaths recently associated with breakthrough of a normally attenuated Flury LEP strain fixed rabies vaccine.[138] Necrotic and inflammatory changes are shown in Figures 3 and 4.

Corresponding to the rapid dissemination of virus antigen in the brain, sequential assays of viral infectivity have revealed very rapid increase in virus titer following onset of brain infection,[23,176] often occurring before the onset of clinical signs.[177] The mechanism of spread of infectivity within the brain is unresolved. Furthermore, there is no explanation for the fact that in terminal infection, especially with street virus, the titer of virus in the brain may be as low as 10^3 infectious units[133] — clearly far less than one infectious virus particle per infected cell! Such low titers may reflect the influence of neutralizing antibody, interferon, or DI particles.

Ultrastructural studies of rabies-infected brain tissue have tended to strongly support the generalization that fixed rabies virus causes more severe cytopathological changes in neurons than does street virus[55,172,178] and thus fail to shed any real light on the cause of death in natural infections. Only very recently have electron microscopic observations begun to illuminate the mechanism of dissemination of rabies within the brain.

Maturation of rabies virions normally requires the coiling of nucleocapsid strands

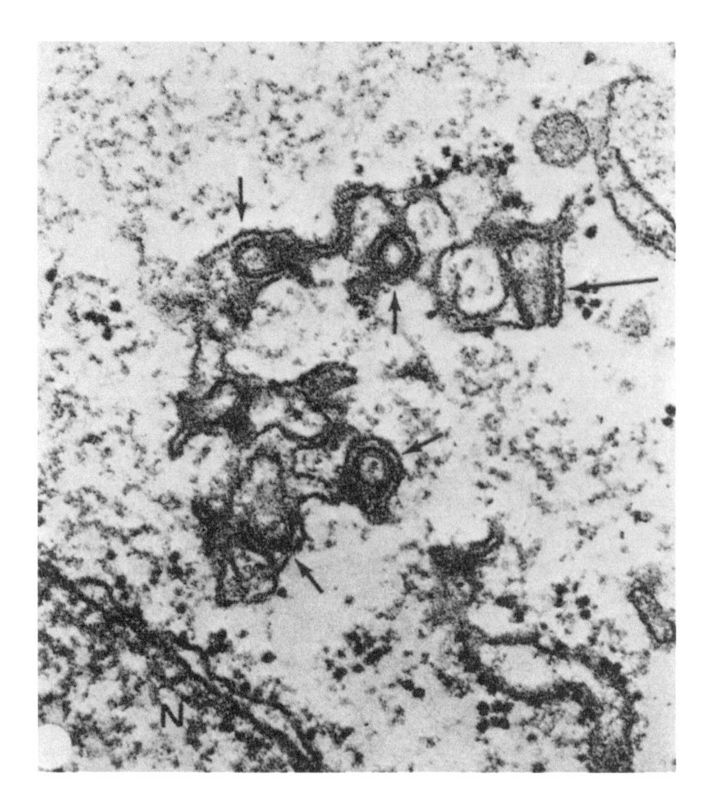

FIGURE 8. Rabies virus maturation within cytoplasm. ERA fixed rabies virus in newborn mice. Endoplasmic reticulum is shown 120 hr postinoculation. N = nucleus. (Magnification × 75,000.) (From Iwasaki, Y. and Clark, H. F., *Lab. Invest.*, 33, 391, 1975. With permission.)

formed in the cytoplasm into budding formations in host cell membranes.[57] As recently as 1975, Matsumoto[179] could report that such maturation was seen in neurons only on intracellular membranes, primarily those of the endoplasmic reticulum (Figure 8). Less frequently (and primarily with street virus), virions appeared to form in or near a matrix of nucleocapsid strands (cytoplasmic inclusions) in the apparent absence of contiguous cell membrane (Figure 9). Bizarre filamentous or branched virus-like structures are often seen in neurons infected with street virus;[180,181] these are presumably not infectious and may partially explain the low titer of infectious virus often found in such brains.

Budding of rabies virus from neuronal plasma membrane in the brain was initially detected only in suckling rodents infected with fixed virus.[98,171] Subsequently, however, Iwasaki and Clark[178] demonstrated plasmalemmal budding of fixed and street rabies viruses in neurons of both newborn and adult mice. In each of these systems, sequential studies revealed budding of virus from plasma membrane and accumulation of virions in intracellular spaces (Figure 10) before the initiation of maturation on intracellular membranes. Apparent direct cell-to-cell transmission of budding virus into an apposed pinocytotic vesicle of an adjacent neuron was also frequently observed (Figure 11). Both cell-to-cell spread and dissemination of virus through cerebrospinal fluid-filled intracellular spaces in the brain were postulated as means of rapid extension of the infection.

Similar observations have also been reported using a much more natural model of skunk-origin street rabies inoculated via a peripheral route into skunks. In brains examined during the terminal stages of encephalitis, virus budding was occasionally ob-

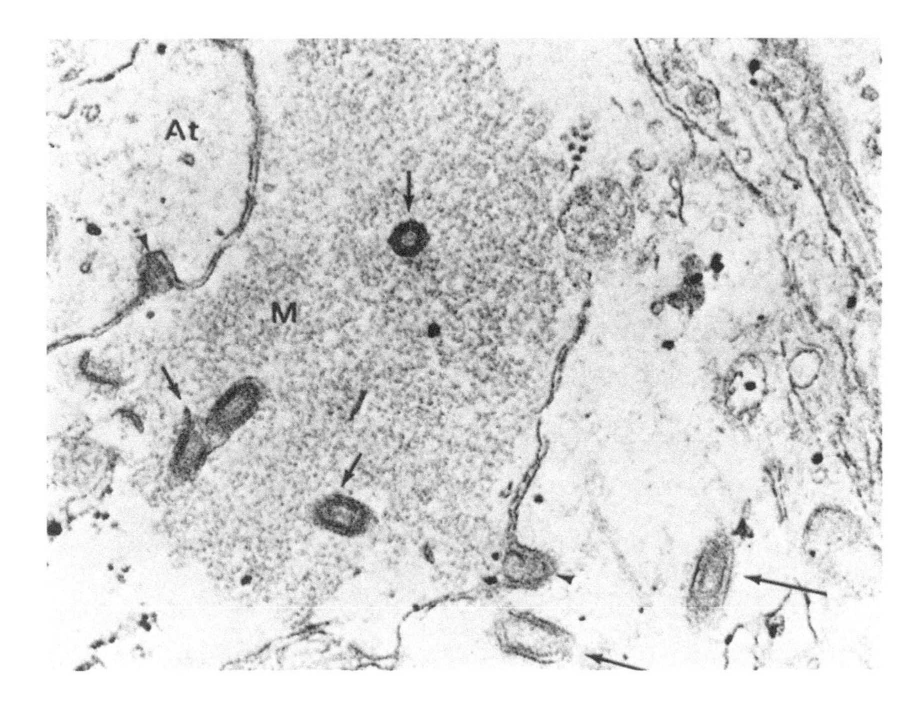

FIGURE 9. Street rabies virus maturation within matrix (M) and from perikaryon of a neuron in the brain of an infected suckling mouse. At = terminal axon. (Magnification × 93,600.) From Iwasaki, Y. and Clark, H. F. *Lab. Invest.*, 33, 391, 1975. With permission.)

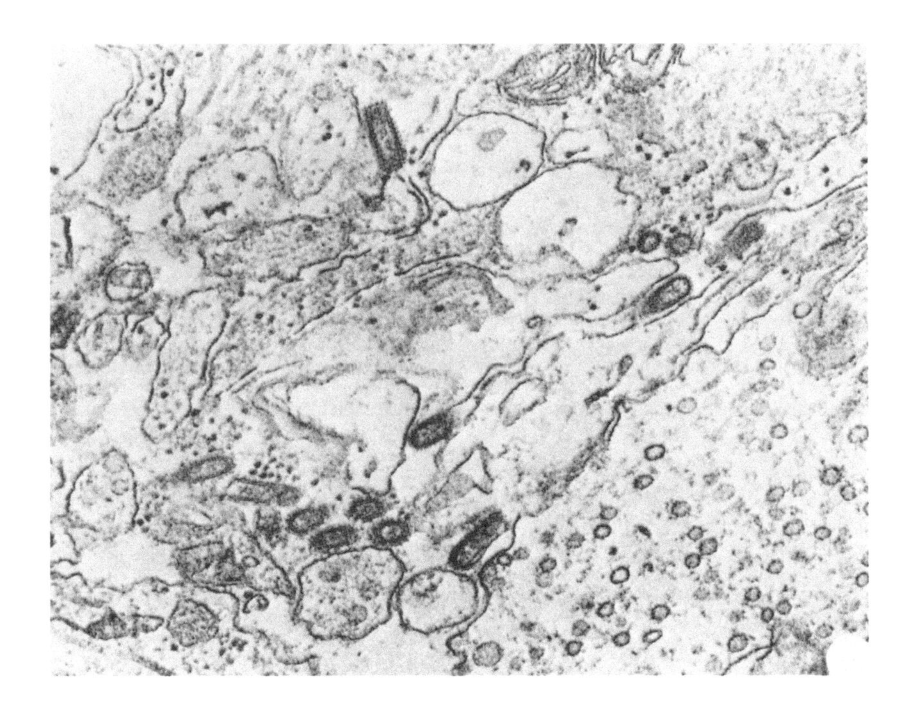

FIGURE 10. Strain ERA fixed rabies virus in newborn mouse brain. Virions in intercellular spaces between neuronal processes. (Magnification × 48,600.) (From Iwasaki, Y. and Clark, H. F., *Lab. Invest.*, 33, 391, 1975. With permission.)

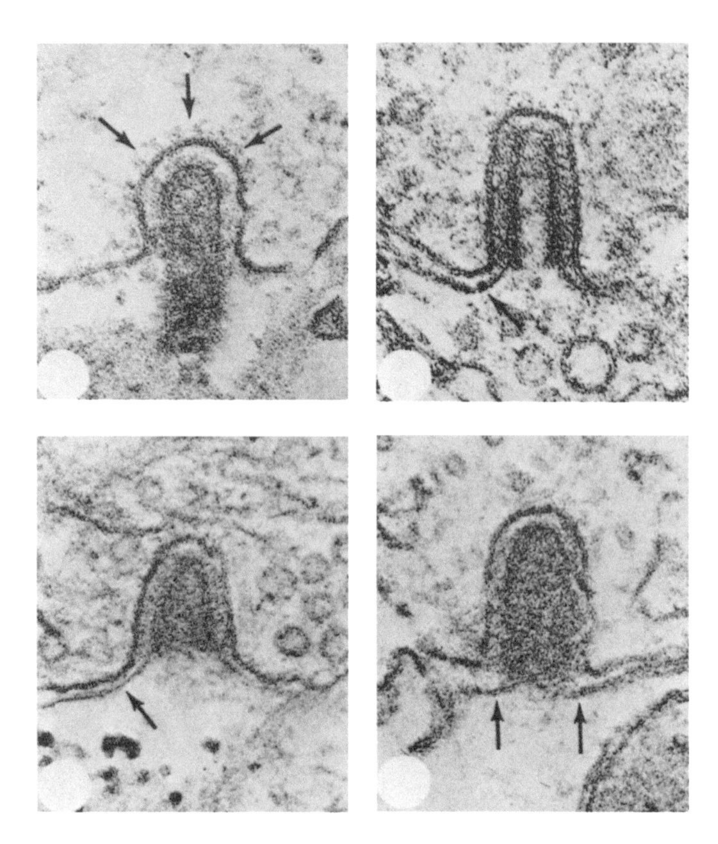

FIGURE 11. Rabies virus spread between contiguous neuronal cell processes and perikarya. Upper left — CVS fixed virus in suckling mouse. Upper right and lower left — street virus in adult mouse. Lower right — street virus in suckling mouse. (Magnification × 100,000.) (From Iwasaki, Y. and Clark, H. F., *Lab. Invest.*, 33, 391, 1975. With permission.)

served from plasma membranes of neurons, primarily at "synaptic or adjacent membranes of dendrites".[165] Budding of virus directly into contiguous neurons occurred more frequently than budding into intracellular spaces.

Recent studies of both natural and experimentally induced rabies encephalitis have shown a predilection for virus budding at synaptic sites in the brain.[182,183] It has been suggested that the relatively low incidence of budding from the membrane of the perikaryon may explain the frequent paucity of obvious neuronal cytopathology.

D. Centrifugal Dissemination of Rabies Virus

Except in human cases of rabies involving heroic therapeutic intervention, there is little suggestion that symptoms other than those directly attributable to encephalitis contribute significantly to rabies fatality. Hence, centrifugal dissemination of virus into nonnervous tissues has been of interest primarily in the case of infection of the salivary gland and occasionally in the case of other tissues examined for diagnostic purposes.

There is clearly little spread of rabies virus in the absence of CNS infection. Sequential studies of the progression of infectious rabies virus in animals inoculated by various parenteral routes early revealed that tissues of contralateral limbs or of parenchymatous organs exposed by i.p. inoculation were not infected until after infection had

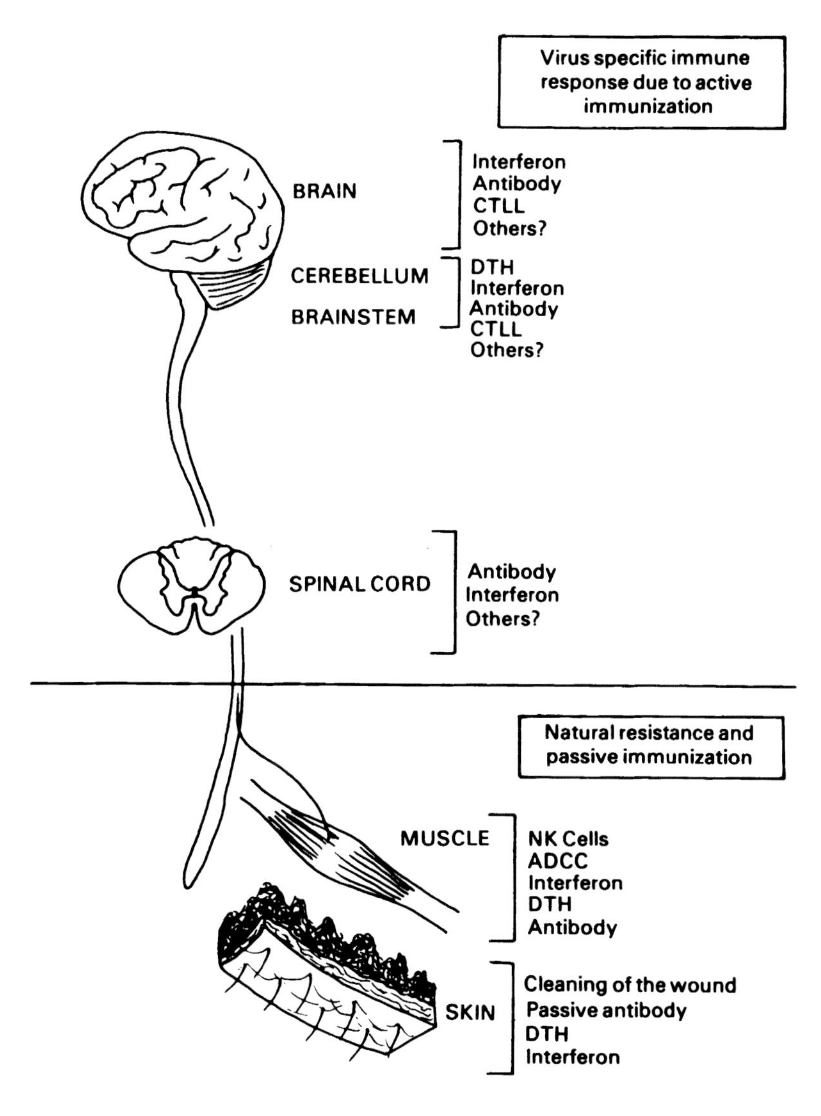

FIGURE 12. The immune response to rabies virus.

reached the cord and/or brain.[176,184] Intravenous inoculation of rabies virus causes no early direct infection of parenchymatous organs in which virus later appears subsequent to infection of the brain.[185] Charlton et al.[186] have very recently demonstrated that even inoculation of street rabies virus directly into the submandibular salivary gland of the skunk did not directly induce infection of the gland: dissemination within the salivary gland was not observed until at least 12 days after inoculation, when infection had also reached the medulla oblongata.

The most direct evidence of neuronal centrifugal spread to the salivary gland was provided by the studies of Dean et al.[7] in which infection of salivary glands of intracerebrally inoculated mice was selectively spared following unilateral section of the lingual nerve and cranial cervical ganglion. Dierks[187] has pointed out that the often florid infections of salivary glands may be established exclusively by neuronal transmission of virus since the nerve networks of such glands provide a direct neural route to each secretory cell. Certainly salivary gland tissue produces infectious rabies virus with an efficiency not approached by any other tissue. The selectively high titers of infectious virus in salivary glands of rabid foxes and skunks have been previously noted in the discussion of rabies epidemiology. The electron microscopic studies of Dierks and colleagues[187,188] revealed production of uniform rabies virions on apical plasma membranes of salivary mucogenic cells and accumulation of uniform virions in salivary gland ducts in a concentration far exceeding that which has been observed in any other tissue, including brain. The morphogenic pattern differed from that characteristic of infected CNS tissue in that: (1) a minority of virions were assembled on intracellular membranes and, (2) branched or otherwise aberrant viral forms were not seen. Concomitant histopathological changes in the salivary gland varied from nonexistent to severe degenerative lesions. Biological characters of salivary gland cells associated with their extraordinarily specific capacity to support efficient replication of street rabies virus have not been identified. Furthermore, it is not known whether the failure of fixed rabies viruses to replicate efficiently in the salivary gland reflects a difference in ability to infect salivary gland cells or simply a failure of centrifugal spread to the salivary glands possibly attributable to the rapid fatal progression of fixed virus encephalitis.

Infection of nonneuronal cells, occasionally accompanied by histopathological lesions, has also been reported in brown fat and cornea,[189] lacrimal gland, kidney, heart muscle,[23] adrenal medulla, exocrine acini of the pancreas,[171] nasal mucosa, lingual papillae, and hair follicles.[185] Murphy et al.[171] have suggested that the intense infection of the nerve end organs close to the epithelial surfaces of the oral and nasal cavities might indicate occasional epidemiologically significant levels of rabies virus shedding from nose and mouth. The frequent infection of corneal epithelium has been exploited in attempts to use fluorescent antibody-stained corneal impressions for antimortem diagnosis of rabies.[154]

E. Abortive and Chronic Infections

"Abortive" and "chronic" infections with rabies virus are of particular interest because of the still commonly accepted dogma that rabies is invariably fatal. Abortive rabies infection may be defined as infection that is accompanied by an immune response (most easily demonstrated by subsequent resistance to challenge) but does not progress to fatal encephalitis. Abortive infections presumably account for the immunity induced by live virus vaccines. Chronic infections are those in which prolonged salivary shedding of virus occurs in the absence of encephalitic disease, or clinical disease progresses over a very prolonged period to an eventual fatal outcome. Although apparently very rare, chronic rabies might occasionally play a role in virus dissemination in nature.

An understanding of the factors leading to abortive infection would be invaluable in designing strategies of immunoprophylaxis. However, while many factors contributing to abortive infection are known, few are understood in terms of mechanism of action. With "attenuated" vaccine strains of rabies virus, both host species and route of inoculation are all-important. Flury HEP rabies virus does not cause disease in peripherally inoculated adult mammals; it is immunogenic in most species of domestic animals, but it is not effective in man.[190] When given intracerebrally, this virus is (1) consistently lethal for newborn mice and adult rhesus monkeys, (2) lethal according to an irregular dose response curve in hamsters and guinea pigs, and (3) almost totally innocuous in adult mice.[120] The widely used veterinary virus strain ERA rabies virus is extremely safe and effective when inoculated peripherally in many species of domestic animals.[191,192] Yet ERA induces fatal rabies in mice inoculated intracerebrally[191] and in several species of rodents inoculated orally.[193,194] ERA virus has induced encephalitis leading to severe permanent brain damage in an adult laboratory worker exposed to a high-titered virus aerosol.[195]

The site of peripheral replication of an attenuated "modified live" rabies virus administered by the standard i.m. route has never been identified. However, it is generally assumed that peripheral replication does occur because the antigenic mass of the virus in a live vaccine is not adequate to induce the observed levels of antibody. Presumably the coursing of such virus to the brain is interrupted because of enhanced ability to elicit and/or respond to specific immune mechanisms. Some modified rabies viruses (such as Flury HEP[120,196]) also fail to induce fatal disease when introduced directly into the brain, despite the fact that virus may replicate in the brain and induce meningoencephalitis with neuronal necrosis.[197] That such attenuated rabies viruses do have the potential to cause fatal dysfunction has been indicated in numerous studies in which typically abortive infections have been made fatal by use of chemical immunosuppression (see Section VII.A.).

The delicacy of the precarious balance between the pathogenic and immunogenic potential of attenuated rabies viruses was shown in our series of experiments consisting of the intracerebral inoculation of serial dilutions of Flury HEP virus into adult mice.[111] In seven different experiments, no deaths occurred in mice given this virus in high concentration, but 15 of 54 mice given between 2 and 200 mouse infectious doses suffered fatal encephalitis. Selective intracerebral pathogenicity caused by minimal infectious doses has been demonstrated with several other rabies virus strains considered to be either partially or fully attenuated.[111]

Abortive rabies infections have also been observed with virulent strains of rabies virus. Many investigators have presented observations of apparently rabies-specific serum antibodies in clinically normal wild or domestic animals as evidence of naturally occurring abortive infections.[198]

Unequivocal evidence has been presented that dogs,[199] mice,[21] and many other species of mammals[198,200] may develop clinically apparent infections with street rabies virus and subsequently survive indefinitely. Survivors may be clinically normal or may exhibit permanently persisting paralytic signs. The latter group have been called survivors *cum sequelae* or "SCS" by Bell.[19]

Dramatic and unpredictable differences in the capacity of different street rabies strains to induce abortive infection have been reported by Baer et al.[21] Rabies virus-infected salivary gland suspensions from 11 different wild animals, all collected from the same region during a single year, were inoculated into mice via the footpad or i.m. route. The incidence of survivors with evidence of infection (resistance to subsequent challenge) varied from none to 75%. Histopathologic changes were limited to the dorsal root ganglia (primarily on the inoculated side) of the surviving infected animals.

These efforts to develop model systems in which abortive rabies infections are consistently induced have usually required either environmental or immunological manipulation of the host animal or the use of atypical variant viruses. Baer et al.[21] reported a greatly enhanced incidence of abortive infections, with up to 33% of SCS mice, following inoculation of mice with several different street rabies strains via the i.p. route. A predominant role of the immune response in survival was suggested by the facts that survival was strongly correlated with relatively long incubation periods prior to appearance of clinical signs and that all survivors had high-titered serum and brain antibody levels.[201]

It was subsequently shown that exposing rabies-infected mice to elevated ambient temperatures (35°C) led to a markedly increased incidence of abortive infections.[202] Several substrains derived from ts mutants of strain CVS fixed rabies virus[22] caused infections in mice that were aborted with very high efficiency under the influence of elevated ambient temperature.[203]

One of the most efficient systems reported for inducing rabies paralysis with survival consisted of the inoculation into mouse footpads of a stable strain ts 2 mutant:revertant virus mixture originally derived from the highly pathogenic CVS strain virus.[116] In one series of mice so inoculated, a paralytic disease with an incidence of 82% was accompanied by a mortality rate of only 29%. In another series of experiments, inoculation of strain ts 2 rabies virus intracerebrally into mice caused a 100% incidence of paralytic disease accompanied by severe histopathologic evidence of inflammation and degeneration in the CNS but only 80% mortality. The same virus inoculated into cyclophosphamide-immunosuppressed mice caused encephalitis accompanied by little paralysis and minimal histopathological signs of brain damage; 100% of these mice died.[204]

Smith[200] induced paralytic disease with survival in mice inoculated in the footpad with high-titered strain ERA rabies virus. A similar inoculation into cyclophosphamide-immunosuppressed or athymic nude mice caused encephalitic death without paralysis. Virus infection was shown to progress to the brain at a similar rate in each of the three groups of mice and to replicate to similar concentrations. However, virus was cleared from the brain after 4 to 5 days only in the immunocompetent mice.

A possible role for defective interfering (DI) rabies particles in mediating abortive infection was indicated by the studies of Wiktor et al.[103] who found that excess DI particles of strain ERA virus could protect mice against an intracerebral challenge with lethal fixed or street rabies virus. We subsequently also demonstrated a small degree of protection against rabies virus-induced fatality with excess ERA virus DI particles.[105] However, when we physically separated DI and standard virus particles from an ERA virus preparation that selectively caused abortive infection when inoculated intracerebrally at high concentrations, the enriched standard particle preparation continued to cause abortive infection. Although DI particles of rabies virus may play a role in the progression of encephalitis, the nature of that role has not yet been effectively characterized.

Coupling of active immunization of animals against rabies with challenge with virulent virus had been utilized to develop models of infection with survival. Wiktor et al.[20] reported that intracerebral inoculation of mice with attenuated Flury HEP virus shortly before or as late as 2 days after intraplantar challenge with virulent virus frequently spared the animals from fatal disease (although persistent paralysis often developed). The effect was thought to be interferon-mediated. Gribencha[205] subsequently reported occasional survival *cum sequelae* in rats and rabbits challenged intracerebrally with virulent virus following a previous i.p. inoculation with a sublethal dose of street virus; no mechanism to explain these observations was suggested. Development of transient paralytic rabies followed by recovery was reported in a dog challenged i.m. with street virus 3 years after vaccination with Flury LEP virus vaccine.[206] This dog differed from asymptomatic survivors in possessing an extremely high level of antibody in the cerebrospinal fluid.

Atypical rabies infections so prolonged as to be designated "chronic infections" have occasionally been reported. Perl et al.[207] reported that a cat inoculated i.p. with street virus developed transient clinical rabies 11 days postinoculation followed by a symptom-free interval of more than 100 weeks prior to reappearance of progressive paralytic disease. The cat showed progressively increasing serum antibody titers to rabies virus during most of the asymptomatic period and exhibited severe inflammatory changes in the brain at necropsy. Rabies virus could not be isolated from necropsy tissue nor visualized by electron microscopic study of the brain.

Murphy et al.[139] reported two additional cases of "chronic rabies" in cats infected with street rabies via simultaneous i.p. and i.m. inoculations. One of these cats survived early symptomatic disease but sickened again at 100 weeks postinoculation; the other was asymptomatic until the onset of disease 120 weeks after inoculation. Both showed progressive increase in serum antibody from inoculation until sacrifice, irrespective of their expression or lack of expression of symptoms. The cat with initial onset of disease at 120 weeks yielded viable virus from the brain (in low concentration) and exhibited several foci of immunofluorescent rabies antigen in the brain and cord. Inflammatory changes in the brain were severe, but neuronal cytopathology was limited. The investigators suggested that neuronal infection may have been characterized by predominantly nucleocapsid replication without maturation of complete virions. It was postulated that in such an infection the absence of viral coat proteins was associated with failure of the host to mount a cytolytic immune defense capable of clearing the infection.

Virus has not been recovered from saliva of most abortively or "chronically" rabies-infected animals, but several instances of persistent salivary shedding of virus have been reported with rabid dogs in Ethiopia and India.[208] In one recent experiment,[208] four dogs were inoculated i.m. with a virus isolated from such a chronic shedder; two of the four recovered completely after developing clinical rabies, but one of these continued to shed virus intermittently in saliva for at least 305 days after recovery.

Clearly the diversity of patterns of both classical and atypical rabies infections is nearly infinite. Hopefully, study of each observed variation may provide some insight towards understanding of the interplay of rabies viruses and host responses that so typically results in clinical encephalitis of a severity not explained by the extent of observable neuronal damage. That study of abortive rabies is not simply academic exercise is indicated by two recent well-documented cases of survival after rabies infection in man, one with complete recovery[13] and one with persisting severe behavioral sequelae.[195] Each observation suggests that the continued search for life-saving therapeutic measures applicable to rabies may not be without hope.

F. Cell-Virus Interaction In Vitro

Given a consensus of opinion that the clinical severity of rabies encephalitis cannot consistently be explained by neuronal cell damage detectable by histopathologic or ultrastructural observations, a concentration of effort on the study of cell physiologic manifestations of rabies virus infection would seem well-justified. Surprisingly, however, the effect of rabies virus on cultured cells has not been explored in depth. Cell:virus interaction studies have more often concentrated on the effect of the host cell upon the virus.

In the preponderance of susceptible cell culture systems, rabies virus replication is not accompanied by characteristic cytopathic effects.[4] In very early cell culture studies, Fernandes et al.[209] associated the failure of division of CVS strain virus-infected BHK/21 and human diploid cells with mechanical damage from very large inclusions. Subsequent studies with these and other cell types have not supported this conclusion. Indeed, Love et al.[210] found that the largest inclusions ("Negri bodies") in neurons in vivo was so regularly correlated with absence of cytopathology that he considered large inclusion formation to be a cell defense mechanism!

Efficiency of cytopathology in vitro as indicated by large plaque size has clearly not been correlated with in vivo virulence of either rabies[114] or rabies-related[115] viruses. Particularly efficient induction of CPE with BHK/21 cells has repeatedly been associated with Flury HEP strain rabies virus, one of the most highly attenuated vaccine strains.[102,105]

Apparently complete compatibility of rabies virus replication with cell growth and division was first shown in CVS virus-infected rabbit endothelial cells; host cell nucleic acid synthesis did not seem to be affected by the virus infection, which was termed "endosymbiotic".[3,211] Numerous investigators subsequently demonstrated that persistently rabies virus-infected cell lines capable of indefinite propagation could be established with ease.[102,212-215] Continued subculture of such cell lines was often characterized by cyclic rise and fall of infectious virus production and/or production of defective interfering particles. Mouse-virulent rabies viruses recovered after prolonged propagation in chronically infected cell systems were frequently found to be greatly attenuated. It has not been possible to demonstrate the existence of integrated "proviral" rabies cDNA in chronically infected cell lines.[213]

In efforts to more closely approximate in vivo infection, rabies virus has recently been propagated with difficulty in organized cultures of mammalian tissues,[181] but very efficiently in neuroblastoma cell lines that express many of the phenotypic characters of mature neurons.[119,121,216] Several (but not all) strains of rabies and rabies-related viruses were highly cytopathic in neuroblastoma cells, but, as in the case of nonneuronal cells, cytopathic potential could not be correlated with mouse virulence. A comparative study of attenuated and virulent substrains of Flury HEP virus revealed selective cytopathic potential in the attenuated substrain.[217]

Cultivation in human or murine neuroblastoma cells had a profound effect on the rabies virus phenotype: attenuated strains of fixed rabies consistently reverted to a mouse-virulent state.[119,121] The most likely explanation for this phenomenon is that neuroblastoma cells selectively support enhanced replication of randomly occurring back-mutants of attenuated rabies virus strains.[121]

Extensive comparative studies of mouse neuroblastoma (C1300) and BHK/21 cells chronically infected in parallel with mouse-virulent or attenuated strains of fixed rabies were performed in our laboratory.[111,215] Particular interest attends the results with neuroblastoma cells, as factors favoring increased virulence (propagation in neuroblastoma cells) and favoring decreased virulence (propagation in a chronically infected cell system) were acting concurrently. In fact, attenuated Flury HEP virus remained attenuated for the first few chronic passages, then expressed a high degree of mouse virulence from the 20th to the 34th passage level, and finally reverted to an attenuated phenotype after the 45th passage. Mouse-pathogenic CVS and PM strain viruses gradually lost pathogenic potential in this system.

After prolonged serial passage (>50 passages) both rabies virus-infected neuroblastoma and BHK/21 cells produced no plaque-forming virus and only low levels of infectivity detectable by inoculation of newborn mice. Defective interfering particles were produced irregularly, particularly by Flury HEP virus-infected cells. The percentage of infected cells detectable by immunofluorescence fluctuated cyclically over a wide range (<1 to 100%) in neuroblastoma cells, but remained constant at nearly 100% in BHK/21 cells.

Rabies virus-infected C1300 neuroblastoma cells retained a normal potential for inducing tumors in syngeneic A/J mice, but cells from such tumors reestablished in culture were free of rabies antigenic expression. These, and other C1300 cell sublines freed of rabies virus-infection following either spontaneous "cure" associated with cell passage at elevated (40.5°C) temperature, exhibited complete susceptibility to reinfection with rabies virus. In contrast, each of three infected BHK/21 sublines cured by passage at 40.5° exhibited an altered cell phenotype. These cells uniformly expressed remarkably increased resistance to infection with either rabies virus or the heterologous rhabdovirus VSV.

It is possible that chronically rabies virus-infected cell systems may provide a system amenable to molecular biological studies shedding light on events occurring in vivo during prolonged incubation periods. Ultrastructural studies of chronically CVS virus-infected C1300 cells have recently revealed selected budding of new virus from cell processes, thus sparing the cell body and providing a possible explanation for the minimal cytocidal effects often characteristic of rabies virus infection of neurons in the CNS.[216]

Very few physiologic studies of rabies-infected cells have been reported. Diaz et al.[218] described activation of lysosomal enzymes following infection of two different hamster cell lines with either CVS or ERA rabies virus. Holian and Clark[261] detected several differences in the response of C1300 neuroblastoma and BHK/21 cells to chronic rabies infection. In comparison with control cells, rabies virus-infected C1300 cells exhibited reduced cell size, normal cell doubling time, and increased rates of oxygen consumption and glucose oxidation coupled with elevated pyruvate concentrations. Chronically infected BHK/21 cells were grossly enlarged (two to threefold) and retarded in cell doubling time; oxygen consumption and glucose oxidation were increased, but pyruvate concentrations were depressed. The results indicated profound cell type-mediated differences in the response to chronic infection with identical strains of rabies virus.

The study of biochemical changes possibly affecting neurophysiological function in rabies virus-infected cells in vitro has only recently been approached. Using a neuroblastoma:glioma rat cell line that expresses several neuronal functions, Koschel and Halbach[219] demonstrated that rabies virus infection led to alteration in membrane receptor-mediated changes of intracellular cyclic AMP levels. The normal action of prostaglandin E, and of isoproterenol (but not that of acetylcholine) upon cyclic AMP levels was impaired following infection with Flury HEP strain virus. The effect upon prostaglandin E activity was later reported to be caused by reduced binding affinity rather than by a reduced number of prostaglandin membrane receptors in infected cells.[220] In contrast to these results, Tsiang[221] presented evidence that acetylcholine receptors were altered in rat neurons infected with rabies virus in vivo.

The possibility that the acetylcholine cell membrane receptor may be the receptor for rabies virus was suggested by Lentz et al.[222] who cited both: (1) a morphologic correspondence of initial sites of virus attachment and acetylcholine receptors in mouse diaphragm and myotube preparations and (2) an inhibition of rabies virus attachment by the acetylcholine antagonists α-bungarotoxin and d-tubocurarine. It was hypothesized that the identity of the acetylcholine receptor and the rabies virus receptor would provide a molecular explanation for the selective entry of rabies virus into the nervous system. It has subsequently been suggested that physiologic studies to determine changes in acetylcholine receptor function after exposure to rabies virus would be particularly interesting.[223]

VII. HOST RESPONSE TO RABIES VIRUS INFECTION

The host's defense against viral infections is a combination of nonspecific and specific immunologic responses. The relative importance of different components of the response may vary. Macrophages form one line of defense against rabies infection by limiting the spread of virus from the primary site of infection to the target organs.[224] Interferon has been shown to play a direct role both in limiting the virus spread[225] and in augmenting the immune response.[226] The important immunologic mechanisms that mediate specific antiviral activity are humoral and cell-mediated responses. The importance of these two factors has been long debated and indeed may vary from case to case. Circulating antibody may be an important factor in neutralizing the virus and limiting its spread. Antibody can also bind to virus on the cell surface and facilitate destruction of the infected cells by either complement or cytotoxic cells. Induction of the cell-mediated immune response is probably also relevant because of its ability to destroy the virus-infected cells. Thus, a multitude of immunological factors can interact in harmony in terminating the viral infection.

A. Role of Humoral Immunity

Many studies of active immunization and more recently of passive immunization have focused our attention on neutralizing antibody as the key factor in postexposure protection of man.[227] The usefulness of passive administration of antibody along with vaccine to individuals exposed to rabies has been demonstrated by Bahmanyar et al.[228] In Iran a number of individuals that were exposed to a rabid wolf were given postexposure prophylaxis. Among those most severely bitten, those who were given serum along with vaccine showed only 8% mortality (1/13), whereas three of five individuals (60%) who were given vaccine alone died. This along with a number of studies in experimental animals strongly points to the importance of antibody in terminating rabies infection (see below).

Recent studies by Miller et al.[229] with mice depleted of bone marrow-derived lymphocytes (B cells) or complement also suggest a role for antibody. Animals were treated with anti-IgM heavy chain antiserum beginning at 24 hr after birth to cause B cell-specific immunosuppression. These animals, when inoculated with a nonvirulent strain of rabies virus (Flury HEP virus) failed to produce antibodies. The death due to HEP virus infection was attributed to the lack of B lymphocytes because these animals had normal T cell populations as indicated by their response to ConA, which is a T cell mitogen. These experiments strongly suggested a role for antibodies in the recovery from experimental rabies virus infection.

There are a number of studies that have employed antibody for passive immunization in a postexposure situation. Animals were inoculated with either pathogenic[230] or nonpathogenic[197,229,231] strains of rabies and subsequently immunosuppressed and then given passive antibody treatment. Survival of a high proportion of animals given passive antibody suggested that antibody can slow the progression of the disease to such an extent that the host can mount an active response adequate to clear the virus.

This protective capacity of passively administered antibody was further confirmed by adoptive cell transfer experiments. Nonlethal HEP virus infection of adult mice was converted to potentially lethal infection by transient immunosuppression. When such animals were treated with either B lymphocytes or hyperimmune serum, approximately 50% of the animals survived.[231] This survival was attributed to the passive immunotherapy in the absence of an active immune response of the host. The protection could be achieved only with enriched B lymphocytes obtained from immune (but not normal syngeneic) donors. Although a weak cell-mediated immune response was noted in the recipients, virus clearance corresponded temporally more closely with increase in antibody titer.

Together these and other studies clearly suggest a very important role for antibody in determining the outcome of rabies virus infection. How antibody actually clears the infection is not yet obvious. The antibody could effectively neutralize the virus that is present in the intercellular spaces or in the body fluids. However, the intracellular virus cannot be eliminated by this mechanism. It has been shown in vitro that antibody can bind to virus expressed on the cell surface and allow complement or antibody-dependent cytotoxic T cells to mediate killing, thus effectively eliminating infected cells and limiting the spread of virus. This possibility may also exist in vivo. Studies by Miller et al.[229] using cobra venom factor, which depletes C_3 component of the complement cascade in the serum for up to 4 days, have shown that the blocking of complement action results in a delayed virus clearance. This observation suggests a role for complement-mediated antibody-dependent clearance of virus from the animal. Yet another way that the antibody can clear the virus is by promoting antibody-dependent cell-mediated cytotoxicity. It has been shown that when nonimmune lymphocytes are added to rabies virus-infected neuroblastoma cells preincubated with antibody, the lymphocytes destroy the virus infected cells in the presence of antibody, but not in its absence. This strongly suggests that antibody can mediate the killing of virus infected cells causing the release of cell-associated virus that can then be neutralized.

B. Role of Cell-Mediated Immune Response

The initial evidence for the role of T lymphocytes in protection against rabies virus infection came from studies by Kaplan et al.[232] BALB/c nude mice, which are deficient in T cells, and their normal littermates were inoculated with the avirulent strain Flury HEP rabies virus. The infection was lethal in athymic BALB/c nude mice although normal littermates showed no signs of the disease. Only 3 out of 18 nude mice developed measurable levels of SN antibody, and these antibody levels were lower than those of controls. Thus, the requirement of T cells for an optimal immune response to rabies antigens was also demonstrated. Turner[233] subsequently showed that nude mice inoculated with vaccine could neither mount an antibody response nor develop resistance to subsequent challenge with a virulent strain of rabies virus. Thus, Turner's studies established the T cell dependence of the immune response to rabies virus.

The T lymphocytes may mediate the clearance of the virus by a number of mechanisms. The T helper cells (Th) can provide the necessary factors for the stimulation of both cytotoxic T cells (Tc) and B cells or may recruit macrophages, which can phagocytose the antigen and thus limit its spread. The experiments discussed earlier suggested that Th cells are necessary for processing rabies virus antigens, but did not provide evidence of the role of other possible mechanisms of T cell action.

Evidence for the presence of rabies virus cytotoxic T cells has been accumulating for some time. Wiktor et al.[234] provided evidence for cell-mediated immunity after exposure to both live and inactivated rabies virus. Virus-specific cytotoxic T cells appeared as early as 3 to 4 days after immunization, reached their peak activity by day 6, and disappeared by the 9th day. These investigators also showed that although the rabies virus-infected syngeneic target cells were destroyed most efficiently, some nonsyngeneic cells were also killed by rabies virus induced cytotoxic T cells. Furthermore, this study demonstrated that the treatment of mice with hyperimmune antirabies serum prior to vaccination with the virus effectively blocked the generation of cytotoxic T cells. This particular observation is of special interest because the currently recommended regimens of postexposure prophylaxis involve administration of serum along with vaccine. Although these studies demonstrated the appearance of cytotoxic T cells, their role in the clearance of virus in vivo was not shown. Studies to address this question were carried out in an experimental system using HEP virus. Animals that were immunologically compromised after infection with HEP virus were given enriched B or T lymphocytes. This showed that enriched T lymphocytes when given in the absence of a humoral immune response could confer protection.[235] Only T cells obtained from donors either 7 days after primary immunization or 3 days after secondary immunization were effective in consistently preventing the disease, suggesting that the effector cells are cytotoxic T cells rather than helper T cells. Cytotoxic T cell response in recipients measured at various times after infection and cell transfer showed a strong correlation between the peak of cytotoxic activity and the clearance of virus. This provided further evidence to suggest that this may be an active mechanism by which the virus is cleared in vivo.

The role of T lymphocytes in eliciting a delayed type hypersensitivity response has only recently been established. Lagrange et al.[224] showed that inactivated rabies virus could induce a DTH response in mice pretreated with BCG or with rabies vaccine. This response was measured as footpad swelling at various times after challenge with the virus. The response was maximal at 24 hr and was accompanied by infiltration of mononuclear cells into the site of inoculation. This phenomenon could be transferred to naive recipients by transfer of cells from immune donors, but not with either normal cells or serum. A similar phenomenon has also been noted using street virus in conjunction with BCG.[236] The demonstration of DTH with street virus raises the possibility that this may be a mechanism by which the virus spread may be limited or eliminated, especially in the early stages of infection.

A fourth mechanism by which cell-mediated immunity can operate during rabies infection has been demonstrated by Harfast et al.[237] These investigators have shown that there is a subpopulation of cells that can destroy rabies infected targets in the presence of antibody. This has been interpreted as antibody-dependent cell-mediated cytotoxicity.[237] This observation again provides evidence for the interplay of both humoral and cellular mechanisms in the clearance of rabies virus infection. Tignor[238] has also shown that CTL as well as ADCC can be effector mechanisms in the destruction of rabies virus infected targets.

Together, there is ample evidence that cell-mediated immune response is very critical for an effective clearance of the virus. Most of these mechanisms have been shown in experimental systems and in vitro, and limited attempts have been made to understand their relative role in the recovery from rabies virus infection.

C. Interferon

The effects of exogenous interferon have been studied using both cell culture systems and experimental animals. Inhibition of rabies virus infection of a number of cultured cells has been noted using exogenous interferon induced in tissue culture using different viruses.[239] Protection in rabbits or guinea pigs can be induced by administering interferon from 24 hr prior to infection to 1 hr after infection. The administration of interferon postinfection at the site of virus inoculation conferred partial protection when the source of interferon was guinea pig serum. However, when interferon of tissue culture origin was used, it was totally ineffective. Whether this was due to differences in the quantity or the source of interferon inoculated is not clear.

One can also administer interferon inducers to generate endogenous interferon that in turn provides protection. Interferon inducers used have included poly IC,[240] polyIC-LC,[241] bovine parainfluenza virus type 3,[242] and Newcastle disease virus.[243] Endogenously induced interferon has been shown to inhibit rabies virus replication in mice, rabbits, and hamsters. The inducers have been shown to be effective when inoculated between 24 hr before and 3 hr after infection. The degree of protection conferred varied from one experimental system to another. For example, poly IC administered prior to infection conferred almost complete protection in mice and rabbits, but the treatment was less effective when given after infection.

The effects of interferon also depend on whether the source is exogenous or endogenous and on the strain of virus used for testing. Different strains of rabies virus are known to have different susceptibilities to the action of interferon. In general, highly pathogenic strains such as CVS are less sensitive to the effects of interferon because of their rapid progression from peripheral tissue to the CNS. However, most street strains tested show some degree of susceptibility to the effects of interferon.[244]

Detailed studies on the effects of interferon administered along with vaccine in a postexposure situation in mice and rhesus monkeys strongly suggest that there is an additive effect of interferon in preventing disease. When mice were given vaccine and exogenous interferon or vaccine and poly IC-LC, 24 hr after infection, Baer et al.[241] noted no mortality as compared to 24% mortality among controls. This treatment was most effective if the interferon or the interferon inducer was inoculated in the same leg into which virus had been injected. A similar observation has been made in monkeys up to 6 hr after infection.[225,226] Vaccine and interferon or interferon inducers each reduced mortality to 9 to 10% as compared to 90% in controls. However, in another study, treatment of monkeys infected with street rabies virus with highly concentrated vaccine resulted in survival of approximately 60% of the monkeys, and addition of interferon inducer did not change the outcome appreciably. This suggested that in a postexposure situation interferon may play a limited role and that other factors may be more important.

Although the higher efficacy of human diploid cell vaccine has been attributed to its ability to induce both a CMI response and high levels of interferon, no definitive proof exists that would indicate an important role for interferon in the postexposure treatment of humans. Interferon may aid in augmenting virus-specific immune responses, thus contributing indirectly to recovery.

D. Complement

One of the several mechanisms by which an infected target cell can be destroyed is through complement-mediated lysis. There is ample evidence that rabies-infected cells in culture that have been pretreated with antibody can be lysed by the addition of complement.[245] This particular mechanism may be quite relevant in rabies viral infection, because the virus buds from the cell membrane and generally does not destroy the infected cell. The virus is probably cleared by this mechanism in vivo also. This particular contention draws its support from experiments in which animals were rendered complement deficient by treatment with cobra venom factor.[229] This factor inactivates C_3 component of the complement cascade, thus effectively blocking the complement pathway. Such a treatment resulted in a delayed virus clearance associated with increased mortality.

Yet another mechanism by which complement could mediate effector function is by direct virolysis. It has been shown with a number of enveloped viruses that addition of complement to antibody-treated virus results in direct destruction of the virus. This antigen-antibody complex is cleared by phagocytes. Complement may also prevent the spread of virus by aiding in the recruitment of nonspecific mononuclear cells to the site of infection or the site of virus replication. Definitive evidence of its relative contribution in human disease, however, has not been established.

E. Immunopathological Aspects of Rabies Virus Infection

Although the immunopathology of rabies infection has been discussed for more than 10 years, the precise mechanism by which the immune response contributes to the disease was not known until recently. In 1972 Koprowski reviewed a number of mechanisms that could cause early death.[246] These included the action of lytic antibody, cytolytic T cells, and immune complexes. Several investigators working with animal models and using a variety of vaccines have observed that animals dying of rabies challenge, despite pre- or postexposure vaccination, tend to die after shorter incubation periods than do unvaccinated controls.[247-249] This phenomenon has been termed "early death". Tignor et al.[250] have demonstrated prolonged survival and decreased paralysis after immunosuppression with cyclophosphamide or thymectomy or sublethal irradiation. Iwasaki et al.[204] have also studied the role of immune response in pathogenesis and pathology of rabies virus infection. These investigators noted that immunocompetent mice infected with rabies were severely paralyzed with marked inflammation and degeneration of the CNS parenchymatous tissue. However, when infected mice were immunosuppressed they developed encephalitic symptoms with degeneration or necrosis of neurons with a mild microglial reaction. A similar outcome was noted after infection of immunodeficient mice. These observations suggest a possible role for the immune system in determining the nature of both the pathological changes and the clinical expression of the disease.

A number of recent reports using different experimental systems have confirmed the existence of an immunopathologic component. Blancou et al.[251] showed that mice vaccinated with a suboptimal dose of vaccine 3 days before challenge died several days earlier than unvaccinated controls. Based upon this observation, studies were conducted to understand the underlying mechanism of the early death. Adoptive transfer of plasma from immune donors to unvaccinated infected mice also resulted in early death of the recipients, suggesting a role for antibody. More definitive evidence as to the mechanism was obtained by adoptive transfer experiments using enriched subpopulations of lymphocytes. In this study, high-egg-passage Flury (HEP) virus was shown to cause a mild inapparent infection of the CNS in adult mice that could be converted into a lethal infection by immunosuppression.[197] The fatal infection could be aborted by transfer of syngeneic immune lymphocytes.[235] However, when secondary immune spleen cells were given to the recipients, although a majority of the animals survived, approximately 30% of the mice died much earlier than the controls. In order to determine the type of cells involved, transfers were made using cells enriched for either T or B lymphocytes. The results indicated that infact, B lymphocytes are involved in early death. To further confirm this conclusion, these animals were given hyperimmune serum that also resulted in early death. These studies established that "early rabies death" is mediated by antibodies.[231]

Further evidence for this comes from studies using street rabies virus infection of a large number of mice. Smith et al.[252] have shown that immunosuppression after street rabies virus infection results in prolonged survival. They noted that these animals showed minimal paralysis. However, the return of immune responses in these suppressed mice corresponded with the onset of paralysis. Furthermore, treatment of infected immunosuppressed mice with hyperimmune serum resulted in early paralysis and death.

Reports in the literature[247,249] also indicate that the "early death" phenomenon is seen in humans who have received postexposure treatment with different types of vaccines with or without accompanying immune serum therapy. This suggests that in rabies, the immune response may play a dual role, either enhancing the disease or favoring survival. It is not yet clear whether immunopathology is a major cause of death in rabies virus infection of humans.

VIII. PERSPECTIVE

The sole effective change in the postexposure immunoprophylaxis of rabies since Louis Pasteur has been the introduction of hyperimmune serum as a part of the postexposure treatment. There is ample evidence that vaccine alone after exposure to rabies virus is of limited value. It was demonstrated by Koprowski and Black[253] that nerve tissue vaccine alone given 24 hr after infection of guinea pigs with street virus was ineffective. This and other observations with similar results have been substantiated by experimental observations that serum with or without vaccine can confer some protection. This was further confirmed by the treatment of humans exposed to rabid wolf bites in Iran with serum and vaccine. Because of the proven efficacy of combined vaccine and serum therapy the World Health Organization has recommended use of hyperimmune equine serum as part of postexposure treatment. Although the addition of serum has reduced the mortality due to rabies, it also has contributed to iatrogenic complications of serum sickness. This, however, has been alleviated by the use of hyperimmune human antirabies serum.

The administration of serum may also affect the host's immune response. There is evidence in both man and experimental animals that the administration of serum results in the suppression of an active immune response.[254,255] This has been noted with serum of both equine and human origin.[255,256] These observations have led to the administration of booster doses of vaccine to overcome the suppressive effects of the serum.

The most impressive improvement in the quality of rabies vaccine since the time of Pasteur is the development of human diploid cell vaccine (HDCV). This new concentrated vaccine has been tested extensively in both animals and humans and has been shown to induce excellent cell-mediated and humoral immune responses. Use of this vaccine in monkeys and humans in postexposure situations has provided excellent protection with a highly reduced risk of allergic responses when compared to previously used vaccines.

Availability of HDCV and purified immunoglobulins from hyperimmune human serum have greatly improved the treatment of rabies. Attempts are underway to produce more potent and less costly vaccines using recombinant DNA technology and artificially synthesized polypeptides. Rabies virus glycoprotein analogs have been biosynthesized in *E. coli*; since glycoprotein is responsible for the induction of neutralizing antibody, these could provide an excellent source of vaccine for future use. This is of particular interest because of reports of vaccine-induced rabies in pet skunks, cats, and dogs[257-260] vaccinated with live attenuated vaccines. This type of accidental conversion of an inapparent infection into an overt disease could be avoided by using subunit vaccines. In the light of the fact that there are a large number of antigenic variants, the spectrum of synthetic polypeptides required for an effective rabies vaccine needs to be established.

The isolation of hybridomas that synthesize monoclonal antibodies to rabies virus has led to the identification of a large number of naturally occurring antigenic variants. This has also provided a new impetus to the understanding of molecular mechanisms involved in the generation of antigenic variants. Monoclonal antibodies are also useful in the differential diagnosis of rabies and rabies-related viruses.

Ironically, the progress in understanding the pathogenesis of rabies and the development of better vaccines has not contributed to alleviating the global problem of rabies. Most of the developing countries still have large numbers of human rabies cases in urban areas because of improper control of the dog population. The problem in developed countries has taken on a new dimension with the recognition of the disease in a large number of wild animals. Efforts are underway in Europe to eliminate rabies in wild foxes, the prime vectors, through vaccination with a live attenuated strain of rabies in bait. Some field successes have been obtained, but unresolved questions about the genetic stability (and safety for all species of wild animals) of the virus strains used justify extreme caution in this approach.

An effective public health program consisting of increased public awareness and control of urban stray dog populations should contribute enormously to reducing the rabies problem in developing countries. However, in developed countries, eradication of rabies requires new innovative approaches to control the disease in wild animals. As long as the virus can persist in the wild, there will be opportunities for spill-over into urban areas. Although rabies is one of the oldest maladies known, it continues to pose new challenges.

ACKNOWLEDGMENTS

This research was supported in part by the Commonwealth of Pennsylvania and by research grant AI-09706 from the National Institutes of Health.

REFERENCES

1. Steele, J. H., History of rabies, in *The Natural History of Rabies*, Vol. 1, Baer, G. M., Ed., Academic Press, New York, 1975, 1.
2. Kissling, R. E., Growth of rabies virus in non-nervous tissue culture, *Proc. Soc. Exp. Biol. Med.*, 98, 223, 1958.
3. Fernandes, M. V., Wiktor, T. J., and Koprowski, H., Endosymbiotic relationship between animal viruses and host cells. A study of rabies virus in tissue culture, *J. Exp. Med.*, 120, 1099, 1964.

4. Wiktor, T. J. and Clark, H. F., Growth of rabies virus in cell culture, in *The Natural History of Rabies*, Vol. 1, Baer, G. M., Ed., Academic Press, New York, 1975, 155.
5. Pasteur, L., Roux, E., Chamberland, C., and Thuillier, L., Note sur la rage, *C. R. Acad. Sci.*, 92, 1259, 1881.
6. Negri, A., Contributo allo studio dell' eziologia della rabia (German transl.), *Z. Hyg. Infektionskr.*, 43, 507, 1903.
7. Dean, D. J., Evans, W. M., and McClure, R. C., Pathogenesis of rabies, *Bull. W. H. O.*, 29, 803, 1963.
8. Pasteur, L., Méthode pour prévenir la rage après mossure, *C. R. Acad. Sci.*, 101, 765, 1885.
9. Clark, H. F., Wiktor, T. J., and Koprowski, H., Human vaccination against rabies, in *The Natural History of Rabies*, Vol. 2, Baer, G. M., Ed., Academic Press, New York, 1975, 343.
10. Murphy. F. A., Rabies: its pathogenesis predicts its ecologic entrenchment, in *Viral Diseases in South-East Asia and the Western Pacific*, Mackenzie, J. S., Ed., Academic Press, New York, 1982, 553.
11. Dupont, J. R. and Earle, K. M., Human rabies encephalitis. A study of forty-nine fatal cases with a review of the literature, *Neurology*, 15, 1023, 1965.
12. Hattwick, M. A. and Gregg, M. B., The disease in man, in *The Natural History of Rabies*, Vol. 2, Baer, G. M., Ed., Academic Press, New York, 1975, 281.
13. Hattwick, M. A., Weis, T. T., Stechschulte, J., Baer, G. M., and Gregg, M. B., Recovery from rabies. A case report, *Ann. Intern. Med.*, 76, 931, 1972.
14. Gode, G. R., Raju, A. V., Jayalakshmi, T. S., and Kaul, H. L., Intensive care in rabies therapy, *Lancet*, II, 6, 1976.
15. Tierkel, E. S., Canine rabies, in *The Natural History of Rabies*, Vol. 2, Baer, G. M., Ed., Academic Press, New York, 1975, 123.
16. Toma, B. and Andral, L., Epidemiology of fox rabies, *Adv. Virus Res.*, 21, 1, 1977.
17. Parker, R. L., Rabies in skunks, in *The Natural History of Rabies*, Vol. 2, Baer, G. M., Ed., Academic Press, New York, 1975, 41.
18. Baer, G. M., Rabies in nonhematophagous bats, in *The Natural History of Rabies*, Vol. 2, Baer, G. M., Ed., Academic Press, New York, 1975, 79.
19. Bell, J. F., Abortive rabies infection. I. Experimental production in white mice and general discussion, *J. Infect. Dis.*, 114, 249, 1964.
20. Wiktor, T. J., Koprowski, H., and Rorke, L. B., Localized rabies infection in mice, *Proc. Soc. Exp. Biol. Med.*, 140, 759, 1972.
21. Baer, G. M., Cleary, W. F., Diaz, A. M., and Perl, D. F., Characteristics of 11 rabies virus isolates in mice: titers and relative invasiveness of virus, incubation period of infection, and survival of mice with sequelae, *J. Infect. Dis.*, 136, 336, 1977.
22. Clark, H. F. and Koprowski, H., Isolation of temperature-sensitive conditional lethal mutants of "fixed" rabies virus, *J. Virol.*, 7, 295, 1971.
23. Johnson, R. T., Experimental rabies. Studies of cellular vulnerability and pathogenesis using fluorescent antibody staining, *J. Neuropathol. Exp. Neurol.*, 24, 662, 1965.
24. Johnson, R. T., The pathogenesis of experimental rabies, in *Rabies*, Nagano, Y. and Davenport, F. M., Eds., University Park Press, Baltimore, 1971, 59.
25. Vaughn, J. B., Gerhardt, P., and Newell, K. W., Excretion of street rabies virus in the saliva of dogs, *JAMA*, 193, 363, 1965.
26. Fekadu, M., Shaddock, J. H., and Baer, G. M., Excretion of rabies virus in the saliva of dogs, *J. Infect. Dis.*, 145, 715, 1982.
27. Preston, E., Computer simulated dynamics of a rabies-controlled fox population, *J. Wildl. Manage.*, 37, 501, 1973.
28. Everard, C. O. R., Race, M. W., Price, J. L., and Baer, G. M., Recent epizootiological findings in wildlife rabies on Grenada, *Symp. Adv. Rabies Res.*, Center for Disease Control, U.S. Department of Health, Education, and Welfare, Atlanta, September 1976, 8.
29. Acha, P. N., Epidemiology of paralytic bovine rabies and bat rabies, *Bull. Off. Int. Epiz.*, 67, 343, 1967.
30. Pawan, J. L., Rabies in the vampire bat of Trinidad, with special reference to the clinical course and the latency of infection, *Ann. Trop. Med. Parasitol.*, 30, 401, 1936.
31. Sodja, L., Lim, D., and Matouch, O., Isolation of rabies-like virus from small wild rodents, *J. Hyg. Epidemiol. (Praha)*, 15, 271, 1971.
32. Schneider, L. G. and Schoop, U., Pathogenesis of rabies and rabies-like viruses, *Ann. Inst. Pasteur*, 123, 469, 1972.
33. Rabies Surveillance Annual Summary, 1980 — 1982, HHS Publ. No. (CDC) 84-8255, Center for Disease Control, U.S. Public Health Service, Atlanta, 1982.
34. McLean, R. G., Raccoon rabies, in *The Natural History of Rabies*, Vol. 2, Baer, G. M., Ed., Academic Press, New York, 1975, 53.

35. Morbidity and Mortality Weekly Report (MMWR), Vol. 32, No. 7, Center for Disease Control, U.S. Department of Health and Human Services, Atlanta, February 25, 1983.
36. Parham, G. L., Rabies in the United States, 1981, Morbidity and Mortality Weekly Report (MMWR), Vol. 32, No. 1, Center for Disease Control, Department of Health and Human Services, Atlanta, February 1983, 33.
37. Constantine, D. G., Rabies transmission by nonbite route, *Publ. Health Rep.*, 77, 287, 1962.
38. Bigler, W. J., Hoff, G. L., and Buff, E. E., Chiropteran rabies in Florida: a twenty-year analysis, 1954 to 1973, *Am. J. Trop. Med. Hyg.*, 24, 347, 1975.
39. Constantine, D. G., Transmission experiments with bat rabies isolates: bite transmission of rabies to foxes and coyote by free-tailed bats, *Am. J. Vet. Res.*, 27, 20, 1966.
40. Constantine, D. G., Transmission experiments with bat rabies isolates: reactions of certain carnivora, opossum, rodents, and bats to rabies virus of red bat origin when exposed by bat bite or by intramuscular inoculation, *Am. J. Vet. Res.*, 27, 24, 1966.
41. Sikes, R. K., Pathogenesis of rabies in wildlife. I. Comparative effect of varying doses of rabies virus inoculated into foxes and skunks, *Am. J. Vet. Res.*, 23, 1041, 1962.
42. Parker, R. L. and Wilsnack, R. E., Pathogenesis of skunk rabies virus: quantitation in skunks and foxes, *Am. J. Vet. Res.*, 27, 33, 1966.
43. Baer, G. M., Bovine paralytic rabies and rabies in the vampire bat, in *The Natural History of Rabies*, Vol. 2, Baer, G. M., Ed., Academic Press, New York, 1975, 155.
44. Torres, S. and de Queiroz Lima, E., A raiva e os morcegos hematophagos, *Rev. Dep. Nac. Prod. Anim. Brazil*, 3, 165, 1936.
45. Moreno, J. A. and Baer, G. M., Experimental rabies in the vampire bat, *Am. J. Trop. Med. Hyg.*, 29, 254, 1980.
46. Johnson, H. N., General epizootiology of rabies, in *Rabies*, Nagano, J. and Davenport, F. M., Eds., University Park Press, Baltimore, 1971, 237.
47. Stamm, D. D., Kissling, R. E., and Eidson, M. E., Experimental rabies infection in insectivorous bats, *J. Infect. Dis.*, 98, 9, 1956.
48. Sulkin, S. E., Allen, R., Sims, R., Krutzsch, P. H., and Kim, C., Studies on the pathogenesis of rabies in insectivorous bats. II. Influence of environmental temperature, *J. Exp. Med.*, 112, 595, 1960.
49. Sulkin, S. E., Krutzsch, P. H., Allen, R., and Wallis, C., Studies on the pathogenesis of rabies in insectivorous bats. I. Role of brown adipose tissue, *J. Exp. Med.*, 110, 369, 1959.
50. Constantine, D. G., Transmission experiments with bat rabies isolates: responses of certain carnivora to rabies virus isolated from animals infected by nonbite route, *Am. J. Vet. Res.*, 27, 13, 1966.
51. Constantine, D. G., Transmission experiments with bat rabies isolates: reaction of certain carnivora, opossum, and bats to intramuscular inoculations of rabies virus isolated from free-tailed bats, *Am. J. Vet. Res.*, 27, 16, 1966.
52. Brown, F. and Crick, J., Natural history of rhabdoviruses of vertebrates and invertebrates, in *Rhabdoviruses*, Vol. 1, Bishop, D. H. L., Ed., CRC Press, Boca Raton, Fla., 1979, 1.
53. Pasteur, L., Chamberland, C., and Roux, E., Nouvelle communication sur la rage, *C. R. Acad. Sci.*, 98, 457, 1884.
54. Clark, H. F. and Wiktor, T. J., Rabies virus, in *Strains of Rabies Viruses*, Majer, M. and Plotkin, S. A., Eds., S. Karger, Basel, 1972, 177.
55. Murphy. F. A., Rabies pathogenesis. Brief review, *Arch. Virol.*, 54, 279, 1977.
56. Clark, H. F., Systems for assay and growth of rhabdoviruses, in *Rhabdoviruses*, Vol. 1, Bishop, D. H. L., Ed., CRC Press, Boca Raton, Fla., 1979, 23.
57. Murphy, F. A., Morphology and morphogenesis, in *The Natural History of Rabies*, Vol. 1, Baer, G. M., Ed., Academic Press, New York, 1975, 33.
58. Murphy. F. A. and Harrison, A. K., Electron microscopy of the rhabdoviruses of animals, in *Rhabdoviruses*, Vol. 1, Bishop, D. H. L., Ed., CRC Press, Boca Raton, Fla., 1979, 65.
59. Matsumoto, S., Morphology of rabies virion and cytopathology of virus infected cells, *Symp. Series Immunobiol. Stand.*, 21, 25, 1974.
60. Sokol, F., Chemical composition and structure of rabies virus, in *The Natural History of Rabies*, Vol. 1, Baer, G. M., Ed., Academic Press, New York, 1975, 79.
61. Sokol, F., Schlumberger, D. H., Wiktor, T. J., and Koprowski, H., Biochemical and biophysical studies on the nucleocapsid and on the RNA of rabies virus, *Virology*, 38, 651, 1969.
62. Madore, H. P. and England, J. M., Rabies virus protein synthesis in infected BHK-21 cells, *J. Virol.*, 22, 102, 1977.
63. Wiktor, T. J., György, E., Schlumberger, D., Sokol, F., and Koprowski, H., Antigenic properties of rabies virus components, *J. Immunol.*, 110, 269, 1973.
64. Cox, J. H., Dietzschold, B., and Schneider, L. G., Rabies virus glycoprotein. II. Biological and serological characterization, *Infect. Immun.*, 16, 754, 1977.

65. Dietzschold, B., Cox, J. H., Schneider, L. G., Wiktor, T. J., and Koprowski, H., Isolation and purification of a polymeric form of the glycoprotein of rabies virus, *J. Gen. Virol.*, 40, 131, 1978.
66. Cox, J. H., Dietzschold, B., and Schneider, L. G., Preparation and characterization of rabies virus hemagglutinin, *Infect. Immun.*, 30, 572, 1980.
67. Lafon, M., Wiktor, T. J., and Macfarlan, R. I., Antigenic sites on the CVS rabies virus glycoprotein: analysis with monoclonal antibodies, *J. Gen. Virol.*, 64, 843, 1983.
68. Dietzschold, B., Wiktor, T. J., Macfarlan, R., and Varrichio, A., Antigenic structure of rabies virus glycoprotein: ordering and immunological characterization of the large CNBr cleavage fragments, *J. Virol.*, 44, 595, 1982.
69. Reagan, K. and Wunner, W. H., personal communication.
70. Murphy, F. A., Halonen, P. E., Gary Jun, G. W., and Reese, D. R., Physical characterization of rabies virus haemagglutinin, *J. Gen. Virol.*, 3, 289, 1968.
71. Schneider, L. G., Horzinek, M., and Novicky, R., Isolation of a hemagglutinating, immunizing, and non-infectious subunit of the rabies virion, *Arch. Ges. Virusforsch.*, 34, 360, 1971.
72. Anilionis, A., Wunner, W. H., and Curtis, P. J., Structure of the glycoprotein gene in rabies virus, *Nature (London)*, 294, 275, 1981.
73. Yelverton, E., Norton, S., Obijeski, J. F., and Goeddel, D. V., Rabies virus glycoprotein analogs: biosynthesis in *Escherichia coli*, *Science*, 219, 614, 1983.
74. Schneider, L. G., Dietzschold, B., Dierks, R. E., Matthaeus, W., Enzmann, P.-J., and Strohmaier, K., Rabies group-specific ribonucleoprotein antigen and a test system for grouping and typing of rhabdoviruses, *J. Virol.*, 11, 748, 1973.
75. Sureau, P., Rollin, P., and Wiktor, T. J., Epidemiologic analysis of antigenic variations of street rabies virus: detection by monoclonal antibodies, *Am. J. Epidemiol.*, 117, 605, 1983.
76. Wiktor, T. J. and Koprowski, H., Monoclonal antibodies against rabies virus produced by somatic cell hybridization: detection of antigenic variants, *Proc. Natl. Acad. Sci. U.S.A.*, 75, 3938, 1978.
77. Prabhakar, B. S., Parks, N. F., and Clark, H. F., Altered expression of rabies virus antigenic determinants associated with chronic infection and virulence, *J. Gen. Virol.*, 64, 251, 1983.
78. Delagneau, J.-F., Perrin, P., and Atanasiu, P., Structure of the rabies virus: spatial relationships of the proteins G, M_1, M_2, and N, *Ann. Virol.*, 132E, 473, 1981.
79. Zaides, V. M., Krotova, L. I., Selimova, L. M., Selimov, M. A., Elbert, L. B., and Zhdanov, V. M., Reevaluation of the proteins in rabies virus particles, *J. Virol.*, 29, 1226, 1979.
80. Wunner, W. H., personal communication.
81. Sokol, F. and Clark, H. F., Phosphoproteins, structural components of rhabdoviruses, *Virology*, 52, 246, 1973.
82. Emerson, S. U. and Yu, Y. H., Both NS and L proteins are required for *in vitro* RNA synthesis by vesicular stomatitis virus, *J. Virol.*, 15, 1348, 1975.
83. Dietzschold, B., Cox, J. H., and Schneider, L. G., Rabies virus strains: a comparison study by polypeptide analysis of vaccine strains with different pathogenic patterns, *Virology*, 98, 63, 1979.
84. Wagner, R. R., Reproduction of rhabdoviruses, in *Comprehensive Virology*, Vol. 4, Frankel-Conrat, H. and Wagner, R. R., Eds., Plenum Press, New York, 1975, 1.
85. Szilágyi, J. F. and Uryvayev, L., Isolation of an infectious ribonucleoprotein from vesicular stomatitis virus containing an active RNA transcriptase, *J. Virol.*, 11, 279, 1973.
86. Cunningham, J., Nicholas, M. J., and Lahiri, B. N., Investigation into the value of etherized vaccine in prophylactic treatment of rabies: action of ether on "street virus" infected brains, *Ind. J. Med. Res.*, 15, 85, 1927.
87. Sokol, F., Clark, H. F., and György, E., Heterogeneity in the phospholipid content of purified rabies virus (ERA strain) particles, *J. Gen. Virol.*, 16, 173, 1972.
88. Blough, H. A., Tiffany, J. M., and Aaslestad, H. G., Lipids of rabies virus and BHK-21 cell membranes, *J. Virol.*, 21, 950, 1977.
89. McSharry, J. J., The lipid envelope and chemical composition of rhabdoviruses, in *Rhabdoviruses*, Vol. 1, Bishop, D. H. L., Ed., CRC Press, Boca Raton, Fla., 1979, 107.
90. Dietzschold, B., Oligosaccharides of the glycoprotein of rabies virus, *J. Virol.*, 23, 286, 1977.
91. Wertz, G. W., RNA replication, in *Rhabdoviruses*, Vol. 2, Bishop, D. H. L., Ed., CRC Press, Boca Raton, Fla., 1979, 75.
92. Iwasaki, Y., Wiktor, T. J., and Koprowski, H., Early events of rabies virus replication in tissue cultures. An electron microscopic study, *Lab. Invest.*, 28, 142, 1973.
93. Kawai, A., Transcriptase activity associated with rabies virion, *J. Virol.*, 24, 826, 1977.
94. Flamand, A., Delagneau, J. F., and Bussereau, F., An RNA polymerase activity in purified rabies virions, *J. Gen. Virol.*, 40, 233, 1978.
95. Holloway, B. P. and Obijeski, J. F., Rabies virus-induced RNA synthesis in BHK-21 cells, *J. Gen. Virol.*, 49, 181, 1980.

96. Coslett, G. D., Holloway, B. P., and Obijeski, J. F., The structural proteins of rabies virus and evidence for their synthesis from separate monocistronic RNA species, *J. Gen. Virol.*, 49, 161, 1980.
97. Hummeler, K., Koprowski, H., and Wiktor, T. J., Structure and development of rabies virus in tissue culture, *J. Virol.*, 1, 152, 1967.
98. Iwasaki, Y., Ohtani, S., and Clark, H. F., Maturation of rabies virus by budding from neuronal cell membrane in suckling mouse brain, *J. Virol.*, 15, 1020, 1975.
99. Lunger, P. D. and Clark, H. F., Comparative morphology of three strains of rabies virions propagated in the reptilian cell line, VSW, 32nd Ann. Proc. Electron Microscopy Soc. Am., Arceneaux, C. J., Ed., St. Louis, Mo., 1974.
100. Dierks, R. E., Murphy, F. A., and Harrison, A. K., Extraneural rabies virus infection, *Am. J. Pathol.*, 54, 251, 1969.
101. Huang, A. S. and Baltimore, D., Defective viral particles and viral disease process, *Nature (London)*, 226, 325, 1970.
102. Kawai, A., Matsumoto, S., and Tanabe, K., Characterization of rabies viruses recovered from persistently infected BHK cells, *Virology*, 67, 520, 1975.
103. Wiktor, T. J., Dietzschold, B., Leamnson, R. N., and Koprowski, H., Induction and biological properties of defective interfering particles of rabies virus, *J. Virol.*, 21, 626, 1977.
104. Crick, J. and Brown, F., An interfering component of rabies virus which contains RNA, *J. Gen. Virol.*, 22, 147, 1974.
105. Clark, H. F., Parks, N. F., and Wunner, W. H., Defective interfering particles of fixed rabies viruses: lack of correlation with attenuation or auto-interference in mice, *J. Gen. Virol.*, 52, 245, 1981.
106. Wunner, W. H. and Clark, H. F., Regeneration of DI particles of virulent and attenuated rabies virus: genome characterization and lack of correlation with virulence phenotype, *J. Gen. Virol.*, 51, 69, 1980.
107. Holland, J. J. and Villarreal, L. P., Purification of defective interfering T particles of vesicular stomatitis and rabies viruses generated *in vivo* in brains of newborn mice, *Virology*, 67, 438, 1975.
108. Bussereau, F., Benejean, J., and Saghi, N., Isolation and study of temperature-sensitive mutants of rabies virus, *J. Gen. Virol.*, 60, 153, 1982.
109. Wright, J. T. and Habel, K., Comparison of antigenicity and certain biological characteristics of 6 substrains of Pasteur fixed rabies virus, *J. Immunol.*, 60, 503, 1948.
110. Habel, K., Factors influencing the efficacy of phenolized rabies vaccines. I. Strains of fixed virus, *Publ. Health Rep.*, 55, 1619, 1940.
111. Clark, H. F., Factors affecting the virulence for mice of fixed rabies virus, in Mechanisms of Viral Pathogenesis and Virulence, Munich Symposia on Microbiology, Bachmann, P. A., Ed., Proc. 4th Munich Symp. on Microbiology, Munich, 1979, 281.
112. Komarov, A. and Hornstein, K., Studies on the pathogenicity of an avianized street rabies virus, *Cornell Vet.*, 43, 344, 1953.
113. Sedwick, W. D. and Wiktor, T. J., Reproducible plaquing system for lymphocytic choriomeningitis, rabies, and other ribonucleic acid viruses in BHK-21/13S agarose suspensions, *J. Virol.*, 1, 1224, 1967.
114. Clark, H. F. and Wiktor, T. J., Temperature-sensitivity characteristics distinguishing substrains of fixed rabies virus: lack of correlation with plaque-size markers or virulence for mice, *J. Infect. Dis.*, 125, 637, 1972.
115. Clark, H. F. and Wiktor, T. J., Plasticity of phenotypic characters of rabies-related viruses: spontaneous variation in the plaque morphology, virulence, and temperature-sensitivity characters of serially propagated Lagos bat and Mokola viruses, *J. Infect. Dis.*, 130, 608, 1974.
116. Clark, H. F. and Ohtani, S., Temperature-sensitive mutants of rabies virus in mice: a mutant (ts 2) revertant mixture selectively pathogenic by the peripheral route of inoculation, *Infect. Immun.*, 13, 1418, 1976.
117. Aubert, M. F. A., Bussereau, F., and Blancou, J., Pathogenic immunogenic and protective powers of ten temperature-sensitive mutants of rabies virus in mice, *Ann. Virol. Inst. Pasteur*, 131, 217, 1980.
118. Bussereau, F. and Flamand, A., Isolation and preliminary characterization of *ts* mutants of rabies virus, in *Negative Strand Viruses and the Host Cell*, Mahy, B. W. J. and Barry, R. D., Eds., Academic Press, New York, 1978, 701.
119. Clark, H. F., Rabies viruses increase in virulence when propagated in neuroblastoma cell culture, *Science*, 199, 1072, 1978.
120. Koprowski, H., Biological modification of rabies virus as a result of its adaptation to chicks and developing chick embryos, *Bull. W. H. O.*, 10, 709, 1954.
121. Clark, H. F., Rabies serogroup viruses in neuroblastoma cells: propagation, "autointerference," and apparently random back-mutation of attenuated viruses to the virulent state, *Infect. Immun.*, 27, 1012, 1980.

122. Crick, J., Brown, F., Fearne, A. J., and Razavi, J. H., Antigenic variation between rabies virus strains and its relevance in vaccine production and potentcy testing, in *The Replication of Negative Strand Viruses*, Bishop, H. L. and Compans, R. W., Eds., Elsevier/North Holland, Amsterdam, 1981, 122.
123. Wiktor, T. J. and Clark, H. F., Comparison of rabies strains by means of the plaque reduction test, *Ann. Microbiol.*, 124A, 283, 1973.
124. Kohler, G. and Milstein, C., Continuous cultures of fused cells secreting antibody of predefined specificity, *Nature (London)*, 256, 495, 1975.
125. Flamand, A., Wiktor, T. J., and Koprowski, H., Use of hybridoma monoclonal antibodies in the detection of antigenic differences between rabies and rabies-related virus proteins. II. The glycoprotein, *J. Gen. Virol.*, 48, 105, 1980.
126. Flammand, A., Wiktor, T. J., and Koprowski, H., Use of hybridoma monoclonal antibody in the detection of antigenic differences between rabies and rabies-related virus proteins. I. The nucleocapsid proteins, *J. Gen. Virol.*, 48, 97, 1980.
127. Wiktor, T. J., Flammand, A., and Koprowski, H., Use of monoclonal antibodies in diagnosis of rabies virus infection and differentiation of rabies and rabies-related viruses, *J. Virol. Meth.*, 1, 33, 1980.
128. Wiktor, T. J. and Koprowski, H., Antigenic variants of rabies virus, *J. Exp. Med.*, 152, 99, 1980.
129. Schneider, L. G. and Meyer, S., Antigenic determinants of rabies virus as demonstrated by monoclonal antibody, in *The Replication of Negative Strand Viruses*, Bishop, D. H. L. and Compans, R. W., Eds., Elsevier/North Holland, Amsterdam, 1981.
130. Coulon, P., Rollin, P., Aubert, M., and Flamand, A., Molecular basis of rabies virus virulence. I. Selection of avirulent mutants of the CVS strain with anti-G monoclonal antibodies, *J. Gen. Virol.*, 61, 97, 1982.
131. Dietszschold, B., Wunner, W. H., Wiktor, T. J., Lopes, A. D., Lafon, M., Smith, C. L., and Koprowski, H., Characterization of an antigenic determinant of the glycoprotein that correlates with pathogenicity of rabies virus, *Proc. Natl. Acad. Sci. U.S.A.*, 80, 70, 1983.
132. Perl, D. P., The pathology of rabies in the central nervous system, in *The Natural History of Rabies*, Vol. 1, Baer, G. M., Ed., Academic Press, New York, 1975, 236.
133. Schneider, L. G., Spread of virus within the central nervous system, in *The Natural History of Rabies*, Vol. 1, Baer, G. M., Ed., Academic Press, New York, 1975, 199.
134. Lapi, A., Davis, C. L., and Anderson, W., The gasserian ganglion in animals dead of rabies, *J. Am. Vet. Med. Assoc.*, 120, 379, 1952.
135. Herzog, E., Histologic diagnosis of rabies, *Arch. Pathol.*, 39, 279, 1965.
136. Hurst, E. W. and Pawan, J. L., A further account of the Trinidad outbreak of acute rabic myelitis: histology of the experimental disease, *J. Pathol. Bacteriol.*, 35, 301, 1932.
137. Leach, C. N. and Johnson, H. N., Human rabies, with special reference to virus distribution and titer, *Am. J. Trop. Med.*, 20, 335, 1940.
138. Murphy, F. A., personal communication.
139. Murphy, F. A., Bell, J. F., Bauer, S. P., Gardner, J. J., Moore, G. J., Harrison, A. K., and Coe, J. E., Experimental chronic rabies in the cat, *Lab. Invest.*, 43, 231, 1980.
140. Atanasiu, P., Animal inoculation and the negri body, in *The Natural History of Rabies*, Vol. 1, Baer, G. M., Ed., Academic Press, New York, 1975, 374.
141. Goodpasture, E. W., A study of rabies, with reference to a neural transmission of the virus in rabbits, and the structure and significance of Negri bodies, *Am. J. Pathol.*, 1, 547, 1925.
142. Goldwasser, R. A. and Kissling, R. E., Fluorescent antibody staining of street and fixed rabies virus antigens, *Proc. Soc. Exp. Biol. Med.*, 98, 219, 1958.
143. Miyamoto, K. and Matsumoto, S., The nature of the Negri body, *J. Cell. Biol.*, 27, 677, 165.
144. Matsumoto, S., Rabies virus, *Adv. Virus Res.*, 6, 257, 1970.
145. Duenas, A., Belsey, M. A., Escobar, J., Medina, P., and Sanmartin, C., Isolation of rabies virus outside the human central nervous system, *J. Infect. Dis.*, 127, 702, 1973.
146. Howard, D. R., Rabies virus tropism in naturally infected skunks *(Mephitis mephitis)*, *Am. J. Vet. Res.*, 42, 2187, 1981.
147. Wandeler, A., Wachendörfer, G., Förster, U., Krekel, H., Müller, J., and Steck, F., Rabies in wild carnivores in Central Europe. II. Virological and serological examinations, *Zentralbl. Vet. Med. B*, 21, 757, 1974.
148. Constantine, D. G., Emmons, R. W., and Woodie, J. D., Rabies virus in nasal mucosa of naturally infected bats, *Science*, 175, 1255, 1972.
149. Winkler, W. G., Baker, E. F., Jr., and Hopkins, C. C., An outbreak of non-bite transmitted rabies in a laboratory animal colony, *Am. J. Epidemiol.*, 95, 267, 1972.
150. Bryceson, A. D. M., Greenwood, B. M., Warrell, D. A., Davidson, N. McD., Pope, H. M., Lawrie, J. H., Barnes, H. J., Bailie, W. E., and Wilcox, G. E., Demonstration during life of rabies antigen in humans, *J. Infect. Dis.*, 131, 71, 1975.

151. Blenden, D. C., Rabies in a litter of skunks predicted and diagnosed by skin biopsy, *J. Am. Vet. Med. Assoc.*, 179, 789, 1981.
152. Baer, G. M., Shaddock, J. H., Houff, S. A., Harrison, A. K., and Gardner, J. J., Human rabies transmitted by corneal transplant, *Arch. Neurol.*, 39, 103, 1982.
153. Roux, F., Bourgeade, A., Salaün, J. J., Bondurand, A., Ette, M., and Bertrand, Ed., L'atteinte cardiaque dans la rage humaine, *Tome*, 15, 37, 1976.
154. Schneider, L. G., Spread of virus from the central nervous system, in *The Natural History of Rabies*, Vol. 1, Baer, G. M., Ed., Academic Press, New York, 1975, 273.
155. Baer, G. M., Pathogenesis to the central nervous system, in *The Natural History of Rabies*, Vol. 1, Baer, G. M., Ed., Academic Press, New York, 1975, 181.
156. Murphy, F. A., Bauer, S. P., Harrison, A. K., and Winn, W. C., Jr., Comparative pathogenesis of rabies and rabies-like viruses: viral infection and transit from inoculation site to the central nervous system, *Lab. Invest.*, 28, 361, 1973.
157. Murphy, F. A. and Bauer, S. P., Early street rabies virus infection in striated muscle and later progression to the central nervous system, *Intervirology*, 3, 256, 1974.
158. Watson, H. D., Tignor, G. H., and Smith, A. L., Entry of rabies virus into the peripheral nerves of mice, *J. Gen. Virol.*, 56, 371, 1981.
159. Lawson, K. F., Johnston, D. H., Patterson, J. M., Black, J. G., Rhodes, A. J., and Zalan, E., Immunization of foxes *Vulpes vulpes* by the oral and intramuscular routes with inactivated rabies vaccines, *Can. J. Comp. Med.*, 46, 382, 1982.
160. Debbie, J. G., Abelseth, M. K., and Baer, G. M., The use of commercially available vaccines for the oral vaccination of foxes against rabies, *Am. J. Epidemiol.*, 96, 231, 1972.
161. Winkler, W. G. and Baer, G. M., Rabies immunization of red foxes *(Vulpes fulva)* with vaccine in sausage baits, *Am. J. Epidemiol.*, 103, 408, 1976.
162. Fischman, H. R. and Ward, F. E., III, Oral transmission of rabies virus in experimental animals, *Am. J. Epidemiol.*, 88, 132, 1968.
163. Correa-Giron, E. P., Allen, R., and Sulkin, S. E., The infectivity and pathogenesis of rabies virus administered orally, *Am. J. Epidemiol.*, 91, 203, 1970.
164. Baer, G. M. and Cleary, W. F., A model in mice for the pathogenesis and treatment of rabies, *J. Infect. Dis.*, 125, 520, 1972.
165. Charlton, K. M. and Casey, G. A., Experimental rabies in skunks: immunofluorescence light and electron microscopic studies, *Lab. Invest.*, 41, 36, 1979.
166. Baer, G. M., Shanthaveerappa, T. R., and Bourne, G. H., Studies on the pathogenesis of fixed rabies virus in rats, *Bull. W. H. O.*, 33, 783, 1965.
167. Jenson, A. B., Rabin, E. R., Bentinck, D. C., and Melnick, J. L., Rabiesvirus neuronitis, *J. Virol.*, 3, 265, 1969.
168. Bijlenga, G. and Heaney, T., Post-exposure local treatment of mice infected with rabies with two axonal flow inhibitors, colchicine and vinblastine, *J. Gen. Virol.*, 39, 381, 1978.
169. Tsiang, H., Evidence for an intraaxonal transport of fixed and street rabies virus, *J. Neuropathol. Exp. Neurol.*, 38, 286, 1979.
170. Leestma, J. E., Velocity measurements of particulate neuroplasmic flow in organized mammalian CNS tissue cultures, *J. Neurobiol.*, 7, 173, 1976.
171. Murphy, F. A., Harrison, A. K., Washington, W. C., and Bauer, S. P., Comparative pathogenesis of rabies and rabies-like viruses. Infection of the central nervous system and centrifugal spread of virus to peripheral tissues, *Lab. Invest.*, 29, 1, 1973.
172. Miyamoto, K. and Matsumoto, S., Comparative studies between pathogenesis of street and fixed rabies infection, *J. Exp. Med.*, 125, 447, 1967.
173. Karasszon, D., Experimental infection of golden hamster with Hogyes' fixed strain of rabies virus, *Acta Vet. Acad. Sci. Hungaricae Tomus*, 20, 251, 1970.
174. Johnson, H. N., Rabies virus, in *Viral and Rickettsial Infections of Man*, 4th ed., Horsfall, F. L., Jr. and Tamm, I., Eds., Lippincott, New York, 1965, 814.
175. Rausch, R. L., Rabies in experimentally infected bears, *Ursus* spp., with epizootiologic notes, *Zentralbl. Vet. Med. B*, 22, 420, 1975.
176. Schindler, R., Studies on the pathogenesis of rabies, *Bull. W.H.O.*, 25, 119, 1961.
177. Johnson, R. T. and Mercer, E. H., The development of fixed rabies virus in mouse brain, *Aust. J. Exp. Biol. Med. Sci.*, 42, 449, 1964.
178. Iwasaki, Y. and Clark, H. F., Cell to cell transmission of virus in the central nervous system. II. Experimental rabies in mouse, *Lab. Invest.*, 33, 391, 1975.
179. Matsumoto, S., Electron microscopy of central nervous system infection, in *The Natural History of Rabies*, Vol. 1, Baer, G. M., Ed., Academic Press, New York, 1975, 217.
180. Matsumoto, S., Electron microscope studies of rabies virus in mouse brain, *J. Cell. Biol.*, 19, 565, 1963.

181. Matsumoto, S. and Yonezawa, T., Replication of rabies virus in organized cultures of mammalian neural tissues, *Infect. Immun.*, 3, 606, 1971.
182. Iwasaki, Y., personal communication.
183. Tignor, G., personal communication.
184. Kligler, I. J. and Bernkopf, H., The path of dissemination of rabies virus in the body of normal and immunized mice, *Br. J. Exp. Pathol.*, 24, 15, 1943.
185. Fischman, H. R. and Schaeffer, M., Pathogenesis of experimental rabies as revealed by immunofluorescence, *Ann. N.Y. Acad. Sci.*, 177, 78, 1971.
186. Charlton, K. M., Casey, G. A., and Campbell, J. B., Experimental rabies in skunks: mechanisms of infection of the salivary glands, *Can. J. Comp. Med.*, 47, 363, 1983.
187. Dierks, R. E., Electron microscopy of extraneural rabies infection, in *The Natural History of Rabies*, Vol. 1, Baer, G. M., Academic Press, New York, 1975, 303.
188. Dierks, R. E., Murphy, F. A., and Harrison, A. K., Extraneural rabies virus infection. Virus development in fox salivary gland, *Am. J. Pathol.*, 54, 251, 1969.
189. Schneider, L. G. and Hamann, I., Die pathogenese der tollwut bei der maus. III. Die zentrifugale Virusausbreitung and die Virus-generalisierung in Organismus, *Zentvalbl. Bakteriol. Parasitol., Infektionskr. Abt. 1 Orig.*, 212, 13, 1969.
190. Fox, J. P., Koprowski, H., Conwell, D. P., Black, J., and Gelfand, H. M., Study of antirabies immunization of man. Observations with HEP Flury and other vaccines, with and without hyperimmune serum, in primary and recall immunizations, *Bull. W.H.O.*, 17, 869, 1957.
191. Abelseth, M. K., An attenuated rabies vaccine for domestic animals produced in tissue culture, *Can. Vet. J.*, 5, 279, 1964.
192. Abelseth, M. K., Further studies on the use of ERA rabies vaccine in domestic animals, *Can. Vet. J.*, 8, 221, 1967.
193. Winkler, W. G., Shaddock, J. H., and Williams, L. W., Oral rabies vaccine: evaluation of its infectivity in three species of rodents, *Am. J. Epidemiol.*, 104, 294, 1976.
194. Wachendörfer, G. and Förster, U., Safety Testing of Attenuated Rabies Vaccines in European Wildlife Species, Symp. Adv. Rabies Research, September 7, 1976, Center for Disease Control, Atlanta, Ga.
195. Tillotson, J. R. and Axelrod, D., Follow-up on rabies — New York, Morbidity and Mortality Weekly Report, 26, No. 31, Center for Disease Control, Atlanta, 1977.
196. Komarov, A. and Hornstein, K., Studies on the pathogenicity of an avianized street rabies virus, *Cornell Vet.*, 43, 344, 1953.
197. Fischman, H. R. and Strandberg, J. D., Inapparent rabies virus infection of the central nervous system, *J. Am. Vet. Med. Assoc.*, 163, 1050, 1973.
198. Bell, J. F., Latency and abortive rabies, in *The Natural History of Rabies*, Vol. 1, Baer, G. M., Ed., Academic Press, 1975, 331.
199. Fekadu, M. and Baer, G. M., Recovery from clinical rabies of 2 dogs inoculated with a rabies virus strain from Ethiopia, *Am. J. Vet. Res.*, 41, 632, 1980.
200. Smith, J. S., Mouse model for abortive rabies infection of the central nervous system, *Infect. Immun.*, 31, 297, 1981.
201. Lodmell, D. L., Bell, J. F., Moore, G. J., and Raymond, G. H., Comparative study of abortive and nonabortive rabies in mice, *J. Infect. Dis.*, 119, 569, 1969.
202. Bell, J. F. and Moore, G. J., Effects of high ambient temperature on various stages of rabies virus infection in mice, *Infect. Immun.*, 10, 510, 1974.
203. Bell, J. F., Clark, H. F., and Moore, G. J., Differences in efficiency of protective effect caused by high ambient temperature in mice infected with diverse substrains of rabies virus, *J. Gen. Virol.*, 36, 307, 1977.
204. Iwasaki, Y., Gerhard, W., and Clark, H. F., Role of host immune response in the development of either encephalitic or paralytic disease after experimental rabies infection in mice, *Infect. Immun.*, 18, 220, 1977.
205. Gribencha, S. V., Abortive rabies in rabbits and white rats infected intracerebrally, *Arch. Virol.*, 49, 317, 1975.
206. Arko, R. J., Schneider, L. G., and Baer, G. M., Nonfatal canine rabies, *J. Vet. Res.*, 34, 937, 1973.
207. Perl, D. P., Bell, J. F., Moore, G. J., and Stewart, S. J., Chronic recrudescent rabies in a cat, *Proc. Soc. Exp. Biol. Med.*, 155, 540, 1977.
208. Fekadu, M., Shaddock, J. H., and Baer, G. M., Intermittent excretion of rabies virus in the saliva of a dog two and six months after it had recovered from experimental rabies, *Am. J. Trop. Med. Hyg.*, 30(5), 1113, 1981.
209. Fernandes, M. V., Wiktor, T. J., and Koprowski, H., Mechanism of the cytopathic effect of rabies virus in tissue culture, *Virology*, 21, 128, 1963.

210. Love, R., Fernandes, M. V., and Wiktor, T. J., The response of the cell to infection with rabies virus, *Rev. Pathol. Comp.*, 781, 533, 1966.
211. Defendi, V. and Wiktor, T. J., Metabolic and autoradiographic studies of rabies virus-infected cells, in *International Symposium on Rabies*, Vol. 1, S. Karger, Basel, 1966, 119.
212. Wiktor, T. J. and Clark, H. F., Chronic rabies virus infection of cell cultures, *Infect. Immun.*, 6, 988, 1972.
213. Holland, J. J., Villarreal, L. P., Welsh, R. M., Oldstone, M. B. A., Kohne, D., Lazzarini, R., and Scolnick, E., *J. Gen. Virol.*, 33, 193, 1976.
214. Andzhaparidze, O. G., Bogomolova, N. N., Boriskin, Y. S., Bektemirova, M. S., and Drynov, I. D., Comparative study of rabies virus persistence in human and hamster cell lines, *J. Virol.*, 37, 1, 1981.
215. Clark, H. F., unpublished data.
216. Iwasaki, Y. and Minamoto, N., Scanning and freeze-fracture electron microscopy of rabies virus infection in murine neuroblastoma cells, *Comp. Immunol. Microbiol. Infect. Dis.*, 5, 1, 1982.
217. Iwasaki, Y. and Clark, H. F., Rabies virus infection in mouse neuroblastoma cells, *Lab. Invest.*, 36, 578, 1977.
218. Diaz, A. M., Diaz, F. J., Yager, P., and Baer, G. M., Activation of lysosomal enzymes in rabies-infected tissue culture cells without accompanying cytopathic effect, *Infect. Immun.*, 4, 549, 1971.
219. Koschel, K. and Halbach, M., Rabies virus infection selectively impairs membrane receptor functions in neuronal model cells, *J. Gen. Virol.*, 42, 627, 1979.
220. Munzel, P. and Koschel, K., Rabies virus decreases agonist binding to opiate receptors of mouse neuroblastoma-rat glioma hybrid cells 108-CC-15, *Biochem. Biophys. Res. Commun.*, 101, 1241, 1981.
221. Tsiang, H., Neuronal function impairment in rabies-infected rat brain, *J. Gen. Virol.*, 61, 277, 1982.
222. Lentz, T. L., Burrage, T. G., Smith, A. L., Crick, J., and Tignor, G. H., Is the acetylcholine receptor a rabies virus receptor?, *Science*, 215, 182, 1982.
223. Wunner, W. H., Is the acetylcholine receptor a rabies-virus receptor?, *Trends Neurosci.*, 5, 413, 1982.
224. Lagrange, R. H., Tsiang, H., Hurtrel, B., and Ravisse, P., Delayed-type hypersensitivity to rabies virus in mice: assay of active or passive sensitization by the foot pad test, *Infect. Immun.*, 21, 931, 1978.
225. Wiktor, T. J., Koprowski, H., Mitchell, J. R., and Merigan, T. C., Role of interferon in prophylaxis of rabies after exposure, *J. Infect. Dis.*, 133, A260, 1976.
226. Hilfenhaus, J., Weinmann, E., Majer, M., Barth, R., and Jaeger, O., Administration of human interferon to rabies-infected monkeys after exposure, *J. Infect. Dis.*, 135, 846, 1977.
227. Baltazard, M., Bahmanyar, M., Ghodssi, M., Sabeti, A., Gajdusek, C., and Rouzbehi, E., Essai pratique du serum antirabique chez les mordus par loups enrages, *Bull. W.H.O.*, 13, 747, 1955.
228. Bahmanyar, M., Fayaz, A., Nour-Salehi, S., Mohammedi, M., and Koprowski, H., Successful protection of humans exposed to rabies infection, *JAMA*, 236, 2751, 1976.
229. Miller, A., Morse, E. H., III, Winkelstein, J., and Nathanson, N., The role of antibody in recovery from experimental rabies, *J. Immunol.*, 121, 321, 1977.
230. Andral, B. and Blancou, J., Study of the mechanisms of early death occurring after vaccination in mice inoculated with street rabies virus, *Ann. Virol.*, 132E, 503, 1981.
231. Prabhakar, B. S. and Nathanson, N., Acute rabies death mediated by antibody, *Nature (London)*, 290, 590, 1981.
232. Kaplan, M. M., Wiktor, T. J., and Koprowski, H., Pathogenesis of rabies in immunodeficient mice, *J. Immunol.*, 114, 1761, 1975.
233. Turner, G. S., Thymus dependence of rabies vaccine, *J. Gen. Virol.*, 33, 535, 1976.
234. Wiktor, T. J., Doherty, P. C., and Koprowski, H., In vitro evidence of cell-mediated immunity after exposure of mice to both live and inactivated rabies virus, *Proc. Natl. Acad. Sci. U.S.A.*, 74, 334, 1977.
235. Prabhakar, B. S., Fishchman, H. R., and Nathanson, N., Recovery from experimental rabies by adoptive transfer of immune cells, *J. Gen. Virol.*, 56, 26, 1981.
236. Tsiang, H. and Lagrange, P. H., In vivo detection of specific cell-mediated immunity in street rabies virus infection in mice, *J. Gen. Virol.*, 47, 183, 1980.
237. Harfast, B., Andersson, T., and Grandien, M., Enhanced cytotoxicity of human lymphocytes against rabies infected cells by rabies-specific antibodies, *Scand. J. Immunol.*, 6, 1107, 1977.
238. Tignor, G., Summary of a workshop on the immunopathology of rabies, *J. Infect. Dis.*, 140, 431, 1979.
239. Sulkin, S. E. and Allen, R., Interferon and rabies virus infection, in *The Natural History of Rabies*, Vol. 1, Baer, G. M., Ed., Academic Press, New York, 1975, 319.
240. Nemes, M. M., Tytell, A. A., Lampson, G. P., Field, A. K., and Hilleman, M. R., Inducers of interferon and host resistance. VI. Antiviral efficacy of poly I:C in animal models, *Proc. Soc. Exp. Biol. Med.*, 132, 776, 1969.

241. Baer, G. M., Shaddock, J. H., Moore, S. A., et al., Successful prophylaxis against rabies in mice and rhesus monkeys: the interferon system and vaccine, *J. Infect. Dis.*, 136, 286, 1977.
242. Fayaz, A., Afshar, A., and Bahmanyar, M., Interference between bovine parainfluenza 3 virus and a street strain of rabies virus in rabbits, *Arch. Gesamte Virusforsch.*, 29, 159, 1970.
243. Atanasiu, P., Barroeta, M., Tsiang, H., and Favre, S., Inhibition *in vivo* de la multiplication du virus rabique par un interfiron endogene, *Ann. Inst. Pasteur Paris*, 119, 767, 1970.
244. Janis, B. and Habel, K., Rabies in rabbits and mice: protective effect of polyribinosinic polyribocytidylic acid, *J. Infect. Dis.*, 125, 342, 1972.
245. Wiktor, T. J., Kuwert, E., and Koprowski, H., Immune lysis of rabies virus-infected cells, *J. Immunol.*, 101, 1271, 1968.
246. Koprowski, H., Immunopathology of rabies virus infection, *Symposia Series in Immunological Standardization*, S. Karger, Basel, 1972, 21, 90.
247. Held, J. R., Tierkel, E. S., and Steele, J. H., Rabies in man and animals in the United States, 1946 — 1965, *Public Health Rep.*, 82, 1009, 1967.
248. Sikes, R. K., Cleary, W. F., Koprowski, H., Wiktor, T. J., and Kaplan, M. M., Effective protection of monkeys against death from street virus by post-exposure administration of tissue culture rabies vaccines, *Bull. W.H.O.*, 45, 1, 1971.
249. Veeraraghavan, N., in *Annual Report, Director, Pasteur Institute of Southern India*, Coonoor Diocesan Press, Madras, 1970.
250. Tignor, G. H., Shope, R. E., Gershon, R. K., and Waksman, B. H., Immunopathologic aspects of infection with Lagos bat virus of the rabies serogroup, *J. Immunol.*, 112, 260, 1974.
251. Blancou, J., Andral, B., and Andral, L., A model in mice for the study of the early death phenomenon after vaccination and challenge with rabies virus, *J. Gen. Virol.*, 50, 433, 1980.
252. Smith, J. S., McClelland, C. L., Reid, R. L., and Baer, G. M., Dual role of the immune response in street rabies virus infection of mice, *Infect. Immun.*, 35, 213, 1982.
253. Koprowski, H. and Black, J., Studies on chick embryo-adapted rabies virus. V. Protection of animals with antiserum and living attenuated virus after exposure to street strain of rabies virus, *J. Immunol.*, 72, 85, 1954.
254. Wiktor, T. J., Lerner, R. A., and Koprowski, H., Inhibitory effect of passive antibody on active immunity induced against rabies by vaccination, *Bull. W.H.O.*, 45, 747, 1971.
255. Atanasiu, P., Dean, D. J., Habel, K., Kaplan, M. M., Koprowski, H., Lepine, P., and Serie, C., Rabies neutralizing antibody response to different schedules of serum and vaccine inoculation in non-exposed persons. IV, *Bull. W.H.O.*, 36, 361, 1967.
256. Hahurick, M. A. W., Rubin, R. H., Music, S., Sikes, R. K., Smith, J. S., and Gregg, M. B., Post-exposure rabies prophylaxis with human rabies immune globulin, *JAMA*, 227, 407, 1974.
257. Bellinger, D. A., Chang, J., Bunn, T. D., Pick, J. R., Murphy, M., and Rahija, R., Rabies induced in a cat by high-egg-passage Flury strain vaccine, *J. Am. Vet. Med. Assoc.*, 183, 997, 1983.
258. Esh, J. B., Cunningham, J. G., and Wiktor, T. J., Vaccine-induced rabies in four cats, *J. Am. Vet. Med. Assoc.*, 180, 1336, 1982.
259. Pedersen, N. C., Emmons, R. W., Selcer, R., Woodie, J. D., Holliday, T. A., and Weiss, M., Rabies virus vaccine infection in three dogs, *J. Am. Vet. Med. Assoc.*, 172, 1092, 1978.
260. Debbie, J. G., Vaccine induced rabies in a pet skunk, *J. Am. Vet. Med. Assoc.*, 175, 376, 1979.
261. Holian, A., and Clark, H. F., unpublished data.

INDEX

A

B

C

D

G

H

I

J

K

L

M

S

T

U

V

W

Z